DARWINISM
& PHILOSOPHY

DARWINISM
& PHILOSOPHY

EDITED BY

Vittorio Hösle and Christian Illies

UNIVERSITY OF NOTRE DAME PRESS

Notre Dame, Indiana

Published in the United States of America

Library of Congress Cataloging-in-Publication Data
Darwinism and philosophy / edited by Vittorio Hösle and Christian Illies.
p.   cm.
Includes bibliographical references and index.
ISBN 0-268-03072-3 (hardcover : alk. paper)
ISBN 0-268-03073-1 (pbk. : alk. paper)
1. Evolution.   2. Evolution (Biology)—Philosophy.
3. Darwin, Charles, 1809–1882—Influence.
I. Hösle, Vittorio, 1960–   II. Illies, Christian.

B818.D26   2005
146'.7—dc22

2005016191

∞ This paper meets the requirements of ANSI/NISO Z39.48-1992
(Permanence of Paper).

CONTENTS

PART IV

Darwinism and the Place of the Human 299

Vittorio Hösle and Christian Illies

Any reasonable theory about the relation between philosophy and the sciences must avoid two extremes. On the one hand, no philosophy can be convincing if it ignores the results of the sciences of its time. For, after all, the results of the sciences are less disputed than those of philosophy, and the claim of the sciences to offer an image of the world is well grounded. On the other hand, philosophy cannot abdicate in the face of scientific theories—for the *interpretation* of the results of the sciences is something that is not as obvious as the results themselves sometimes are. This is true both on a special and on a more general level. A good example of the former is quantum mechanics: even if there is an agreement on the formalism to be used, the real meaning of it remains controversial, including the question of whether quantum mechanics is a deterministic or an indeterministic theory. On the more general level, realist as well as instrumentalist interpretations of scientific theories remain possible even after a scientific theory has been accepted. Even less does the acceptance of modern physics, for example, entail physicalism, as the founders of modern physics prove— neither Descartes nor Leibniz, to name only two of them, was a physicalist.

It is characteristic of the early twenty-first century that the philosophically most challenging science is no longer physics but biology. And, of course, modern biology is essentially Darwinian. It is hardly an exaggeration to state that Charles Darwin has shaped biology more significantly than anyone else before or since.

His importance can be compared only with that of Aristotle, the founder of biology as a science dedicated to the systematic exploration of the phenomena of life and the classification of organisms. Aristotle's ideas—including his tenet that species are eternal—dominated biology for more than two thousand years. This ended with the paradigm shift toward modern biology—that is, evolutionary biology—in the nineteenth century. Of course, this paradigm shift, like almost all paradigm shifts, was not sudden and abrupt—it occurred as gradually as biological evolution does according to Darwin. Evolutionary ideas had been cherished since the eighteenth century, and even the mechanism of natural selection had been mentioned before Wallace and Darwin. The reception of Darwin's *On the Origin of Species by Means of Natural Selection, or the Preservation of Favoured Races in the Struggle for Life* (1859) was clearly facilitated by the fact that scientists had been alerted to the possibility of descent with modification for many decades; after all, Charles was Erasmus Darwin's grandson, and he had himself participated in 1847 in the session of the British Association for the Advancement of Science in which Robert Chambers's transformistic *Vestiges of the Natural History of Creation* was subjected to sharp criticism. (He sat silently, while carrying with him a long manuscript on his own theory.) Nevertheless, the logical rigor and the range of Darwin's work raised the discussion to a new level, and it was thus fair enough that the new theory was very soon called "Darwinism."

Although the history of the reception of Darwin's theory is highly complex and by no means linear, differing also according to the various countries, one can say that, despite much resistance, his theory was successful in a relatively short time. There are some reasons—both scientific and nonscientific—that might explain this astonishingly rapid response. One of the most striking ones is probably that Darwin could explain with a very simple argument an abundance of different facts and also show the internal connection of various phenomena that had hitherto been regarded as separate. Further, the years proved his research program to be extremely fruitful, since it could be linked to and find support from new results in almost all areas of biological research—from paleontology, embryology, and biogeography to the modern disciplines of genetics and molecular biology (the "synthetic theory" of evolution being a good example of this). To be sure, the theory of natural selection has gone through many modifications, and substantial additions have been made. What used to be a simple idea that could be expressed in purely qualitative terms is now a complex group of (mostly mathematical) models, designed for many explanatory purposes. But Darwin's insight is still so essential to them that it continues to make sense to refer to them as "Darwin's theory" (or "Darwinism") in the broadest sense.

From the point of view of the history of ideas, another reason seems even more important for Darwin's spectacular reception (which contrasts so strongly with the neglect of the other genius of nineteenth-century biology, Gregor Mendel): natural selection offered a causal, nonteleological explanation of the admirable order in organisms. The possibility of explaining complexity by a very simple structure — the mechanism of natural selection — is indeed the core of Darwinism and is the basis of both its attractiveness and the hostility that it has provoked. By using this mechanism, Darwin made biology compatible with the type of explanation peculiar to physics, the paradigm of a proper science since Galileo and Newton. By changing the way we think about teleology in nature as well as our concept of species, Darwin's intellectual revolution had profound philosophical consequences that are very close to some general ideas of the late nineteenth century. (It is noteworthy that he himself was aware of the consequences and had been particularly interested in them immediately after the conception of his theory. The metaphysical speculations present in the M and N Notebooks demonstrate this.) And from the beginning many of his contemporaries observed that the significance of his theory went far beyond the limits of biology. The theory could serve as a cornerstone of a general weltanschauung that encapsulated the main ideas of its time. (See, e.g., the fifteenth chapter of *The Education of Henry Adams,* where we find Henry Adams's half-melancholic, half-ironic description of how he became a Darwinist, although he lacked the most rudimentary knowledge necessary for evaluating the merits of Darwin's theory.)

To show in more detail how Darwin's theory fits into the intellectually dominant ideas of the nineteenth century would require a very careful analysis. For the present purpose it is sufficient to point to some particularly strong affinities between his theory and two sciences that were very important for the self-understanding of the time: history and economics.

History, in particular the ideas of change and of a continuous development in time, had proven to be very fertile for the understanding of human culture achieved by the contemporary humanities and had been applied, for example, to social (e.g., legal and political) institutions, the arts, religion, and philosophy. It could now be extended to nature. Thus the gap between nature and culture was diminished: for many, Darwinism promised to be the basis of a naturalistic interpretation of the world and thus to be the perfect replacement for the official theological metaphysics. However, few of the numerous enthusiastic Darwinians understood that the main point of Darwin's theory (not always recognized by Darwin himself, particularly in his later years) was the rejection of a guaranteed telos of the evolution — an idea dear to the nineteenth century. Darwin's approach to evolution was in truth very different from that of Comte and Marx, far more revolutionary than theirs and much closer to Nietzsche's nonteleological conception of history.

Economics was the other science to which the Darwinian approach to biology was obviously related. This cannot surprise us if we remember that the final idea that Darwin needed for his theory stemmed from Thomas R. Malthus, the first professor of political economy in the British Isles. The reflection on the consequences of scarcity, sharpened by the simultaneous growth of the users of scarce resources, is the common denominator of economics and biology since Darwin. In the twentieth century these links were elaborated in more detail. Michael Ghiselin even regards economics and biology as branches of a larger endeavor, called "general economy," that encompasses as well the economy of social and cultural institutions (see his seminal bioeconomical work *The Economy of Nature and the Evolution of Sex*).

As Ghiselin proves, many aspects of Darwin's theory not only explain why it was so attractive to his contemporaries but still deserve serious philosophical attention today. Especially the last decades have shown that the implications of Darwin go far beyond the explanation of speciation and that his general idea is much more fundamental. It has even been argued that Darwinism is itself a philosophical theory, since it seems to be about "being in general" (to put it in very traditional terms): wherever we have a plurality of entities reproducing themselves and vying for scarce resources, we have a Darwinian situation—the entities may be organisms competing for food, languages competing for speakers, or scientific theories competing for intellectuals supporting them (as Stephen Toulmin has convincingly argued in his epistemology). This generalization might suffice to show that Darwinism is more than a biological theory; it may not apply to all entities (atoms do not reproduce), but it certainly does apply to far more than genes and organisms. It is therefore also a theory about being in general, an ontological (or metaphysical) theory.

The fundamental character—and philosophical importance—of Darwin's theory is also due to the inclusion of humans in its general account. Darwin hinted at this already in the *Origin*, and this application to humans had been a major concern for him from the very beginning. But only after others had dealt with the natural evolution of man was he willing to write and publish *The Descent of Man* and *The Expression of the Emotions in Man and Animals*. These works show profound insights into human language, sentiments, and morality. By including cognitive activities, social behavior, and the aesthetic taste in natural history, evolutionary biology began to play a pivotal role for philosophy, more and more replacing physics, which had been of highest importance for philosophy since the seventeenth century. Of course, physics remains more fundamental than biology (it is possible to develop physics without any knowledge of biology, but not vice versa), but evolutionary biology is more akin to philosophy: it is (among other things) about us humans, and Kant had already noted that philosophy's central concern can be summed up in the question "What is man?"

Biology is the science of life in a double sense. On the one hand, it is about life and us as living organisms. On the other hand, it is an intellectual activity performed by life—that is, by humans as living organisms. And this activity itself can become the subject of biological research (and even more of nonbiological research guided by Darwinian principles). Evolutionary epistemology as an important branch of evolutionary biology aspires to explain the development and mechanisms of animal and human understanding as a process of adaptation. Thus biology reveals a kind of reflexive structure, something that is a traditional feature of philosophy rather than of the natural sciences. (Physics, for example, does not have it, not even in quantum mechanics, for the measurement problem does not deal with human interiority: the Heisenberg cuts have nothing to do with Descartes's cut.) Biology also studies the ways in which organisms interact or express themselves—applied to humans, it analyzes the forms in which we interact with nature (e.g., in order to feed ourselves) as well as with other human beings (e.g., in order to reproduce). Thus evolutionary biology is of great significance for the social sciences but also for the corresponding branches of philosophy—ethics and political philosophy. Since consciousness, as far as we know, is connected to neuronal structures, philosophy of mind (and the mind-body-problem) can hardly be addressed outside the philosophy of biology.

These are just some (very roughly sketched) instances of the profound importance of evolutionary biology, and thus of Darwinism, for philosophy. Given this importance, it can hardly be surprising that philosophers and scientists with philosophic ambitions espoused Darwin's theory from the very start, most notoriously Herbert Spencer in Great Britain and Ernst Haeckel, David Friedrich Strauss, and Friedrich Nietzsche in Germany. We must give them credit for having been the first to realize Darwin's enormous philosophical potential and to try to expand his insight beyond the limits of biology. Although there are serious, perhaps even cogent, objections against the concrete way in which they utilized his ideas and against the rather rash conclusions they drew from evolutionary theory, they were right in sensing the far-reaching explanatory power of "Darwin's Dangerous Idea" (to use the title of Daniel Dennett's book).

However, there is no consensus whatsoever about what these philosophical consequences might be. For a long time most authors regarded Darwinism as being strongly supportive of some kind of naturalism, if not identical with it. (And indeed, while physicalism was never a very plausible form of naturalism, biologism is: the increasing invasion of the domain of the social sciences and of the humanities by biological explanations is an expression of the attraction as well as the remarkable success of biologism.) Nevertheless, in the last decades this view has been challenged by historians of science and of philosophy as well as by systematic philosophers. In this context the work of historians is important, for it is generally wrong to believe in an automatic progress in the history of philosophy, and particularly misleading in the case of the interpretation of Darwinism. The richness of philosophical ideas to be found in the notebooks of the young Darwin is far from having been fully exploited and brought to the general attention of philosophers. The more we know about the intellectual background of Darwin, the more we recognize that his worldview was not naturalistic (and thus was "metaphysical" in one of the many senses of the word); it was not even explicitly atheistic. On the contrary, even after the loss of his Anglican faith Darwin shared the conviction of the fathers of modern physics that a rational concept of God was a proper foundation for the scientific endeavor. Darwin continued to define himself as a theist and only in the last phase of his life as an agnostic—in sharp contrast to the dominant atheistic interpretation given to his theory by many of his friends and foes alike. But also among his contemporaries were scientists able to grasp the theory and at the same time willing to integrate it into a religious worldview—St. George Mivart being the best known.

The twentieth century has brought forth different attempts to give a non-naturalist reading of Darwinism, Hans Jonas being one of the most promising. (Jonas's ontological analysis of the organic mode of existence was a generalization of Heidegger's existential analytic of *Dasein*.) It is probably not an accident that particularly in German-speaking countries, where the influence of Kant and German idealism's *Naturphilosophie* is still felt, a non-naturalistic philosophy of biology is more present than in the Anglo-American world. The tradition of critical epistemology in the aftermath of Kant, the defense of the irreducibility of the normative realm—particularly of the moral law, but also of epistemic claims—to the physical world, the emphasis on development and more holistic models over genetic reductionism, and the recognition of intentionality and agency in organisms and its consequences for the mind-body problem are some of the insights of the German philosophy of biology that necessitate further discussion and seemed to make an encounter between representatives of the German and the Anglo-American tradition of philosophy of biology an important task. (The authors of this introduction do not want to hide that they, too, have been educated in this German tradition.)

Of particular relevance is the fact that, although the German tradition of philosophy of biology was more prone to be linked to the program of a rational theology than the philosophies of biology of other countries, already in the early nineteenth century its idea of rational theology was more modern than that of most other European countries, including Britain. On the one hand, the philosophical theology of German idealism is markedly distinct from the tradition of physico-theology that Darwin so convincingly destroyed. Kant's criticism of physico-theology was far more influential in Germany than Hume's criticism had been in Britain, and after Kant no serious German philosopher ever again used physico-theological arguments. Even those who, like Hegel, rejected Kant's subjectivism in the *Critique of Judgment* were grateful to him for having eliminated forever the pettiness of eighteenth-century physico-theology. One will look in vain for an equivalent to the Bridgewater Treatises in the Germany of the nineteenth century, so Darwin's discoveries were less disruptive in Germany than they were in Britain. (Already the young Schelling and later Schopenhauer had liked to quote Erasmus Darwin.) It is true that in the second half of the nineteenth century German idealism had been given up by most German philosophers, but not because of Darwin's discoveries; and when neo-Kantianism and later the various neo-Hegelianisms emerged, they could connect with Kant and Hegel without feeling that going back to the corresponding tradition had become an impossibility due to Darwin.

On the other hand, post-Kantian German rational theology was not threatened by forms of biblicism so typical of the American reaction to Darwinism. A high level of sophistication in hermeneutics, based on an awareness of profound historical changes in the development of humankind, prevented such a reaction. Unfortunately, however, in popular American culture often enough the rejection of the naturalistic interpretation of Darwinism is linked to a general polemic against the scientific theory as such. Even if there are various open scientific questions related to Darwin's theory, such a polemic rarely shows the competence necessary to be taken seriously by scientists. Due more to a peculiar biblical hermeneutics than to sound theological principles, in some U.S. states (recently in Kansas), although in no European country, the question of the scientific recognition of Darwin's theory is still a religious and even a political issue. This certainly does not contribute to a sober analysis of the philosophical implications and the correct philosophical interpretation of Darwinism.

The aim of the essays collected in this book is to start a new and more comprehensive discussion about the philosophical implications of Darwinism. The term *implications* is used in a broad sense, including both the presuppositions of the theory and its consequences: to be true, Darwinism entails certain tenets regarding the structure of reality—that is, a certain "ontology"—and if it is true, certain philosophical consequences follow from it.

The obvious divergence of philosophical styles and approaches to this question was the greatest challenge to this enterprise. Philosophers as much as scientists have very different ideas about what these implications are in detail, where to look for them, and even what questions one should ask. However, we see this diversity not as a threat but as an opportunity. We have thus consciously tried to integrate very different approaches into this volume to make the wide range of possibilities obvious. As in the case of biodiversity, a plurality of backgrounds is a fundamental value that enriches the final consensus one has to aim at.

The choice of the persons invited to contribute to this volume was intended to include the following groups. First, we decided that philosophers as well as scientists (ideally, persons who were both) should be present. Even if interdisciplinary discussion is a demanding task, it goes without saying that a convincing discussion of the philosophical implications of Darwinism presupposes a conversation between the two groups. Second, we decided that systematic reflections should be complemented by historic ones: if Darwin's contribution to the interpretation of his own theory is not inferior to later ones, and if the genesis of modern science in the seventeenth century was already accompanied by a consciousness of its philosophical presuppositions rarely achieved afterward, it is indispensable to reconstruct these early reflections. Historians of philosophy, geology, biology, and psychology—the disciplines to which Darwin contributed most—wrote chapters for this volume.

Third, we decided that philosophers from different traditions should meet, namely those from the continental and the analytic schools of philosophy, mainly from German- and English-speaking countries (one colleague comes from France). Very often, the two have seemed to have such disparate understandings of what philosophy is that no communication appeared to be possible. Most recently, however, things have begun to change—slowly, too slowly, but still one can feel an increasing reciprocal curiosity that is one of the most hopeful events in the philosophy of the last decades. For until now in the Anglo-American world the contribution of post-Kantian continental philosophers of biology (Schelling, Hegel, Schopenhauer, Scheler, Driesch, Bergson, Gehlen, Plessner, Portmann, Jonas) has not been really recognized—it is absent from the syllabi taught as well as from the ongoing discussions (some of the texts have not even been translated into English). In a highly influential article with the title "What Philosophy of Biology Is Not" (*Synthese* 20 [1969], 157–84), David Hull discussed the criteria that books and articles in the philosophy of biology have to satisfy in order to be regarded as useful contributions to the field—a test that according to him only few texts can pass. Even if the rigor of a discipline depends on what it excludes, the danger of strict boundaries is that more is eliminated than is good for the proliferation of ideas (which is certainly not a sufficient but still a necessary condition for the development of both science and philosophy). One of the main objects of Hull's criticism was a formalist way of proceeding, which often enough expresses "more or less commonplace ideas . . . in tiresome exactitude" (173). But he tended to reject also approaches that defended a more traditional philosophy of nature—Hans Jonas's important work *The Phenomenon of Life: Toward a Philosophical Biology* was mentioned only once in a dismissive footnote. There is still a long way to go before the different traditions will meet, as this collection shows. But a willingness to learn from each other seems to develop in the philosophy of biology, as this collection can also show.

The split between "continental" and "analytical" philosophy is, by the way, not linked to the more or less critical appraisal of Darwin's theory: Jonas had great admiration for as well as a full grasp of the theory, while analytical philosophy of religion is sometimes inimical to it. Since intellectual and geographic boundaries fortunately do not coincide, it was an explicit intention of the editors to invite German philosophers of biology who work in the analytical tradition as well as American colleagues familiar with the tradition of German *Naturphilosophie*. Finally, both older and famous scholars who marked the field decades ago and younger philosophers were brought together at this conference.

It would be naive to expect that a common agreement among all speakers could be achieved regarding the questions mentioned. But this volume has succeeded in mapping the problems, detecting many of the implications and consequences, both scientific and interpretive, of our construction of Darwinism, and showing that there are very different ways of making philosophical sense of Darwinism. Naturalism may be a very tempting interpretation of Darwinism, but it is certainly not the only possible one—as Charles Darwin himself already understood with impressive clarity.

A final word of thanks has to be directed to Mark Roche, the dean of the College of Arts and Letters of the University of Notre Dame, who graciously funded the conference that was the basis for this volume; to Phillip Sloan, who gave us invaluable advice with regard to both the organization of the conference and the publication of this volume with an uncommon generosity of ideas and time; to Cornelius Delaney, Rev. Ernan McMullin, and the late Phil Quinn from the Department of Philosophy of the University of Notre Dame for very helpful contributions in the discussions after the lectures; and of course to all the contributors to this volume for their dedication to the enterprise.

PART I

What Kind of Science Is Darwinian Biology, and What Are Its Ontological Presuppositions?

We have already mentioned in the general introduction that one of the reasons for the success of Darwinism was its structural affinity to the type of causal explanation to be found in modern physics. Final causes became superfluous, and explanations took place according to the pattern of what was later called the Hempel-Oppenheim scheme (which had actually been presented already much earlier, in the first book of Spinoza's *Ethics*). Despite the evidence of this scheme, modern science is not as free from presuppositions as some scientists believe. One of the most important presuppositions is that of the constancy of natural laws. Charles Lyell's actualism was a special case of a general belief in constant laws of nature. Can this belief be justified? Is it simply one of the constituents of modern science? How was it connected at the beginning of modern science to theological assumptions? How is it related to materialism? And what is the status of this principle? Is it only a methodological principle, or does it apply to nature as such? Of course, this question is linked to the controversy over whether an instrumentalist or a realist interpretation of science is appropriate. Such problems are dealt with by Peter McLaughlin in his introductory essay.

Since Darwin acquired his fame first as a geologist, the second essay is by a historian of geology, David Oldroyd. After all, few books influenced Darwin as strongly as Lyell's *Principles of Geology;* in a letter to L. Horner (August 29, 1844), Darwin wrote that his own books seemed to him to come partly out of Lyell's brain.

Without Lyell's actualism, *the* alternative to Georges Cuvier's catastrophism in geology, Darwin would have lacked a basic presupposition of his theory—the great amount of time he needed to explain the gradual evolution of species. What are the common presuppositions of geology and paleontology? How do the two disciplines relate to each other? What are the principles of taxonomy, and how were they influenced by the geological discoveries of the eighteenth, nineteenth, and twentieth centuries? And what is the distinctive approach of cladification, the new challenge to traditional classification?

One of the unresolved issues in the discussions of theory of science dealing with Darwinism is the status of the formula of the survival of the fittest. What in Darwin's argument is based on empirical facts, what is deduced from them, and is there anything that can be known independently from experience? Ever since Darwin's own time we find people pointing out that his essential formula can be known a priori—mostly, however, on the basis of the assumption that it is ultimately a tautology. But what if we do not consider Darwin's explanation to be tautological? Is it still possible to grasp some important element of natural selection through mere reflection? This is what Christian Illies tries to explore in his essay by analyzing the structure of an explanation through natural selection.

Michael Ruse continues in his essay the analysis of some of the topics mentioned by McLaughlin. Is modern science, and is Darwinism, that form of biology that finally corresponds to the ideal of modern science, necessarily materialistic? Is atheism then a corollary of Darwinism? Are miracles excluded by Darwinism as such? Is the belief in a Creator God incompatible with the rejection of special creationism? Has the insistence on the struggle for life rendered the old problem of theodicy more difficult or perhaps easier to solve? And could the central Christian concept of an organism created in God's image be instantiated also by an organism very different from *Homo sapiens?*

It can hardly come as surprise for Darwinians that Darwinism itself has also developed and that it has changed in the course of its history. The common ground of these different avatars of Darwinism is certainly the theory of natural selection. But what are the differences, both in the scientific substance and in the popular image formed of it—the iconic Darwinism, as one could call it? Do the different Darwinisms presuppose different ontologies? Is there in general a variety of ontological interpretations behind each form of Darwinism? Do interpretive frameworks and conceptual schemes, far more than the scientific facts themselves, determine the philosophical conclusions? To name a particularly important question: What is the place of agency within a Darwinian theory? Do organisms only adapt to their environment, or do they change it? How strongly is the choice of an ontology determined by extrascientific factors, such as either the support or the rejection of a social Darwinist theory of society? What is the role of cultural and political factors in the increasing appeal of iconic Darwinism in the last two decades? Is it linked to the demise of Marxism and Freudianism, to a general discontent with the cultural revolution that tried to change society, and to a political resignation accepting human nature as it inalterably seems to be? All these are the type of questions David Depew deals with in the final essay of this section.

CHAPTER 1

Materialism, Actualism, and Science
What's Modern about Modern Science?

Peter McLaughlin

> Order and purposiveness in nature must themselves be explained from natural grounds
> and according to natural laws; and the wildest hypotheses, if only they are physical, are
> here more tolerable than a hyperphysical hypothesis, such as the appeal to a divine
> Author, assumed simply in order that we may have an explanation.
>
> Immanuel Kant, *Critique of Pure Reason*

The general theme of the conference is the relation of Darwinism and metaphysics, and what that relation is depends on what is included under the heading of Darwinism, which can mean many things to many people. For some it is an explanation of organic change, for others a doctrine of progress; for some it has been a general monistic weltanschauung, for others a social philosophy. I shall take Darwinism in a somewhat narrow sense covering the mechanism of directed organic change and the historical product of its action over time. In this narrower sense Darwinism has as much and as little to do with metaphysics as does plate tectonics. Neither the scientific *theory of evolution* by means of variation and natural selection nor the scientific *fact of phylogeny* or common descent has any metaphysical implications per se. Only in combination with metaphysical presuppositions and commitments can an empirical theory or a historical reconstruction have such implications. Just as no number of descriptive premises will support a normative conclusion unless some normative premises are added, so too no empirical scientific propositions will support a metaphysical conclusion unless some metaphysics is implicitly slipped in. This does not of course mean that scientific theories cannot be metaphysically consequential, nor does it mean that there is anything wrong or even superfluous about metaphysical commitments. Given certain metaphysical presuppositions, the consequences of a particular theory can be very great. For instance, the consequences of the theory of evolution given a metaphysical position like scientific realism are quite different from those it has given, say, critical idealism. But we are always talking about the consequences of a *set* of propositions including metaphysical commitments. Furthermore, since most scientific theories don't raise the kind of metaphysical questions that Darwinism does, we can restrict our focus of attention to one particular aspect: where evolution plays a relevant role in worldview conflicts.

To illustrate the problem, take a trivial case: someone might object to natural selection, which inevitably involves some contingency in its results, because he or she doesn't see how God could simply make matter and let things evolve and still be said intentionally to create man in his image and likeness (I assume that man is being considered here as a physical object, not just as a moral agent). But even here there is no direct contradiction between the theory of evolution and such speculative assertions of historical fact. Only in conjunction with a particular metaphysical theory of the will and of the nature of God and his actions does such a contradiction arise, always presupposing a commitment to some form of scientific realism. Thus our choice given such a conflict is not between belief in science and belief in a revealed historical fact but rather (on the assumption of certain historical facts) between giving up a belief in science and giving up a belief in a particular metaphysical theory—of the will, the nature of God, or the nature of religious and scientific knowledge.

None of this I take to be outrageously original or even controversial. I mention it only to make explicit what we should be looking for when we consider the implications or the operative metaphysical assumptions of Darwinism.

I am going to talk about materialism, actualism, and the nature of science both from a systematic and from a historical perspective. Scientific materialism and actualism in this context are methodological prescriptions taken to be constitutive of modern science. Materialism (or naturalism), as the name indicates, demands that material phenomena be explained in terms of material objects and events only. Actualism is a corollary to materialism stating that in historical reconstructions—in particular in geology and paleontology—no kinds of causes or laws that are not empirically accessible to current science are to be adduced to explain past events. This doesn't mean you can't use, say, the crash of an asteroid or some other unusual or even unique event to explain the extinction of dinosaurs, it just means you don't use different laws of nature to get the asteroid to cross our path: laws of nature don't change or expire. In science you may speculate as much as you wish about exotic boundary conditions but not about the arbitrary suspension of the laws of nature.

I am going to argue that these two principles dependably distinguish modern science from various other cognitive practices. Darwinism presupposes them both but little more, and it therefore has few, if any, metaphysical implications of its own. In fact, as we shall see, Darwinism is not even incompatible with special creation unless we bring in additional commitments to some particular metaphysical position—as we normally do.

DEMARCATION

Now, the relation of science—or whatever we want to call a particular society's form of epistemically distinguished knowledge—to other cognitive practices is a theme that has accompanied philosophy from its beginnings. According to some this is the question with which philosophy actually began. And indeed one of the oldest formal definitions of philosophy that has come down to us deals with the examination of cognitive claims. According to the *Horoi,* or definitions of the Platonic school, philosophy is the habit of examining the true as to how true it is—or perhaps more fashionably, philosophy takes an analytical stance toward the truthfulness of the true (ἕξις θεωρητικὴ τοῦ ἀληθοῦς πῶς ἀληθές).[1] One twentieth-century version of this habit familiar from logical empiricism and especially from Popper's critical rationalism is called the demarcation problem.

The demarcation problem has for some time deservedly had a rather bad reputation. All attempts at finding universally applicable, sufficient, and necessary conditions for the proper application of the term *science* have failed. And according to the original schemes of the mid–twentieth century such conditions were merely conventions, free stipulations that could not be given an unassailable justification. There was no specifiable reason that could compel assent to them, and they could be motivated only by some intuitive connection between formal logic and rationality and by vague hand waving at something called fertility. It is thus hard to imagine what really would have been gained if the project had succeeded. Furthermore, demarcationists, especially those of the Popperian stripe, are notorious for slipping into ad hominem argument and, instead of attacking certain propositions as unscientific, attacking certain persons or their attitudes as unscientific. This state of affairs is quite unfortunate, since there is a genuine philosophical question about the nature and aims of science.

About twenty years ago the demarcation problem was revivified in a new form when—possibly for the first time since the Galileo trial—a philosopher was called to testify in court as an expert witness and to offer a judgment on a particular claimant to the title of science. In the case *McLean v. Arkansas Board of Education,* the philosopher Michael Ruse asserted that we do in fact have criteria that allow us to distinguish between science and religion and that these tell us that so-called "scientific creationism" is not science.[2] I bring up this example because it allows us to address two questions in a rather clear-cut way. Precisely what is the area of expertise of the philosopher with respect to competing truth claims about the origin of species? And what exactly is the relation between the claim to being scientific and the claim to being true?

Ruse's testimony led to a minor controversy among philosophers of science on the issue of whether we have any demarcation criteria between science and nonscience or, more precisely, between science and religion. Larry Laudan, for instance, argued that we just do not have any demarcation criteria and therefore that the philosopher's argument against creationism can only be that it is *bad* science.[3] But this alternative is not very promising. Even abstracting from its lack of justification, it puts philosophers of science in the position of stating *in the area of their expertise* that creationism is in principle science and then of stating in an area where they have no *professional* credentials that it is no good. When asked as professionals to what genre creationism belongs, we are supposed to say, "For all we know, it is science." If asked as laymen how good that science is, we would reply, "Bad!"— but then, who asks a layman? Philosophers who take this position do not take a stand on the philosophical question of the nature and aims of science; rather, they argue that the real issue before us is whether the empirical evidence better supports evolutionary theory or creationism. But the question of which particular proposition is best supported by the available evidence is something that philosophers ought to leave to the empirical scientists.

Philosophers who take this antidemarcationist position must claim that they can establish an ordering relation between better and worse science without specifying the dimension ordered. It would seem that they in fact share an implicit understanding of science: science is the pursuit of objective truth, and thus any honest pursuit of truth can be called science.[4] They criticize so-called scientific creationism not as being unscientific but as being empirically wrong. And insofar as they are not just playing amateur scientist but also really arguing as philosophers, they are basically merely asserting that naturalism is a stronger metaphysical position than supernaturalism—an assertion with which I have no quarrel. I think, however, it would be better to separate three different questions: (1) whether creationism is or is not science; (2) whether one particular scientific hypothesis is better confirmed than another; and (3) whether creationism is metaphysically wrong because it is incompatible with philosophical naturalism. I shall be concerned only with the first question—and of course with the task of keeping the questions separate.

The antidemarcationist position also has the politically unpleasant consequence that the question of whether the assertion of special creation is true or false must be settled before we can decide whether it ought to be included in science. This seems to admit that someone who believes it to be true is not only entitled but perhaps even morally obliged to insist that it be taught in science class. But the question that concerns me here is not whether creationism should be taught in biology class but whether the *only* reason not to teach it in biology class is that it is less well confirmed than evolution according to the standards of Bayesian confirmation theory. I don't believe that we have to agree on substantial truths about God and the universe before we can reach minimal political agreement on the nature and aims of science. We do not have to know the true answer to a number of existentially important questions in order to be able to say what a proper biological explanation looks like. And philosophy ought to have something to say about this.

If we are going to be able to settle this kind of worldview-linked question in a consensual manner—that is, independent of our prior commitments to the substantial truth of metaphysical propositions—then it must be at least imaginable for each side to say that a particular doctrine should or should not be recognized as science without ruling on its truth or falsity. (For instance, if you want a consensual agreement that creationism is not science, you must be prepared to admit that it is at least conceivable that the true story may be told not by science but by some other intellectual endeavor and that it is at least conceivable that what we call science might aim at something other than the truth.)

Now the basic problem with demarcation is that you can in principle define science any way you want. Popper defined it by completely conventional traits such as rationality, logic, bold hypothesis, and rigorous testing; Feyerabend used his freedom somewhat more creatively, introducing some rather unconventional traits. And even if we assume that science has to be a particular form of knowledge that is in some way socially distinguished, there are different societies and different views of what distinguishes scientific knowledge. In Germany most of the humanities bear the title of *Wissenschaft,* and the only really ecumenical definition of that term is, in Hugo Dingler's phrase, anything for which a chair exists at a German university. On the other side there is a much narrower definition of science derived from early Enlightenment philosophy and characterized by the two principles mentioned in my title: materialism and actualism. This happens for contingent historical reasons also to be the notion of science that is incorporated (if anything is) into American constitutional law. This is indeed only one of many possible conceptions of science, but both history and reason do speak for it.

We can safely ignore the general question of demarcation, since we are not in fact looking for the ultimate necessary and sufficient conditions for the application of the term *science* such that everybody with a Ph.D. will agree to them. We don't need to find one set of criteria that can infallibly separate science from landscape architecture, plumbing, and transcendental meditation. The issue at hand is not to separate science from everything else but to separate science from religion or other worldview commitments. The practically relevant question is whether we have rational grounds to argue that we can distinguish between scientific and religious questions.

MATERIALISM

There has been a great deal of discussion about this newer form of demarcation in recent years, and the position I have been advocating has come to be called methodological naturalism. This term is something of a misnomer: it can be misleading because it seems to suggest that naturalism or materialism is being used heuristically, as an assumption to be legitimated by its producing fruitful hypotheses and successful research programs. But naturalism is not a heuristic, it is a presupposition. It is not what Kant called a regulative principle but rather what he called a transcendental presupposition. For Kant it was a regulative principle, for instance, that we should try to order things into classes and subclasses. But this ordering activity also makes a "transcendental presupposition," namely the assumption that the physical world is orderly or at least orderable. Scientific naturalism is not a heuristic justified by its utility but the presupposition of certain kinds of intellectual activity—which activity itself may of course be justified by its utility. This doesn't make naturalism as a metaphysical position right; it only makes the assumption of naturalism inevitable in certain kinds of practice. I shall be fairly brief and historical about materialism because I think it is relatively unproblematic. The real conflict arises with the less well-understood principle of actualism.

In the *Principles of Philosophy* (1644) René Descartes asserted the causal closure of the material world. The conservation of matter and force was introduced not as a law of nature but as a prerequisite for the laws of nature that could then be formulated. It was given, not a physical justification, but a metaphysical one, namely as something that a reasonable God would guarantee. The possibility envisioned by Descartes that the notion of a conservative or dynamically isolated system might be instantiated by the actual world in which we live was as radical a break with the past as the notion that technical construction could generate natural knowledge.

Any human action depends on reliable and invariant aspects of the world around us. Remember what happened to Alice in Wonderland when she tried to play croquet with flamingos as mallets and hedgehogs as balls. If everything had a will of its own, systematic human action would be impossible. We all have an intimate interest in knowing which aspects of the world around us behave in deterministic (or stochastically reliable) fashion according to their own natures and which aspects are subject to the arbitrary will of various other agents. Natural knowledge need not be the only kind of knowledge, or even the most important kind, but it is important to have it and to be able to distinguish it. And certain forms of this natural knowledge are what we now call science.

Even if we were to admit divine or otherwise non-natural intervention in the material world on an occasional or even a regular basis, we would still want to distinguish between phenomena that depend only on the natural properties of things and those that depend additionally on the special will of the creator or some other being. Seventeenth- and eighteenth-century philosophers distinguished between the "ordinary" and the "extraordinary" concourse of God.[5] In the first case we need only natural knowledge to manipulate or predict the behavior of natural objects; in the second case we need additional knowledge about the volitional states or at least the behavioral dispositions of an immaterial being to be able to act appropriately. It is of great advantage to understand the natural course of events, even if we believe that there is sometimes a great difference between what things do of their own natures and what they actually do empirically. For many purposes, purely natural knowledge is sufficient. We do not have to conjure the deity every time we build a bridge.

The conflict over the competing cognitive claims of science and religion is a significant part of modern intellectual history, and it accompanied the history of modern science from the start. It was also one of the central questions of modern philosophy, so that if the contemporary philosopher of science doesn't know how to tell the difference, he might want to ask Descartes or Leibniz, because they certainly thought a lot about it and thought they knew the difference. And even though you can arbitrarily call anything you want "science," if you want the enterprise to which you are referring to have some significant overlap with that enterprise initiated by Galileo, Descartes, and their ilk—which is generally called science—then there will have to be some constraints placed on what you call "science." That is, there has to be some hermeneutical give-and-take between what we want to consider important today in science and what actually distinguishes *modern* science from the views of Aristotle's medieval followers. And early modern philosophy did in fact reflect on such distinctions and formulated at least two distinguishing characteristics of science: materialism and actualism.

Some of the clearest statements on the principles that distinguish modern science and theology are to be found in Leibniz. In his fifth letter to Samuel Clarke he argues that it is important both for science and for theology to distinguish between natural and supernatural events: "In good philosophy and sound theology, we ought to distinguish between what is explicable by the natures and powers of creatures, and what is explicable only by the powers of the infinite substance."[6] And in an exchange with Pierre Bayle he characterizes more closely the kind of materialism that science presupposes. Bayle was fairly adamant about the importance of natural knowledge, insisting that it was not the business of the philosopher to have recourse to God in explanation: *Non est philosophi recurrere ad deum.*[7] Leibniz was somewhat more consilient, asserting: "In a word, so far as the details of phenomena are concerned, everything takes place in the body as if the evil doctrine of those who believe, with Epicurus and Hobbes, that the soul is material were true, or as if man himself were only a body or an automaton. . . . But in addition to general principles which establish the monads of which compound things are the result, internal experience refutes the Epicurean doctrines."[8] In this particular context the subject under discussion is human conduct, but the position taken is more general. As a general rule, Leibniz tells us, science must proceed as if the evil doctrine of the materialists were true. He believes that even though we have metaphysical grounds to reject materialism as a fundamentally true doctrine, nonetheless for the description of natural phenomena it is indispensable. In his debate with Newton and Clarke, Leibniz insisted that extraordinary interventions of God, namely miracles, occur only for the purposes of faith and morals, not for the purposes of nature. The lesson to be learned from this is that certain forms of knowledge, namely natural knowledge or science, presuppose materialism. You may take this materialism methodologically or ontologically, but you have to take it with the form of knowledge. Leibniz accepts the methodological version of materialism as constitutive of natural science, even though he explicitly repudiates the ontological version.

ACTUALISM

The second postulate relevant to the question is actualism, which may be somewhat less familiar than materialism. I shall use two examples to help make clear what its content is. In the case of materialism, classical modern philosophy sought to distinguish between *metaphysical* truth and what natural knowledge presupposes. In the case of actualism, we may want to distinguish between *historical* truth and what science presupposes or seeks.

I want to try to illustrate in an intuitively plausible way the possibility that truth and science may diverge. Perhaps we may occasionally want to choose science over truth or truth over science. I will try with two thought experiments to present an example in which "truth" and the aims of science diverge. I want to argue that modern science, at least sometimes, has aims better characterized as explanation than as descriptive truth.

Now, I take it to be trivial that some false propositions can be scientific and some nonscientific propositions can be true. For instance, the proposition that proteins are the bearers of hereditary information was a scientific conjecture but turned out to be false. And my belief that I missed two planes yesterday is true but hardly in any reasonable sense scientific. Caesar either crossed the Rubicon or he didn't, but physics doesn't take a stand on this. The genuinely problematical case would be one in which a false scientific proposition and a true nonscientific proposition each made reference to the same individual entity—for example, the planet Earth.

In a popular science fiction spoof, *The Hitchhiker's Guide to the Universe,* a group of space-time travelers visit the deserted planet Magrathea. Magrathea, now closed down, once was an important workshop for manufacturing luxury and vacation planets—and also specialty planets used for scientific experiments. However, in the scene described, it has been temporarily called back into service to make a second copy of the planet Earth according to the original blueprints, since the original

Earth was unfortunately demolished to make way for an intergalactic freeway. (The owners—who had originally had the earth constructed in order, among other things, to perform a few minor experiments on the evolution of primates—want a replacement.) The space-time travelers are given a guided tour of the workshop by the engineer who was responsible for manufacturing the fjords along the coast of Norway. Because of various interruptions, their guide cannot go into very great detail about exactly what he did, but I will embellish a bit and imagine that he tells them how he put the fossils in the rocks and the dinosaur bones in the swamps and then gouged out the glacial valleys. Now, let us suppose that the earthling among space-time travelers is a geologist or a paleontologist and upon his return has to teach a graduate course on the Precambrium. Should he now, due to his visit to this factory, teach the course differently than he taught it before? Concretely, should he continue to teach the historical reconstruction offered by the best science available, or should he rather teach what he now knows to be the historical truth? Or, assuming he teaches *both* stories, we can reformulate the question as one in the empirical sociology of science: In which case would the returned space-time traveler be speaking as an expert professional and in which case as an informed layman? It would seem that the layman would be in possession of the truth, while the expert scientist could only offer warrants for a false belief. Here, the *literally true* story would be told by a *historical revelation,* not by science.

Let me drop this science fiction example, which is good for illustration of a point but, like most examples of this kind, encourages equivocation. Since the master technician or evil scientist of such science fiction examples is finite, it is in principle possible that we could find some flaw or some difference that would make a difference between the real thing and the fake, between nature and the artifact. A traditional, theologically formulated example can exclude this sort of possibility: if an omnipotent God had created the world six thousand years ago so that it would look exactly the same as if it had arisen the way geologists tell us it did, then there would in principle be no (scientific) way of providing evidence for or against this fact. Theologians might have difficulty explaining the rationality and goodness of God's action in such a case (since there seems to be at least a hint of duplicity involved), but if we had enough reason to assent to the contingent historical truth of the matter, then the theologians would have to get to work on an explanation. So let me take up just such an example, the one on which early modern philosophy confronted the issue.

In the third part of the *Principles of Philosophy* Descartes envisions precisely this kind of situation:[9]

There is no doubt that the world was created with all its perfection from the beginning; so that the sun, the earth, the moon, and the stars existed in it; and also that not only the seeds of plants were on the earth but the plants themselves; nor were Adam and Eve born as infants but were made as adult people. The Christian faith teaches us this, and natural reason convinces us that this is true; because, taking into account the immense power of God, we cannot judge him to have made anything that was not wholly perfect of its kind [*omnibus suis numeris absolutum*]. But nevertheless, just as for an understanding of the nature of plants or men it is better by far to consider how they can gradually grow from seeds than how they were created by God at the very beginning of the world; so too, if we can devise some principles which are very simple and easily known and from which we might demonstrate that the stars and the earth, and indeed everything which we perceive in this visible world, could have arisen as if from certain seeds, even though we know well that they did not arise in this manner, we shall in that way explain their nature much better than if we were merely to describe them as they are now. And because I think I have discovered some principles of this kind, I shall here briefly explain them.

Thus Descartes points out a difference between two kinds of knowledge: first, the *true description* of the history of the world as seen by God (or from a God's-eye point of view) and revealed to us by him; and second, a *reconstruction by us* that *explains* how it could have happened, based on our knowledge of certain fundamental principles about the workings of nature. This is to my knowledge the first explicit formulation of the principle of actualism. God's decision to populate the completed earth with a full complement of adult organisms cannot be simulated in the lab or tested in the wild. Had Descartes been the space-time traveling paleontologist, he would not have altered his lecture notes at all. As a scientist he would have continued to teach the best reconstruction that science could provide, and as a Christian—assuming he was one—he would have continued to believe the revealed truth without letting it interfere with his science. In such a case it is not difficult to distinguish between the scientist's reconstructive story and what might happen to be the historical truth.

The point at issue here is the role of primarily *experimental* knowledge in historical disciplines. Experimental knowledge is different in kind from historical. With his famous simile of two differently constructed clocks that nonetheless display the same empirical phenomena, Descartes maintains that we can never say unequivocally from our experiments that the particular mechanism we have suggested is the only one possible for explaining a particular individual system.[10] This borderline case between natural science and natural history can be used to illustrate a difference in goals and means between two different enterprises: accurate descriptions of how particular things really are or were—that is, historical knowledge—and dependable instructions for the production of functional equivalents of a particular kind of thing—experimental science. Thus the specifically new element in modern science (experiment) gives us the means to explain, not how the earth (a particular individual entity) was *really* formed, but how a world of this kind (a functional equivalent of this world) could be produced again. As Descartes says immediately before the long passage I have just cited: "But even though this (hypothesis) may be thought to be false, I shall consider that I have achieved a great deal if all the things which are deduced from it are in conformity with experiments: for thus we will receive from it just as much usefulness for life as from a knowledge of the truth itself [because we shall be able to use it just as effectively to manipulate natural causes so as to produce the effects we desire]."[11]

By this excursion into Descartes's quasi-instrumentalist philosophy of science, I don't want to try to convince you that Descartes was not an epistemological realist. He was. It is difficult to read the *Meditations* in any other sense: there are real objects that cause our ideas of them, and that's that. Once Descartes has overcome his "hyperbolic" methodological doubt, there is no question that the external world is real and can be known by us. I only want to assert that Descartes isn't and that one needn't be a *scientific* realist. Descartes is unwilling to allow his philosophical analysis of scientific knowledge and its claims to depend on any particular metaphysics—even his own realistic metaphysics. He decouples philosophy of science from metaphysics, and thus his own philosophy of science is neither realist nor antirealist; it is merely nonmetaphysical. This illustrates that it is possible to be an epistemological realist like Descartes without conflating historical and experimental knowledge and without contaminating the philosophy of science with prior metaphysical commitments, say to realism.

To return to my point of departure: independent of one's commitment to the truth or falsity of some particular story about the history of a particular individual entity, the planet Earth, it is possible to discuss the kinds of stories that science can or cannot support. Both Descartes, on the question of actualism, and Leibniz, on the question of materialism, advocated scientific explanations of phenomena even when these were based on presuppositions that they considered *false*. Thus the demand that only one of the legitimate scientific reconstructions can claim to be true in some framework-independent sense is not a demand that need be made in the name of science. It may be demanded by the realist metaphysics that the scientist adheres to, but that is a different question.

Let me add a slight historical caveat here before I turn to my conclusion. Before ascribing too strong an instrumentalism to Descartes, we should remember the context. The teachings of science and religion in this case are incompatible if both of them subscribe to appropriate forms of metaphysical realism. As soon as one of the parties drops this postulate of realism, the contradiction disappears. We should expect that the socially or institutionally weaker party to the conflict drops realism. Descartes dropped literal truth and opted for a kind of pragmatic truth for his explanation of the history of the earth. Many theologians today abandon literal truth for the biblical story and opt for a kind of moral truth.

CONCLUSION

Returning to the original question of Darwinism and metaphysics, I have argued that even the assertion of the special creation of organisms is not incompatible with the scientific theory of evolution, which explains how organisms evolve now, and that it is incompatible with the scientific fact of phylogeny only under the assumption that actualism is not just normative for scientific reconstructions but true of the real world. There may just be a sense in which scientific facts are not historical facts. Most of the worldview implications of Darwinism come from taking materialism and actualism not just methodologically but metaphysically. I certainly do not want to dissuade anyone from doing this; and indeed narrow Darwinism's lack of worldview implications may even provide a motive to take one's naturalism metaphysically. Materialism is certainly a legitimate metaphysical position, and it is arguably better than any of the available alternatives; but it is not without alternatives and it is not identical with science. And while scientific naturalism and philosophical naturalism fit well together, they are not the same thing.

On the other hand, in a somewhat looser sense of *implication,* where we are not talking about logical entailments of a set of propositions, even Darwinism in the very narrow sense has some very relevant worldview implications. To the extent that Darwin's theory has made it possible to give scientific explanations of phenomena that were long explained by supernatural causes, it can indeed strengthen the position and enlarge the scope of naturalism as a weltanschauung. For instance, a naturalist like David Hume once capitulated before the phenomenon of adaptation and accepted one particular form of the traditional argument from design because the science of his time had no resources adequate for explaining this phenomenon. Hume was unable to conceive of the eye as an adaptation for seeing without conceiving it as something designed by some agent for seeing; he therefore, albeit in extremely restricted form, admitted the existence of design in nature and allowed a supernatural explanation of this design.[12] Darwinism, however, has made it possible to envision adaptation without design. That is, since the effective scope of a naturalist view of the world is to a large extent dependent on the results of natural science, and since the effective alternative in popular culture (as opposed to the better theological seminaries) is in fact the God-of-the-gaps strategy of proposing supernatural explanations wherever natural explanations are unavailable, every closed gap will of course strengthen naturalism. But this has more to do with the causality of social interactions than with the logic of argumentation.

NOTES

1. Plato, *Sämtliche Werke* (Greek and German), ed. Karl-Heinz Hülser (Frankfurt am Main: Insel, 1991), 10:242.

2. Michael Ruse, "Creation Science Is Not Science," *Science, Technology, and Human Values* 7, no. 40 (1982): 72–78. Ruse's own criteria were unfortunately somewhat Popperian, for which he rightly took a bit of flak, but that is not important in this context.

3. Larry Laudan, "Commentary: Science at the Bar—Causes for Concern," in *But Is It Science? The Philosophical Question in the Creation/Evolution Controversy,* ed. Michael Ruse (Buffalo, N.Y.: Prometheus, 1988), 351–66; Philip P. Quinn, "The Philosopher as Expert Witness," in Ruse, *But Is It Science?* 367–85.

4. Thus the tendency to slip into ad hominem argument: someone who consistently fails at achieving truths that others recognize must be using the wrong methods, and someone who dogmatically sticks by these methods in the face of failure is probably not *really* pursuing truth at all.

5. René Descartes, *Principia philosophiae* (Amsterdam, 1644), pt. 2, § 36, in *Oeuvres de Descartes,* ed. C. Adam and P. Tannery (Paris: Vrin, 1964–74), 8:183. Robert Boyle, *The Works of the Honourable Robert Boyle* (London, 1772), 5:163–64, 179, spoke of God's "ordinary and preserving concourse." See J. E. McGuire, "Force, Active Principles, and Newton's Invisible Realm," *Ambix* 15 (1968): 154–208, for examples from other English writers of the period. See also Francis Oakley, *Omnipotence, Covenant and Order: An Excursion in the History of Ideas from Abelard to Leibniz* (Ithaca, N.Y.: Cornell University Press, 1984), 72–92.

6. G. W. Leibniz, *Die philosophischen Schriften* (Berlin, 1875–90), 7:417; *The Leibniz-Clarke Correspondence,* ed. H. G. Alexander (Manchester: Manchester University Press, 1956), fifth letter, § 112.

7. Pierre Bayle, "Buridan" (note C), in *Dictionnaire historique et critique* (1740; reprint, Geneva: Slatkine, 1995); see also "Sennert" (note C): "Recourir à Dieu comme à la cause immédiate, ce n'est point philosopher."

8. "Reply to Bayle's Article 'Rorarius,'" in Leibniz, *Die philosophischen Schriften,* 4:559.

9. Descartes, *Principia philosophiae,* pt. 3, § 45, 8:99–100. There is a short version in the *Discourse on the Method,* pt. 5.

10. Ibid., pt. 4, § 204, 8:327.

11. Ibid., pt. 3, § 44, 8:99; French translation approved by Descartes, 9:123. The passage in square brackets was presumably added by Descartes in the French translation.

12. David Hume, *Dialogues concerning Natural Religion,* ed. Norman Kemp Smith (Indianapolis: Bobbs-Merrill, 1947), pts. 2 and 12.

CHAPTER 2

Evolution, Paleontology, and Metaphysics

David Oldroyd

Evolution, Darwinism, and paleontology evidently have conceptual connections, and the relationship between Darwinism and metaphysics has long been thought important. So one should anticipate connections between metaphysics and paleontology. But what are they? I want to discuss this question, from a historical point of view, with reference to empirical and scientific issues. I shall only say a little about Darwin.

I should first state my metaphysical position. I am an atheist. The arguments for theism appear so preposterous that their finding any favor can be accounted for, in my view, only by sociological means. I suppose that there are potential "Darwinian" scientific explanations of mind, as adumbrated by Dennett (1995). The nature of matter can, I think, be tackled sufficiently by physics and chemistry. There is no need to agonize over the purpose of life or the universe. Neither has purpose per se, I submit. Our own purposes are our own business and can be understood sufficiently by introspection, biological or Darwinian–psychological theories of the mind or sociological theory. Ethical problems are intransigent but are not clarified by invoking metaphysical entities like "souls," divine beings, or "absolute" truth, goodness, or evil. There *are*, to be sure, big questions to worry about, of which the most important in my view are overpopulation, exhaustion of fossil fuels, desertification, and so on. But such questions are metaphysical only from a rather expanded view of the nature of metaphysics.

For me, the one big metaphysical question has to do with the origin of the universe. There are apparently three possibilities: a universe with no beginning or end (or perhaps one that is cyclic); a "big-bang" universe, somehow coming into being in a naturalistic fashion; or a God-produced or God-sustained universe, which might conceivably pertain to a universe of infinite age or, more likely, one that started with a "big bang." I don't know which of these options is correct. Most modern cosmologists favor a big bang (Rees 2001), but we may yet see a resurgence of the idea of some kind of infinite universe, the "bang" we think we know about being only the last of an endless series of expansions and contractions. None of the theories is comprehensible to me except in the vaguest way. So I put them in the too-hard basket and get on with life. This may be intellectual laziness; but it is no lazier than the theistic option, and less unsatisfactory. For if the origin of the universe is a mystery, it is not clarified by introducing a greater mystery—God. At least we know the universe exists.

All this is rather knockabout philosophizing. But I thought it best to state my atheism (which is probably the majority position among Australian academics, and certainly among philosophers) at the outset. Of course, my naturalism (and rejection of "saltations" in nature) may be regarded as a form of metaphysics, for I cannot demonstrate *its* philosophical warrant. I agree, however, with the first of Gerhard Vollmer's (1994) twelve theses of naturalism, namely "Nur so viel Metaphysik wie nötig." My metaphysics *is* (I hope) minimalist, as Vollmer recommends.

What, then, if anything, is the connection between metaphysics—as traditionally construed—and evolution and paleontology? In some societies, such as Japan, the connection is tenuous. In the Western world, however, metaphysics, evolution, taxonomy, and paleontology have become contingently intertwined. The contingencies have to do with the Aristotelian tradition as regards taxonomy and with the rise of Protestantism—with its emphasis on the Bible as the foundation of faith and the one-time presumption that the Bible was the "word of God"—even concerning the geosciences and evolution.

As is well known, biblical literalism led to interaction between metaphysics and paleontology. (What were fossils doing on mountaintops?) But interaction was not forged by Christianity alone. It also had to do with Plato and Aristotle's ideas on the nature of being, the question of essences (nominal or real), and then the Thomist synthesis of Aristotle and Christianity. The Swedish botanist Carl von Linné (1707–78) also had a hand in the matter. His "invention" of binomial nomenclature (Linné 1758) was scientifically important but had roots in such ventures for providing essential definitions as the "tree of Porphyry" (Warren 1975).

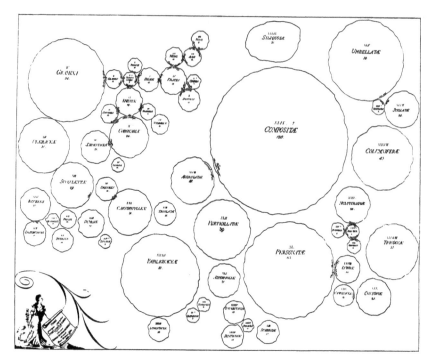

FIGURE 2.1. "Linnean" Representation of Relationships of the Vegetable Kingdom, According to a Plate Accompanying Linné and Giseke (1792). *Source:* By permission of the British Library. 8712.d.1/21.

Linné (1744, 1972) also proposed a creation myth: there was an equatorial island, where the first pairs of species were located, which subsequently reproduced and spread round the globe. So, without going into details, by the beginning of the nineteenth century there was a belief that species could be suitably named in terms of genus and species (as Aristotle and Porphyry had supposed), that nature had "joints" bounding each species, and that there was a "natural" way to carve nature at her joints.

It is interesting that at one stage, Linné (via one of his students, Paul Giseke, whose thesis of 1747 was published in 1767, and again in 1792) began to arrange his plant groupings like a map (Linné and Giseke 1792) (see figure 2.1). But Linné himself was not greatly interested in fossil forms. Though the Linneans found an ever-increasing number of species, the general view in the early 1800s was that the number of species was finite, that they had been created by divine fiat, and that such creation was relatively recent—a few thousand years, according to biblical estimates (Fuller 2001).

But there were also practical men who recognized that fossils could be relevant to engineers, agriculturalists, miners, and so on. Here we need only mention William Smith (1769–1839), civil engineer, land surveyor, and cartographer, whose work took him to all parts of southern Britain (Torrens 2001, 2002). There were obviously different "kinds" of strata (clays, chalk, coal deposits, etc.), but Smith made an issue of the fact that the different rock types contained particular suites of fossils, which he recognized by eye—using his "*nous.*" He devoted himself to arranging the strata of southern Britain in ascending geometrical order, guided by the fossil contents of each kind of stratum, and he prepared maps and sections of the strata. (He was, fortunately, dealing with what geologists call a "layer-cake stratigraphy.") He published his famous geological map of England and Wales (Smith 1815a, 1815b, 1815c). Its purpose was practical and economic, though Smith made it a labor of love to bring it to completion, and it was an aesthetically fine production.

In particular, Smith used fossils to map Upper and Lower Oolitic limestones separately, even though they *looked* similar. He advised against prospecting for coal at certain sites because fossils showed that the rocks there were well above the Coal Measures in the stratigraphic succession. Smith became known as the "Father of English Geology" for his principle that "each stratum is . . . possessed of properties peculiar to itself, has the same exterior characters and chemical qualities, and the same extraneous fossils throughout its course" (Smith 1815c, 2). This was an empirical and commercially significant discovery

Analogous work was conducted in France by two professors and rivals at the Musée d'histoire naturelle in Paris: Jean-Baptiste Lamarck (1744–1829) and Georges Cuvier (1769–1832). They were "natural philosophers," not essentially "practical men," and of different metaphysical dispositions. Lamarck has been called a "Heraclitean," with his metaphysical preference for continuities and flux— evolution or transformism. He imposed his metaphysics on his and others' observations. The fossils in the stratigraphic column were like points on a graph, which his imagination connected. He could not conceive how extinctions could occur.

Cuvier, by contrast, was more of an empiricist. Examining the strata, he found what appeared to be radical stratigraphic discontinuities, indicating radical discontinuities amongst organisms over time. Everything did not blend into everything else. Cuvier (1796a) reached this conclusion early in his career. He investigated African and Indian elephant skeletons and noted differences indicating that they were different species and were different again from the remains of a creature found in Ohio, which Cuvier named the mastodon. The following year he examined a large skeleton from Paraguay, which he named *Megatherium americanum* (Cuvier 1796b). These American animals differed from anything alive today. Seemingly, extinctions had occurred.

Vertebrate remains in France led to the same conclusion. Such facts suggested to Cuvier that there had been a *monde* previous to the one that we know today, which had been destroyed by some kind of grand catastrophe. Thus he stated what came to be his trademark: "catastrophism." To the question "How were the successor types formed?" Cuvier replied that as a metaphysical question this was no more difficult or simple than the production of *anything* (such as the universe, I might suggest). So he brushed the difficulty aside and continued his empirical work, building up evidence for the former existence of a whole succession of previous *mondes,* as suggested by the fossil data. As stated in his lectures to the Collège de France in 1805, slow, continuous causes could not account for the formation of distinct strata (Grandchamp 1994). Contra Lamarck, Cuvier (1812) did not think that the evidence suggested mutations or transformations. Cuvier's ideas can be represented graphically as shown in figure 2.2 (cf. figures 2.1 and 2.3).

Cuvier's geology influenced the next major development of Smithian stratigraphic principles: the advent of the concepts of "stage" and "zone," advanced in France by Alcide d'Orbigny (1802–57) and in Germany by Albert Oppel (1831–65). D'Orbigny listed some 200,000 species, including 40,000 fossil species, which, after weeding synonyms, yielded 18,286 species and 1,440 genera (d'Orbigny 1849–52; Monty 1968; Rioult 1969). His researches suggested there were stable stratigraphic units, across countries, with lines of demarcation between the characteristic fauna above and below marking global upheavals. So we have the *terrains* (d'Orbigny's term), such as Jurassic and Cretaceous (names not coined by d'Orbigny). There were also faunal breaks within the Jurassic (say), which marked *étages*. There might be an unconformity at the break, an erosion surface, a conglomerate, or a condensed deposit such as a bone breccia. There would be facies changes, marked by faunal discontinuities. Sudden rises or falls of the land surface under the seas—successive catastrophes—supposedly produced these effects. Considering the first appearances, maxima, and disappearances of major animal groups from "*Amorphozoires*" to mammals, d'Orbigny proposed twenty-seven stages for his first five periods or *terrains* from Silurian to Tertiary; and there was also the *Terrain contemporain*. Thus organic forms had supposedly been created or recreated twenty-eight times, though there was some continuity from one stage to

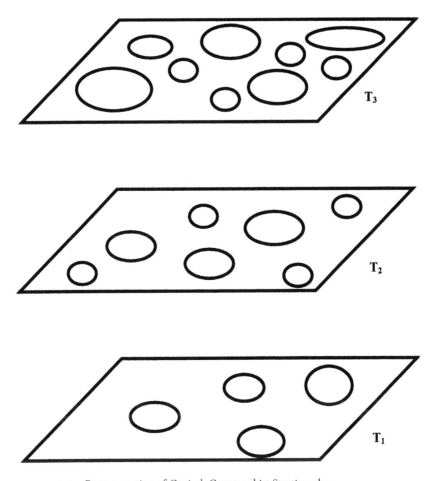

FIGURE 2.2. Representation of Cuvier's Catastrophist Stratigraphy

the next (d'Orbigny 1849–52). D'Orbigny was a catastrophist but to an extent a uniformitarian also. Fossils were presumed to be formed by processes observable today. The catastrophes were produced by the operation of physical laws. But (like Cuvier) he avoided the question of the process(es) involved in the origin of new forms.

So within the Jurassic, for example, d'Orbigny proposed ten stages, each with its characteristic ammonite fauna. Like Smith, but in greater detail, d'Orbigny selected characteristic fossils for each stage that were widely distributed over different facies, abundant, and largely limited to the stage in which they occurred. The faunal content of each stage defined the strata's chronological succession. Thus he supposed that there were faunal "zones," but these approximated to his stages and were coarser entities than Oppel's. For d'Orbigny, a "stage" was something real, resulting from the earth's tectonic activity or associated "water movements" such as the "Bajocian transgression" in the Jurassic. It is interesting that d'Orbigny recognized that there might be some filiation of species within stages (such as Lamarck would have favored) and also that there might be what we would call convergence. So: slow "improvement" within stages; accelerated "improvement" between stages; but no transitions between successive stages.

Finer subdivision of the stages was undertaken by the paleontologist and stratigrapher Oppel, developing the idea of "zones." Working in the Jurassic of Württemburg and Schwabia and across France and England also, Oppel (1856–58) surveyed the literature on Jurassic fossils and determined their stratigraphic ranges in as many localities as possible, building up a composite series of stratigraphic sections for different regions. From the assembled ranges of fossils, he found that there were groups of strata characterized by similar aggregates of fossils or that within a stage there were "clusterings" or "assemblages" of species at different stratigraphic horizons. A species was selected as an "index fossil," its name being used to designate the relevant zone. Fossils having narrow time ranges and wide geographical distributions were preferred. However, if one took an assemblage to delineate a zone, some of its constituent species might occur above or below that zone. By Oppel's method, a zone's lower boundary was determined by the first appearance of a chosen species; the upper boundary was distinguished by the last appearance of some other species. Oppel's investigations did not require any particular theory of biological change or metaphysical speculations. His zone concept is still, in its essentials, used by biostratigraphers.

However, Oppel zone usage has been somewhat flexible and subjective and has been much debated, for the procedure depends in no small degree on "tacit knowledge." In fact, the establishment of zones and other subdivisions of the stratigraphic column is now a formalized social process, undertaken by the relevant subcommissions of the International Commission on Stratigraphy, the decisions of which are then ratified by the International Union of Geological Sciences. In these deliberations, issues are ultimately decided by *vote*.

The boundaries between the geological systems, subsystems (series), and stages are given "Global Stratotype Sections and Points" (GSSPs). Particular sections are chosen as exemplars, rather as species were defined by Linné. Even now, the biostratigraphic definitional work is unfinished. According to a chart produced by the International Commission on Stratigraphy (Gradstein et al. 2004), all GSSPs (also known among geologists as "golden spikes") are settled for the Silurian, Devonian, and Pliocene, but for the Triassic, Jurassic, and Cretaceous, where all these games began, things are far from being finalized. The decisions as to where to locate the "spikes" are taken by commissions of the International Union of Geological Sciences and have to be ratified by that body. Politics can enter the story. Different countries have vied with one another to "host" stratotype sections. And in any case, the decisions are ultimately decided by voting — or achieving consensual agreement. Thus the ontology of stages and zones falls within the purview of the sociology of knowledge. Stratigraphic subdivisions are evidently not produced by emanation from some Platonic heaven — they are clearly "in" the rocks in some sense; but they are also "in" the geological community, which changes them from time to time. Their ontological status seems ambiguous, in fact more so than that of biospecies, for stratotypes cannot be treated cladistically (see below), and though they may change they do not reproduce themselves like living organisms! (Moreover, some stratigraphers think that "golden spikes" are of doubtful use.)

But zonal subdivision is certainly not arbitrary and susceptible to purely nominalist interpretation. Consider the work of Charles Lapworth (1842–1920) in the Southern Uplands of Scotland in the 1870s. Mostly, the terrain is poorly exposed moorland, but there are some isolated stream sections, and Lapworth used one at Dob's Lin near Moffatt to establish a scheme of zonal stratigraphy. He found an anticline of thin shales that gave clear evidence of the "way-upness" of the strata. A tradition in his family reveals that he had a special waistcoat, with multiple pockets. Graptolites were collected, inch by inch, from each stratum, and placed in separate pockets. Then:

> Having already finished our arrangement of the shales into five successive groups, we next busy ourselves by collecting as many as possible of these fossil graptolites. In our cabinet we have five corresponding sets of drawers. We place all the graptolites we procure from the highest grey shales in the first set — those from the second set of beds in the second set of drawers, and so on. In our spare time . . . we name and identify these graptolites but we are always careful to replace them in the proper drawer. Long before we have collected our hundred specimens, we shall have made . . . a startling discovery. We find that our drawers contain three distinct sets of graptolites. The two upper sets of rocks contain one set of graptolites. Those answering to the two middle sets of shales contain a second set and those answering to the flinty beds at the bottom of the section contain a third set. None of these species of graptolites found in one set passes into another.[1]

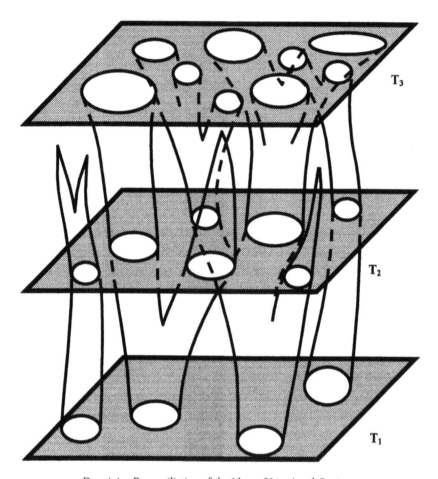

FIGURE 2.3. Darwinian Reconciliation of the Ideas of Linné and Cuvier

Thus Lapworth was able to do, on a more refined scale, what Smith had done for younger rocks. Lapworth could recognize the relative ages of shales by their fossil contents and apply the succession for Dob's Lin to other localities. We think his technique worked because his graptolites were evolving but also were being sorted into identifiable groups by the vagaries of sedimentary processes. Nothing metaphysical here, so far as the stratigraphy was concerned.

In principle, Darwin's theory provided a rapprochement between the ideas of Linné and Cuvier, as shown in figure 2.3 (cf. figures 2.1 and 2.2). The observed "jerkiness" of the stratigraphic record was explained by its supposed incompleteness. Darwin realized, however, that it would be desirable to find examples of smooth evolutionary transitions in the fossil record. In his *Origin* he referred to the occurrence of "graduated forms" of the fresh-water gastropod *Planorbis multiformis* described by Franz Hilgendorf (Darwin 1869, 362). The actual location of the site was at Steinheim in southern Germany, but Darwin thought this place was in Switzerland, and presumably he did not see the relevant publications (Hilgendorf 1866). If he had, he probably would not have thought the evidence suited to his purpose.[2]

The issue of the existence or otherwise of smooth transitions in the stratigraphic record could be said to be of metaphysical importance, since, it would seem, it has a bearing on whether species come into being by some naturalistic evolutionary process or from time to time in some mysterious way, perhaps guided by a "divine hand." For the most part, there *appears* to be jerkiness in the stratigraphic record, which can cast doubt on the Darwinian picture (figure 2.3). The difficulty has been seized on by creationists (e.g., Gish 1978).

To maintain our naturalism, we can proceed in various ways in the face of stratigraphic gaps: by finding empirical evidence to "fill the gaps" so that species can be seen to blend into one another over time—as we know they do over space in "ring species" (Cain 1954); by invoking the incompleteness of the record (as did Darwin); by developing biological theories to provide explanations of the sudden formation of new species (e.g., Schwartz 1999); or perhaps by invoking extraterrestrial causes to explain the record's jerkiness (e.g., Frankel 1999). To give even a summary account of the efforts to explain the seeming jerkiness of the stratigraphic record is a task beyond the scope of the present chapter and would take us into developmental biology and genetics. Here let us just consider some paleontological and stratigraphic examples.

Stephen Gould (1999) has discussed the interesting case of the Russian paleontologist Vladimir (or Woldemar) Kovalevsky (Kovalevskii) (1850–91), who described what he took to be a gradual evolutionary sequence of fossil horses in Europe, with special reference to *Anchitherium* (a three-toed pony-sized animal), but with the sequence extending from *Palaeotherium* in the Eocene, through *Anchitherium* and *Hipparion* in the Miocene, to the modern horse. Moreover, he related the anatomical trends to environmental changes, notably a change from browsing to grazing, associated with the increase of grasses on open plains (Kovalevsky 1876, 1980). It seemed a triumphant illustration of Darwinian expectations, or, as Kovalevsky (1873, 1980) put it, it had a "*charme irrésistible*" for the transformationist. But the evidence was soon upset as the new forms were found to have originated in the Americas and to have successively migrated to Europe.

At the end of the nineteenth century, the English geologist Arthur Rowe (1899) found—by sampling the chalk exposures of the cliffs of southern England—what he thought was evidence for what we would call the "gradualistic" evolution of sea urchins. I was told about this work as a student. But examination of Rowe's plates does not suggest the smooth trends in fossil shape that I had been led to expect. Rather, Darwin's critics might appear to have a point. (However, Rowe examined anatomical details other than overall shape, and these seemed to evidence evolutionary trends.)

Students of fossil plankton have, however, found evidence of evolutionary continuity and speciation in microfossils extracted from drill cores (e.g., Cifelli 1969; Lipps 1970; Malmgren and Kennett 1981; Scott 1982; Tappan and Loeblich 1988; Kucera and Kennett 2000). For larger organisms, Peter Sheldon, now of the Open University, UK, collected some fifteen thousand well-localized Welsh trilobites. He measured the thicknesses of their pygidia ("tails") and their rib numbers, for eight species (Sheldon 1987, 1988), finding that they displayed evolutionary continuity (see figure 2.4). Like the Foraminifera, the trilobites exhibited a kind of "drunken walk" as far as changes in size were concerned, but the transitions seemed to be relatively smooth. Moreover, the directions of "walk" seemed to be roughly correlated among the different species. Sheldon speculated that the size changes might be driven by extraterrestrial causes such as those responsible for Milankovich climatic cycles. Perhaps?

Sheldon also remarked that specimens found at different ends of his connected lineages had been regarded as *different species* by earlier workers such as Christopher Hughes (1969, 1971, 1979). Sheldon (1987) made the important point that the earlier additional species identifications could have been due to "the requirement to apply binomial taxonomy to fossils and the practice of lumping together specimens collected from different horizons in order to amass enough material for full 'species' description." In other words, Linnean naming, together with the insufficiency of collected specimens, might have led paleontologists to suppose there were distinct and separable species within what was actually a continuum. The species were thus figments of a (metaphysical?) preference that "evaporated" in the face of additional data.

The other approach to the problem of apparent absence of missing links and the sudden appearance of new lineages—so useful for stratigraphers working within the d'Orbigny-Oppel tradition—was to give a more careful analysis of what happens during speciation. Such analysis was undertaken in the first half of the twentieth century by the likes of the American George Gaylord Simpson (1902–84), and

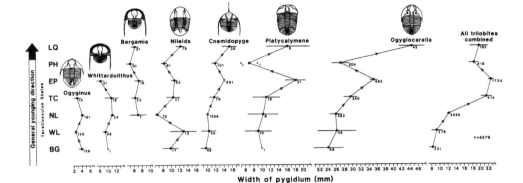

FIGURE 2.4. Evolutionary Trends in Trilobites, According to Peter Sheldon.
Source: Reprinted from "Trilobite Size Frequency Distributions, Recognition of
Instars, and Phyletic Size Change," by P. R. Sheldon, *Lethaia,* www.tandf.no/leth,
1988, 21, 305, by permission of Taylor & Francis AS.

in postwar years we have had the doctrine of "punctuated equilibrium" (PE), which
its authors, Stephen Gould and Niles Eldredge (Eldredge and Gould 1972; Gould
and Eldredge 1977), thought had special significance in that it challenged the supposedly metaphysical preference for gradualism. Actually, I doubt that PE theory
was or is metaphysically different from ordinary Simpsonian evolutionary theory
and stratigraphy.[3] Certainly some studies favor PE, with, for example, what appears
to be punctuated speciation events among conodonts[4] in Argentinian deposits (Albanesi and Barnes 2000), but PE theory seems to have been absorbed into general
evolutionary theory without metaphysical indigestion or consequences.

Simpson was primarily a vertebrate paleontologist, his contribution to the neo-
Darwinian synthesis being to show how statistically based genetics could be compatible with the stratigraphic record (Laporte 2000). New types would be few and
geographically restricted, moving into new environments and subject to severe selection and adaptation. So the *seeming* stratigraphic "gaps" and anatomical saltations
might in a sense be real without there having been saltatory evolution. Simpson
(1944) also envisaged certain ecological or environmental "adaptive zones" where
species could do well and others where they could not. Movement from one zone
to another might occur in short bursts, in what he called "quantum evolution."

In line with such thinking, Simpson (1944, 213) hypothetically represented
evolution as a kind of tree, as so many have done before and since. But his tree was
shaped more like an espaliered apple tree than like an ordinary deciduous tree in
a forest. That is, he envisaged a number of branches separating from the trunk

near the same place, each then growing upwards in approximately the same direction. Thus he envisaged an "explosive phase" when "quantum evolution [was] predominant" and there was an "opening of new adaptive zones." This was followed by a "normal phase" of evolution of the newly formed types in which "phyletic evolution [was] predominant," variation was reduced, and the new forms were apart from one another and without intermediaries.

But all this was manifestly hypothetical or schematic. The successful construction of a real tree diagram, for real organisms, would necessarily depend on knowledge of which items in the tree were older and which younger, and which were ancestral to which. That is, we should require knowledge, at least in a general way, of the historical order of strata. While that might be feasible in areas of "layer-cake" stratigraphy, it is not so easy in tectonically complicated areas or over wide distances, or where the continents and oceans have undergone significant relative movements, as entertained by continental drift or plate-tectonic theorists. Moreover, one may argue in a circle or invoke ad hoc hypotheses to explain away uncomfortable findings: that is, one may use the theory to guide the construction of the tree.

A "classic" example of mammalian evolution is that given in Simpson's book *Horses* (1951), which goes back to his early studies in Texas and New Mexico in the 1920s and earlier to the nineteenth-century work of O. C. Marsh. However, the horses did not come from one site, with a layer-cake stratigraphy. They were from localities distributed over North America, and their arrangement was accomplished to some extent by hypothesizing a tidy evolutionary sequence, in which the number of toes decreased as hooves evolved. Nevertheless, it seemed that they could be situated on the branches of an evolutionary tree (see figure 2.5).

Figure 2.5 shows what one might hope to see according to a Simpsonian/Darwinian view of the world, but it might be contrary to what the stratigraphic evidence warranted. Simpson's cartoons could be called "just-so" stories, though his model for horse evolution was better constrained empirically than that for many scenarios based solely on the paleontological record and recognition of evolutionary changes "by eye." As we shall see, some scientists—cladistic aficionados—take issue with such evolutionary histories. Sheldon's work filled the gaps for a particular series of macrofossils but did not reveal a nice spreading tree such as Darwinian-Simpsonian theory might anticipate.

In fact, well before Sheldon, a quite different, non-Simpsonian but ostensibly Darwinian system for paleontological analysis was already in the marketplace: "phylogenetic systematics" or cladistics, the procedure proposed by the East German

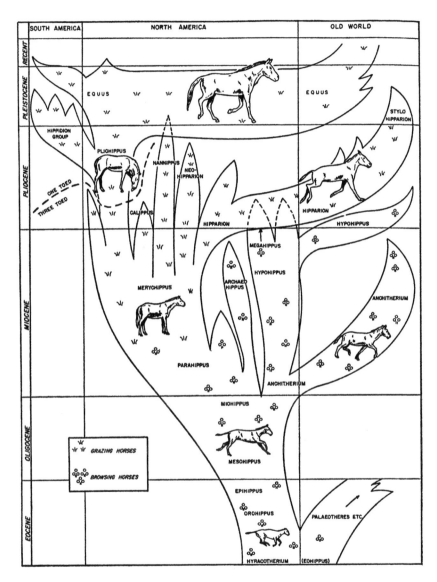

FIGURE 2.5. Lineages of the Horse Family, According to Simpson (1951, 114).
Source: Horses, by George Gaylord Simpson, copyright 1951 by Oxford University Press. Inc. Used by permission of Oxford University Press, Inc.

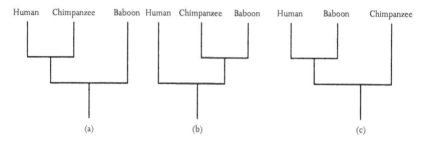

FIGURE 2.6. Possible Cladograms for Humans, Chimpanzees, and Baboons

biologist Willi Hennig (1950, 1965). There are variants of cladistics (Hull 1988), but all claim to be essentially positivist and empirical. The taxonomist takes a group of organisms and examines their biochemical and/or anatomical characters. One might, for example, be interested in the taxonomic and evolutionary relationships of humans, baboons, and chimpanzees. They could be variously grouped, as shown in figure 2.6.

The animals have similarities and differences. We can see that some characters have been *derived* from other states by evolution, as recognized by comparison with other animals. For example, baboons and chimpanzees are hairy; humans (mostly!) are not. But most other mammals have hair. So hairlessness is a derived character. Chimps and humans lack tails; baboons and most other animals have tails. Therefore, absence of a tail is a derived character. Thus humans and chimps share a *derived* character (being tailless), whereas chimps and baboons share a "primitive" character (being hairy). Therefore, in a "cladogram" we should represent humans and chimps as being the most closely related (see figure 2.6a).

The matter can be pursued further by considering other characters: possession of frontal sinuses, short canine teeth, appendix, position of scapula, posture, and so on. Examining as many anatomical features as possible, one can (if dealing with a larger number of species) consider the number of *shared derived characters* identifiable in the several species under consideration and then (usually using a computer) deduce the evolutionary relationships, such that the simplest pattern (most parsimonious) compatible with the data is obtained. All branchings are assumed to be dichotomous. The assumption of parsimony, deployed by the computer in constructing its cladograms, minimizes reversals of characters—such as becoming hairless and then reverting to being hirsute (though such a change *might* occur with alterations in climate). Cladistic technique can also be used by consideration

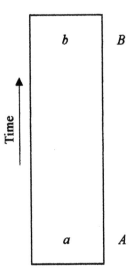

FIGURE 2.7. Representation of Fossils *a* and *b* in Stratigraphic Succession

of biomolecules. For an example of the method deployed on horses, zebras, and donkeys, see Groves and Ryder (2000).

Cladograms do *not* reveal anything about time or the actual course of evolution, as revealed in the fossil record—or at least to one group of cladists they do not. The cladogram bears rather the same relation to the evolutionary tree as does the London Underground map to the actual lines and stations in London. As Henry Gee (2000) has argued (expressing ideas developed by others during the 1970s), one should understand that, for evolution, time (or what he and others call "Deep Time") is, to all intents and purposes, infinite. So the incompletely sampled fossil record does not necessarily reveal what happened in evolutionary history. Because fossil *b* occurs above fossil *a* in a stratigraphic succession, and resembles it, we cannot infer that *b* is *derived* from *a*.

As Gee explains, suppose we have fossil *a* of species *A* found at a stratigraphic horizon below fossil *b* of species *B* (see figure 2.7 and Gee 2000, 146). By the principle of superposition, fossil specimen *b* is younger than fossil specimen *a*. This is compatible with species *B* having evolved directly from species *A*, in what is called "anagenesis" (perhaps as exemplified by Sheldon's trilobites). But it is also possible that *B* could be ancestral to *A*. Alternatively, *A* and *B* could have an unknown common ancestor *C*. So *A* could be older than *B;* or *B* could be older than *A;* or *A* and *B* could have come into being at the same time by division of species *C* into two

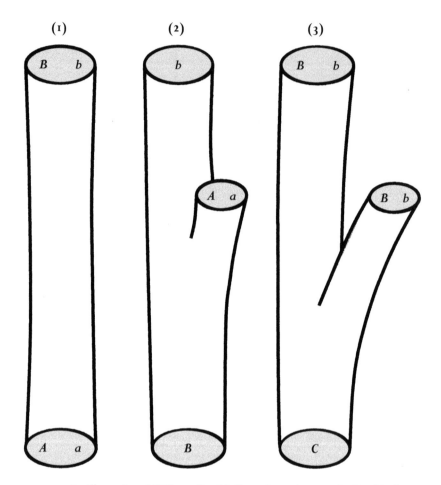

FIGURE 2.8. Illustration of Different Possible Scenarios to Account for Fossil *b* of Species *B* Occurring Higher in the Stratigraphic Column Than Fossil *a* of Species *A*. *Source:* Based on Gee (2000, 146).

new species. The different possibilities are represented in figure 2.8. The possibilities are captured by means of the following cladogram (figure 2.9), which allows (1) *A* to be ancestral to *B* by anagenesis; (2) *B* to be ancestral to *A;* or (3) an unknown species *C* to be ancestral to both *A* and *B*.

If we accept such arguments, it would, according to cladists, put a brake on "evolutionary scenarios" such as those that biologists like Simpson[5] or his colleague Alfred Romer sometimes contemplated. For example, though qualifying his argu-

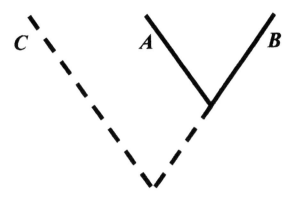

FIGURE 2.9. Cladogram Representing the Three Possibilities Depicted in Figure 2.8

ment with remarks such as "it is possible that" or "it seems not improbable that," Romer (1933) told a gratifying tale of the competition between armored fish and predatory eurypterids in the Silurian and Devonian (over a period of nearly eighty million years). Both types declined during the Devonian, and the armored fish were succeeded by unarmed types. Perhaps, Romer suggested, as the eurypterids declined, the need for fish armor declined concomitantly. Fast-moving carnivorous fish emerged, chasing and eating the eurypterids.

Gee complains, however, that this scenario is speculative and untestable. (And it ignores the existence of the extraordinary giant armored fish such as the Upper Devonian *Dunkleosteus,* which resembles a submarine "warhorse" in its armor.) Gee suggests that Romer's observations could have other explanations—for example, that bony "armor" served to store phosphorus in a phosphorus-deficient environment, the "armor" being merely a side effect. Romer's theory seemed attractive, suggests Gee in his *Deep Time,* because it had similarities to human battles and the rise and fall of empires. But Romer's adaptationist story was untestable and, from a cladistic perspective, unscientific, telling a pseudohistory without the information required for such an historical reconstruction. It was a "cartoon" of the past, illustrating what may happen when one takes Darwinian theory too far, mixing it with paleontology and ignoring the scientific significance of "Deep Time." (But it is hard to interpret *Dunkleosteus* as anything other than an immensely heavily armed fish.)

All this is not to say that cladistics is antievolutionary. On the contrary, cladists think that evolution has occurred in a Darwinian fashion. But the stories that can be told with its assistance should be treated cautiously, with the depth of time in mind. Gee is right, I think, to emphasize the significance of "Deep Time" and the limitations it imposes on our understanding. We must not transcend the evidence, elaborating historical "just-so" stories as we please.

But in the view of Robert Richards (2000), reviewing *Deep Time,* Gee's claim that "no science can be historical" is misguided and incompatible with what sciences such as evolutionary paleontology and stratigraphy actually may accomplish. Richards holds that evolutionary systematists (should) use cladistics as a tool to help them understand genealogical patterns and hypothesize undiscovered ancestral forms. But he commends the formulation of historical scenarios: the methods of historical geology are "logically no different" from those of the historian studying the French Revolution, or whatever.

Here, perhaps, I differ from Richards, in that the historian's documents do have meanings "written on them"—even if they are in code, deliberately mendacious, fragmentary, or muddled up in the archives. But for Gee, fossils, situated as they are in "Deep Time," are (ontologically?) different. They just *are.* In "reading" a history from them, one transcends the evidence: one makes imputations about fossils, as might happen if a feather were found associated with a skeleton and were taken to indicate a capacity for flight, or if what seemed to be "armor" were found on a fish. Of course, a historian imputes meaning to texts, but the connection between meaning and object is much closer when one is dealing with documents, as compared with fossils. I don't think a paleontologist has abandoned naturalism or "gone all metaphysical" if he or she attempts to "read a history" from the rocks. On the other hand, fossils are not one and the same as "medals," though the long use of the metaphor of them as the "medals of creation" has perhaps beguiled us into thinking that the work of the historical geologist *is* essentially the same as that of the general historian, whose documents do "speak" more or less directly.

Though cladistics (in its so-called "reformed" mode, such as Gee espouses) is condemned by many as purely empirical or positivistic, it has, I think, contributed substantially to the development of science, and certainly its techniques are widely used. But some cherished beliefs have been upset in the process. For example, by distinguishing between "clades" and "grades," cladistics also has implications for what we think are natural groupings.[6] We used to smile at Linné having "beasts of burden" as a division of the animal kingdom: horses and camels don't belong to the same "natural kind." But "fishes," and perhaps bats, are not, it seems, natural groupings or "kinds"—each a part of a single clade—either. Fishes don't all have a common lineage. They just happen to live and swim in water. Cladistics shows that "legged" fish like the coelacanth and lungfish are ancestors of tetrapods and belong to a different clade from, say, that which contains the cod and salmon. As tetrapods, we are more closely related to the coelacanth than is the "modern" codfish.

While cladistic critique cautions against Simpsonian–Romerish activities in the matter of the development of evolutionary trees, it has positive features, enabling scientists to do things that would otherwise be impossible. Referring to another of Gee's examples, the paleontologist can't investigate the evolutionary history of roundworm parasites, as they have left no fossil traces. But one can perform a cladistic analysis on the modern roundworm biomolecules, from which it appears that roundworms are not members of a single clade. They are a "grade," for the "invention" of roundworm parasitism has apparently occurred more than once.

Most interesting is what cladistics has to say about birds, another of Gee's examples. What *are* birds? They are animals that have two legs and two wings, have feathers and beaks, can mostly fly, lay eggs, and have lightweight skeletons. Possession of feathers is perhaps the most important trait in people's minds, and when discovered, *Archaeopteryx* was interpreted as a primitive bird. So a Simpsonian would see it as a missing link—a reptile on its progressive way to becoming a bird. In fact, the creation scientist Duane Gish *wanted* evolutionists to read it that way! Thus he pointed out that the discovery of a true bird from the same geological period as the *Archaeopteryx*, reported in 1977, showed that *Archaeopteryx* could not be a bird ancestor (Gish 1978, 87). The famous German fossils were apparently anything but knockdown pro-evolutionary evidence. But "Deep Time" protects evolutionists: as we've seen, they need not suppose that observed stratigraphic order is the same as the evolutionary order.

In fact, *Archaeopteryx* is similar to the theropod dinosaurs (bipedal carnivores), and especially the subset called dromaeosaurs (a fast-running, large-brained group), which, like birds, had air sacs in their bones. The dromaeosaurs occurred in the late Cretaceous, too young to be ancestors of the late Jurassic *Archaeopteryx*. But perhaps there were earlier dromaeosaurs, as yet unfound. This would be compatible with both cladistics and evolutionary theory, but not with a "Simpsonian" evolutionary scenario of progressive evolution from dinosaurs to birds via *Archaeopteryx*.

Recent discoveries from the early Cretaceous in China have revealed primitive feathered and beaked creatures named *Confuciusornis* (Hou et al. 1995) and also a feathered theropod, *Sinosauropteryx*, which, however, is not a dromaeosaur (Hecht 1996; Chen, Dong, and Zhen 1998). This creature had "integumentary structures" seemingly intermediate between hairs and feathers. Another creature found in China, *Caudipteryx* (Qiang et al. 1998), had quasi-feathers on its forelimbs, which, however, were otherwise unadapted for flight. So feathers may have existed before the evolution of birdlike flight. This being so, we can envisage how these ground-runners might have evolved into flying creatures with the help of their preexisting feathers. Thus one of the criticisms of evolutionary theory by creationists begins to lose force: "What use is half a wing or a half-formed feather?" they ask. Of course, we don't *know* what use feathered dinosaurs made of their feathers before they took to the air. Perhaps they kept their eggs warm with their feathered limbs. Perhaps they used them for sexual display. We can (usefully and interestingly) speculate and perhaps test, but we cannot know. There are other birdlike features in the thero-pod dinosaurs. The process of the emergence of flying birds need not have been linear and progressive (or Simpsonian). Perhaps the invention of feathered flight occurred more than once. Perhaps . . . ? The cladist also should beware of attrac-tive "just-so" stories.

So what *is* a bird? Is it a member of the clade that includes all modern birds? If so, it would exclude the flying, feathered *Confuciusornis*—which, however, we would presumably think was a bird if we met one. Or is a bird anything in the clade that includes all creatures down to the nearest modern relatives of birds, namely crocodiles? That would include dinosaurs as birds, which is an unpalatable decision. We might choose our bird clade to include all birds, *Archaeopteryx,* and the most recent common ancestors of all birds. But we don't know what that com-mon ancestor is; and the definition of birds would thus be arbitrary. The ontology of birds is evidently unsteady in the face of cladistic analysis.

Thus cladistics raises perplexing ontological questions. It suggests that a predi-lection for progressive-adaptationist "just-so" stories is not always warranted or helpful, though the search for evolutionary sequences linking isolated fossil discov-eries continues and would appear to be the natural activity of the Darwinian pale-ontologist. Cladistics changes our ontological notions about the nature of kinds, but it does so in a naturalistic manner. It does not imply that there is no such thing as evo-lution or adaptation. However, the methodology of using "the present as the key to the past" (e.g., using *our* concept of what a bird *is* as the basis for understanding avian evolution) becomes suspect. The punctuated equilibrium theorists claimed that they were onto something big when they spoke against metaphysical preferences for phyletic gradualism: no evolution by jerks, please! But cladistic study of the past throws up something of greater methodological-ontological-metaphysical(?) signifi-cance and interest, I think. It can modify ideas about the "being" of natural groups. However, it does not bring God into the picture—rather the reverse. It has to do with the kind of empirical—or some might say hyperempirical—evidence we look for, and what we think we can or cannot do with it.

We have seen that cladistics suggests that because a fossil occurs in an earlier stratum it is not necessarily ancestral to some later, somewhat similar, form. But this does not undermine the methods of stratigraphy based on Oppel zones. The zone concept still holds as a practical tool. With it, stratigraphers can subdivide a succession and effect correlations. They can know the "way-upness" of rocks. In a structurally complex area it is of prime importance to know the stratigraphic succession; then the structure can be elucidated and the geological history told. One can offer complicated "histories"—"just-so" stories if you will—about the movements of plates, collision of terranes, episodes of mountain building, opening and closing of seaways, filling of sediment basins, and so on. We can do geology. But this is not a metaphysical enterprise, except insofar as it maintains a naturalistic posit.

I have assumed here the worth of cladistics for commenting on what can or can not legitimately be done in paleontology. But cladistics has problems too. For example, it regards a species as "that set of organisms between two speciation events, or between one speciation event and one extinction event or that are descended from a speciation event" (Ridley 1989). Or, if you like, it is the organisms that occur between two nodes in a cladogram. This is conceptually clear but not altogether useful from a practical standpoint. One can rarely locate speciation events in the stratigraphic record, though some are claimed (Albanesi and Barnes 2000), and there may be considerable gradual morphological change (anagenesis) between speciations, as the work of Sheldon (1987, 1988) suggests. Consequently, from the point of cladistics two substantially different creatures may have to be regarded as belonging to the same species. So discussions continue about the nature of species.

One final point. There was the question with which we began—the apparent occurrence of catastrophes during earth history, which for d'Orbigny, for example, seemed to separate his stages from one another. The idea that the division of the Cretaceous from the Tertiary was due to a meteorite hitting the earth in Mexico now has many adherents. Indeed, we find authors suggesting that all the major breaks in the stratigraphic column—at least to the level of systems if not stages—were due to extraterrestrial causes (Frankel 1999). Whether this is or is not correct (and it is an idea that the older generation of paleontologists still contests) is an issue for science, not metaphysics (unless you are still worried as to whether we live in a closed Aristotelian cosmos rather than something more open, such as a Newtonian universe). But I don't think the question of whether the breaks in the stratigraphic column are or are not due to catastrophes is a metaphysical issue, despite the efforts of creationists to make it so.

Leaving aside here the ontological problems of microphysics, and assuming a naturalistic perspective in science, the major unresolved (and perhaps unresolvable) metaphysical question is, I submit, the origin or cause of the universe. The nature of biological species is something that is best discussed by biologists or by philosophers who seek to clarify the conceptual issues of the life sciences, not metaphysicians. Even cosmology is best studied scientifically rather than metaphysically, though at present the complexities of that subject seem so great as still to be in the domain that Aristotle thought lay "beyond physics." Paleontology, despite its difficulties and disputes, seems to me to be less problematic and need not be regarded as a metaphysical inquiry.

NOTES

I thank Graham MacKenna (British Geological Survey) for making copies of parts of Oppel's text available to me; Colin Groves for his information on the recent classification of equids and his suggestion for a nice simple example to illustrate cladistic "thinking"; Michiko Yajima for information about Hilgendorf; numerous correspondents who supplied me with references on evidences for evolutionary trends; Robert Richards for discussion of the merits or otherwise of Henry Gee's *Deep Time;* Gerhard Vollmer for supplying me with his list of "twelve theses"; Elisabeth Magnus for valued editorial assistance; and Vittorio Hösle for his invitation to attend an excellent symposium at the University of Notre Dame.

1. Charles Lapworth, MS of lecture, "The Silurian Age in Scotland" (n.d.), 6. Lapworth Archive, Birmingham University.

2. I am informed by Dr Michiko Yajima, Tokyo, who has made a study of Hilgendorf's introduction of Darwin's ideas into Japan and has visited the Steinheim locality near Nördlingen, that the site consists of Miocene freshwater sediments deposited in an old impact crater. I gather that the sequence is highly condensed and that some imagination would be required to "see" smooth transitions in the fossils over time. Even Hilgendorf's own figure (1866, figure between 502 and 503) does not depict an uninterrupted transition of forms. His "tree" for the evolution of *Planorbis* has a markedly conical form in the middle of a sequence of otherwise rather flat coiled shells. On Hilgendorf, see Reif (1983).

3. One might regard the questions of randomness or pattern in evolution, or whether long-term large-scale effects flow from small contingent events, as metaphysical issues, analogous in a way to the old problem of chance and necessity (or even free will and determinism). *Is* the future course of evolution mapped out as a result of what has already happened? Though this question cannot be answered with certainty, so long as the biologist or geologist is open to the different possibilities and does not assume one style of

change a priori, it need not be a metaphysical issue and can be studied empirically with the help of stratigraphy.

4. Conodonts are regarded as the surviving phosphatic teeth from Cambrian-to-Triassic fishlike organisms, which otherwise lacked hard parts. The fish connection is suggested by certain rarely preserved indications of the creatures' soft parts, occasionally found associated with the "toothlike" conodont entities.

5. But Simpson's evidence for horse evolution, based on fossil evidence, was relatively good. The construction of evolutionary scenarios based on the evidence of isolated fossils can be compared with joining up dots to make a picture. However, in the horse family the dots are large, or even dashlike, and it is not unreasonable to join them up to generate working hypotheses. (I am indebted to Colin Groves for this analogy and for the suggestion of using the human/chimp/baboon example to illustrate the cladistic approach in this chapter.)

6. Organisms in a "grade" resemble one another because of evolutionary convergence or parallel evolution rather than common descent.

REFERENCES

Albanesi, G. L., and C. R. Barnes. 2000. Subspeciation within a punctuated equilibrium evolutionary event: Phylogenetic history of the Lower-Middle Ordovician *Paroistodus originalis–P. horridus* complex (Conodonta). *Journal of Paleontology* 74:492–502.

Cain, A. J. 1954. *Animal species and their evolution.* London: Hutchinson University Library.

Chen Pei-Ji, Dong Zhi-Ming, and Zhen Shuo-Nan. 1998. An exceptionally well-preserved theropod dinosaur from the Yixian formation of China. *Nature* 391:147–52.

Cifelli, R. 1969. Radiation of Cenozoic planktonic Foraminifera. *Systematic Zoology* 18:154–68.

Cuvier, G. 1796a. Mémoire sur les éspèces d'élephans tant vivantes que de fossiles, lu à la séance de publique de l'Institut national le 15 Germinal, an IV. *Magasin Encyclopédique* 2 (3): 440–45.

———. 1796b. Notice sur la squelette d'une très-grande éspèce de quadrupède inconnue jusqu'à présent, trouvé au Paraguay, et déposé au Cabinet d'histoire naturelle de Madrid, redigée par G. Cuvier. *Magasin Encyclopédique* 2 (1): 303–10.

———. 1812. *Recherches sur les ossemens fossiles de quadrupèdes, où l'on rétablit le caractères de plusieurs éspèces d'animaux que les révolutions du globe paroissent avoir détruites.* 4 vols. Paris: Déterville.

Darwin, C. 1869. *On the origin of species by means of natural selection, or the preservation of the favoured races in the struggle for life.* 5th ed. London: John Murray.

Dennett, D. C. 1995. *Darwin's dangerous idea: Evolution and the meaning of life.* New York: Simon and Schuster.

D'Orbigny, A. 1849–52. *Cours élémentaire de paléontologie et de géologie stratigraphiques.* 2 vols. in 3, plus *Tableaux.* Paris: Masson.

Eldredge, N., and S. J. Gould. 1972. Punctuated equilibrium: An alternative to phyletic gradualism. In *Models in paleobiology,* ed. T. J. M. Schopf, 82–115. San Francisco: Freeman, Cooper.

Frankel, C. 1999. *The end of the dinosaurs: Chicxulub Crater and mass extinctions.* Cambridge: Cambridge University Press.

Fuller, J. C. M. 2001. Before the hills in order stood: The beginning of the geology of time in England. In Lewis and Knell 2001, 15–23.

Gee, H. 2000. *Deep time: Cladistics, the revolution in evolution.* London: 4th Estate.

Gish, D. T. 1978. *Evolution: The fossils say no!* San Diego: Creation-Life.

Gould, S. J. 1999. Mr Sophia's pony. In *Leonardo's mountain of clams and a diet of worms,* 141–58. New York: Vintage Books.

Gould, S. J., and N. Eldredge. 1977. Punctuated equilibrium: The tempo and mode of evolution reconsidered. *Paleobiology* 3:115–51.

Gradstein, F. M., J. G. Ogg, A. G. Smith, W. Bleeker, and L. J. Lourens. 2004. A new geological time scale, with special reference to Precambrian and Neogene. *Episodes* 27:83–100.

Grandchamp, P. 1994. Deux exposés des doctrines de Cuvier antérieurs au "Discours préliminaire": Les cours de géologie professés au Collège de France en 1805 et 1808. *Travaux du Comité Français d'Histoire de la Géologie,* 3rd ser., 8:13–26.

Groves, C. P., and O. A. Ryder. 2000. Systematics and phylogeny of the horse. In *The Genetics of the Horse,* ed. A. T. Bowling and A. Ruvinsky, 1–24. Oxford: CABI.

Hecht, J. 1996. Downy "dragon" shakes up theories of bird evolution. *New Scientist* 152 (2052): 7.

Hennig, W. 1950. *Grundzüge einer Theorie der phylogenetischen Systematik.* Berlin: Deutscher Zentralverlag.

———. 1965. Phylogenetic systematics. *Annual Review of Entomology* 10:97–116.

Hilgendorf, F. 1866. Über Planorbis Multiformis im Steinheimer Süsswasserkalk: Ein Beispiel von Gestaltveränderung im Laufe. *Monatsbericht der Königlichen Akademie der Wissenschaften zu Berlin,* July, 474–504.

Hou Lian-hai, Zhou Zhonghe, L. D. Martin, and A. Feduccia. 1995. A beaked bird from the Jurassic of China. *Nature* 377:616–18.

Hughes, C. P. 1969. The Ordovician trilobite faunas of the Builth-Llandrindod Inlier, Central Wales. Part I. *Bulletin of the British Museum (Natural History), Geology Series* 18:39–103.

———. 1971. The Ordovician trilobite faunas of the Builth-Llandrindod Inlier, Central Wales. Part II. *Bulletin of the British Museum (Natural History), Geology Series* 20:115–82.

———. 1979. The Ordovician trilobite faunas of the Builth-Llandrindod Inlier, Central Wales. Part III. *Bulletin of the British Museum (Natural History), Geology Series* 32:109–81.

Hull, D. L. 1988. *Science as a process: An evolutionary account of the social and conceptual development of science.* Chicago: University of Chicago Press.

Kovalevsky [Kovalevskii], W. 1873. Sur l'Anchitherium aurelianense Cuv. et sur l'histoire paléontologique des chevaux. Lu le 5 septembre 1872. *Mémoires de l'Académie Impériale des Sciences de St.-Petersbourg,* ser. 7, 20 (5): 1–73.

————. 1876. Monographie der Gattung Anthracotherium Cuv. und Versuch einer näturlich Classification der fossilien Hufthiere, and Osteologie des Genus Anthracotherium Cuv. *Palaeontographica* 22:131–347.

————. 1980. *The complete works of Vladimir Kovalevsky.* Ed. Stephen Jay Gould. New York: Arno Press.

Kucera, M., and J. P. Kennett. 2000. Biochronology and evolutionary implications of late Neogene California margin planktonic foraminiferal events. *Marine Micropaleontology* 40:67–81.

Laporte, L. 2000. *George Gaylord Simpson: Paleontologist and evolutionist.* New York: Columbia University Press.

Lewis, C. E. L., and S. J. Knell, eds. 2001. *The age of the earth: From 4004 BC to 2002 AD.* Special Publication No. 190. London: Geological Society.

Linné, C. von (Carolus Linnaeus). 1744. *Oratio de telluribus habitabilis incremento. Et Andreae Celsii . . . Oratio de mutationibus generalibus quae in superficie corporum coelestium contingunt.* Leyden: Cornelium Haak.

————. 1758. *Systema naturae per regna tria naturae, secundum classes, ordines, genera, species, cum characteribus, differentiis, synonymis, locis.* 10th, rev. ed. Stockholm: Impensis Direct. Laurentii Salvii.

————. 1972. *L'équilibre de la nature: I Oratio de telluris habitabilis incremento (1744) Oeconomia naturae (1749) Politia naturae (1760) II Curiosita naturalis (1748) Cui bono (1752) . . .* Ed. C. Limoges. Paris: Vrin.

Linné, C. von, and P. D. Giseke. 1792. *Caroli a Linné . . . Praelectiones in ordines naturales plantarum: e proprio J. C. Fabricii . . . Msto. edidit P. D. Giseke, . . . Accessit uberior palmarum et scitaminum expositio praeter plurum novorum generum reductiones.* Hamburg: B. G. Hoffmann.

Lipps, J. H. 1970. Planktonic evolution. *Evolution* 24:1–22.

Malmgren, B. A., and J. P. Kennett. 1981. Phyletic gradualism in a late Cenozoic planktonic foraminiferal lineage: D[eep] S[ea] D[rilling] P[roject]. *Paleobiology* 7:230–40.

Monty, C. L. V. 1968. D'Orbigny's concepts of stage and zone. *Journal of Paleontology* 42:689–701.

Oppel, C. A. 1856–58. *Die Juraformation Englands, Frankreichs und des südwestlichen Deutschlands: Nach ihren Einzelnen Gliedern Eingetheilt und Verglichen von Dr. Albert Oppel. Mit einer geognostischen Karte.* Stuttgart: Von Ebner & Seubert.

Qiang Ji, P. J. Currie, M. A. Norell, and Shu-an Ji. 1998. Two feathered dinosaurs from northeastern China. *Nature* 393:753–61.

Rees, M. J. 2001. Understanding the beginning and the end. In Lewis and Knell 2001, 275–83.

Reif, W.-E. 1983. Hilgendorf's (1863) dissertation on the Steinheim planorbids (Gastropoda; Miocene): The development of a phylogenetic research program for paleontology. *Paläontologische Zeitschrift* 57:7–20.

Richards, R. J. 2000. You can't get there from here: How much can we really know about evolutionary processes we weren't around to see? Review of Gee (2000). *New York Times Book Review*, Feb. 27, 32.

Ridley, M. 1989. The cladistic solution to the species problem. *Biology and Philosophy* 4:1–16.

Rioult, M. 1969. Alcide d'Orbigny and the stages of the Jurassic, translated from the French by William A. S. Sarjeant, assisted by Anne Margaret Sarjeant. *Mercian Geologist* 3:1–30.

Romer, A. S. 1933. Eurypterid influence on vertebrate history. *Science* 78:114–17.

Rowe, A. W. 1899. An analysis of the genus *Micraster,* as determined by rigid zonal collecting from the zone of *Rhynconella cuvieri* to that of *Miocraster cor-anguinum. Quarterly Journal of the Geological Society of London* 55:494–547.

Schwartz, J. H. 1999. *Sudden origins: Fossils, genes, and the emergence of species.* New York: John Wiley.

Scott, G. H. 1982. Tempo and stratigraphic record of speciation in *Globorotalia punticulata. Journal of Foraminiferal Research* 12:1–12.

Sheldon, P. R. 1987. Parallel gradualistic evolution of Ordovician trilobites. *Nature* 330:561–63.

———. 1988. Trilobite size–frequency distributions, recognition of instars, and phyletic size changes. *Lethaia* 21:293–306.

Simpson, G. G. 1944. *Tempo and mode in evolution.* New York: Columbia University Press. Facsimile, New York: Hafner, 1965.

———. 1951. *Horses: The story of the horse family in the modern world and through sixty million years of history.* New York: Oxford University Press.

Smith, W. 1815a. *A delineation of the strata of England and Wales, with part of Scotland; exhibiting the collieries and mines, the marshes and fen lands originally overflowed by the sea, and the varieties of soil according to the variations in the substrata. Illustrated by the most descriptive names by W. Smith. To the Right Honble. Sir Joseph Banks, Bart., P.R.S. This map is by permission most respectfully dedicated by his much obliged servant W. Smith, Augst. 1, 1815.* London: J. Cary.

———. 1815b. *A map of the strata of England and Wales, with part of Scotland; exhibiting the collieries, mines, and canals, marshes and fen-lands originally overflowed by the sea, and the varieties of soil, according to the variations in the substrata: Illustrated by the most descriptive names of places and of local districts; showing, also, the rivers, sites of parks, and principal seats of the nobility and gentry, the opposite coast of France, and the lines of strata neatly coloured.* London: J. Cary.

———. 1815c. *A memoir to the map and delineation of the strata of England and Wales, with part of Scotland.* London: John Cary.

Tappan, H., and A. R. Loeblich. 1988. Foraminiferal evolution, diversification and extinction. *Journal of Paleontology* 62:695–714.

Torrens, H. S. 2001. Timeless order: William Smith (1769–1839) and the search for raw materials, 1800–1820. In Lewis and Knell 2001, 61–83.

———. 2002. *The practice of British geology, 1750–1850.* Ashgate: Aldershot.

Vollmer, G. 1994. Was ist Naturalismus? *Logos,* n.s., 1:200–219.

Warren, E. W. 1975. *Porphyry the Phoenician: Isagoge.* Toronto: Pontifical Institute of Mediaeval Studies.

CHAPTER 3

Darwin's A Priori Insight

The Structure and Status of
the Principle of Natural Selection

Christian Illies

Once the nature [of natural selection] has been understood, this principle proves to
be self-evident. It is not an empirical judgement, but a truly *a priori* insight.

Nicolai Hartmann, *Philosophie der Natur*

"**H**ow extremely stupid not to have thought of that," T. H. Huxley
exclaimed once he read the *Origin of Species* (qtd. in Oldroyd
1980, 195). And it is indeed rather surprising that it took such a
long time for the cause of evolution, namely natural selection, to be discovered.
Surprising, because its principle might be found without much empirical research
by mere thinking. At least this is the view of several authors from Darwin's time
onwards. The agriculturist Patrick Matthew, for example, who was credited
by Darwin for having anticipated his and Wallace's theory, considered himself
as having reached the central insight of natural selection in an a priori manner:
"Mr Darwin . . . seems to have worked it out by inductive reason, slowly with due
caution to have made his way synthetically from fact to fact onwards; while with
me it was by a general glance at the scheme of Nature that I estimated this select

production of species as an *a priori* recognizable fact—an axiom requiring only to be pointed out to be admitted by unprejudiced minds of sufficient grasp."[1] A similar point about the principle being knowable a priori has been made by Herbert Spencer, Nicolai Hartmann, and other philosophers since then—mostly, however, based on the assumption that the explanation through natural selection is a mere tautology.

Today, there is a general agreement among philosophers of biology that they are not right and that *natural selection* is not referring to some tautological truism. But what is the structure of natural selection as an explanation? And if this principle is not considered to be tautological, can we still grasp it—or something about it—in an a priori way? These are the two questions I wish to address in my chapter.[2]

To ask this requires a brief clarification of the employed term *a priori*. In what follows, I use the term in the epistemological sense. It marks a kind of knowledge that does not depend upon any evidence or warrant from sense impressions—it must not be the result of an abstraction or of inductive reasoning. (Traditional examples of a priori truths are the truths of mathematics, while examples of a posteriori truths are, of course, the results of scientific research.) For the present purpose, this knowledge-oriented characterization of a priori will suffice. We do not have to enter into metaphysical debates, such as in which sense a priori truths are necessary.

In this chapter I will argue that we can indeed "grasp" the principle of natural selection in an a priori fashion. To do so, I will proceed as follows. First, I will investigate Darwin's critique of special creationism to show the role nonempirical arguments can play in science. In this critique, Darwin himself employs certain a priori principles and demands concerning what a scientific explanation amounts to (like the principle of sufficient reason and the subsequent demand for providing a *vera causa*). Second, I will analyze the structure of natural selection. I will suggest that a core principle carries its main explanatory weight. According to this principle, the persistence of an entity depends upon its environment and upon the properties of this entity. I aim to show that this core principle is not tautological but genuinely explanatory. The final part of this chapter will address the status of natural selection and the core principle. Here I will argue that we can indeed know something about the mechanism of natural selection a priori. This is so because its core principle follows from the principle of sufficient reason once certain conditions are fulfilled. However, we know only empirically whether natural selection is taking place in a certain situation.

DARWIN'S CRITIQUE OF SPECIAL CREATIONISM

It is well known that Charles Darwin returned from his long voyage with two fundamental questions. First, did evolution happen? And, second, if it happened, how? By this time, it was apparent to him that special creationism, the standard nineteenth-century explanation of the variety of species, was not satisfying for the phenomena and relics of the living world. Darwin harbored concrete scientific objections against creationism. Some examples arose from anatomical resemblance between animals of different species, the so-called metamorphic changes seen in vertebrae or in the pistils of flowers and other facts of taxonomy; the geographical distribution of species; homologies and developmental sequences of embryos in the same class; details of the fossil record; and rudimentary organs and other structural and specific puzzles of nature. All these phenomena seemed to be explained much better by descent with modification and migration. They remained obscure on the basis of creationism.

However, Darwin's critique and rejection of creationism as an adequate explanation were also nurtured by the application of some more fundamental and nonempirical principles. First and most important was the principle of sufficient reason. According to this principle, nothing happens without a proper (i.e., sufficient) reason explaining why it does (see, e.g., Darwin 1988a, 123).[3] For natural science that means that every event in nature must be part of a regular, and thus lawlike, order of events. As Darwin remarks, the view of God as a cause of biological phenomena does not follow this principle: instead of giving an explanation, it merely restates the facts in a more dignified way (see Darwin 1988a, 133). For Darwin, scientists put the principle of sufficient reason into practice by looking for the *vera causa* of events, as he called it. (Thus Darwin was following a tradition that started with Newton and found a very clear expression in the work of Thomas Reid and that he got to know mainly through the works of John Herschel and Charles Lyell; see Hodge 1977, 1987.) A *vera causa* is an explanation that satisfies two demands. First, a scientific explanation must provide sufficient cause for the effect in the sense described above, and, second, it must be real or "true" *(verum)*—that is, known to exist even separately from the effect that it is supposed to explain. We must have independent evidence for its existence—namely other observations that show its presence in nature. Newton's gravitational force, for example, was known to exist from observation of the earth and could therefore be a *vera causa* for the planetary orbits. Darwin regarded the demand for a *vera causa* as a crucial constraint and essential standard for all scientific theories and found it not satisfied by special creationism. God is not a creative power observable independent from the creation that it is supposed to explain, nor is he a sufficient cause in the common scientific sense. Immediate acts of creation do not have the strict regularity in time that we expect from a cause in a law-bound system. This explanation neither allows us to make any predictions (because we cannot know God's will) nor adds anything to our knowledge. And, as we might add, *if* God as a special "cause" could be accounted for in a way that did match the criteria for a causal explanation, then he (or this cause) would probably be indistinguishable from a natural law. Special creationism would lose its point.

It is important to stress that the principle of sufficient reason (and the wider demand for a *vera causa*) is in general regarded as a priori and that this is the way Darwin uses it. Within the context of a scientific investigation, this principle is a priori because it is basic and constitutive for the fundamental framework of an empirical science. It is not derived as a scientific result; rather, it states what counts as a result. Science starts from the assumption that we find (causal) order and intelligibility in the world—thus the principle of sufficient reason. This is obvious, since no failure to detect causes in a given case would lead scientists to abandon this principle; rather, it would encourage them to look more carefully for the causes. Moreover, as has been argued by G. J. Warnock (1953), it is impossible even to imagine finding any empirical evidence against this principle.[4] In this sense, we can reformulate Darwin's first criticism: creationism is not a science.

There are two further nonempirical principles underlying Darwin's critique, which are also guidelines for science rather than results from experience or merely "best hypotheses." One of them is that all natural developments are continuous—the law of continuity, known to Darwin as the principle "*natura non facit saltus*" (literally, "nature does not take a jump"). This principle had already been crucial for Charles Lyell, through whom Darwin came to understand its importance.[5] In the way it is used by Lyell and Darwin, it says that all changes in nature are gradual and not sudden occurrences. While Lyell used it to reject Cuvier's catastrophism and to found his uniformist theory of geological change, it also served Darwin as a tool for criticizing creationism in the *Origin of Species*. The thesis that newly created species suddenly enter the stage of the world presupposes exactly the kind of break of continuity that is excluded by the principle. However, the meaning of this principle in the realm of biological evolution cannot be specified as precisely as the principle of sufficient reason (which does not mean that a definition of the latter principle is without difficulties). The law of continuity seems to say, on the one hand, that nature is regularly ordered and governed by laws that operate continuously—in which case it comes very close to the principle of sufficient reason. On the other hand, it can be used to mean that all changes that happen in nature are continuous processes in small steps. There are no major, revolutionary turnovers like Cuvier's sudden geological catastrophes. In the latter case, the focus of the principle is not the demand for a causal order (since even catastrophes might find a full causal explanation) but rather the demand for gradual changes. In this sense Darwin uses it for criticizing creationism and also as a guiding principle for his own research (which means that it is regarded as prior to scientific research and not derived from it).

Another principle behind Darwin's critique is the demand for the simplicity of an explanation. Although the reference to God as a creator makes a single instance responsible for every phenomenon in the world, it requires a rather complex story on many occasions. Darwin asks whether we can, "in truth, feel satisfied by saying each Orchid was created, exactly as we now see it, on a certain 'ideal type.'" It would then seem that God, "therefore, made the same organ to perform diverse functions—often of trifling importance compared with their proper functioning—converted other organs into mere purposeless rudiments, and arranged all as if they had to stand separate and then made them cohere. . . . Is it not a *more simple* and intelligible view that all Orchids owe what they have in common to descent from some monocotyledonous plant . . . ?" (Darwin 1862, 306–7; emphasis mine). Darwin's demand for the most economical explanation is, again, prior to any scientific investigation and not the result of it. And it is a demand that is not satisfied by creationism.

There are two respects in which this principle and *natura non facit saltus* differ from the principle of sufficient reason. First, although they are also regarded as prior to science, they are not indispensable for research in the same way that the latter is. (For the law of continuity, A. G. Kaestner, arguing against Leibniz, already pointed this out in 1751; see Breidert 1976, 1043.) We can, for example, imagine science finding empirical evidence against them. This means that it is a much stronger claim to presuppose the validity of these two principles. Second, they leave much more room for interpretation. What, exactly, is a "gradual" process in comparison to development in jumps? Similarly, how simple must an explanation be to count as such? These questions are profound and Darwin did not consider them in much detail. For our investigation it is enough to mention them and to go on.[6]

THE STRUCTURE OF DARWIN'S EXPLANATION

The Deductive Core of Darwin's Argument

Let us turn to Darwin's explanation of evolution through natural selection.[7] He writes in his *Autobiography* that the *Origin of Species* is but "one long argument" (Darwin 1993, 140). And although Darwin was well known for his noble, Victorian character, this claim is not merely modest but correct: his ingenious and enormously productive insight is centrally one argument in four steps. Roughly, we can state them as follows: (1) Animals or plants produce more offspring than can survive, since the resources they need are limited. The consequence is the struggle for existence. (2) The members of a species always differ in their features to a certain degree; some are better adapted to their environment than others. These will have advantages and thus be positively selected in the struggle for existence. They will gradually replace less adapted ones. (3) Over time the selection of minor variations will accumulate and a new species will develop. (4) If the selection happens in different ways, because the species inhabits an area with a divergent environment, then a parental species can branch into different new ones.

The exact reconstruction of the derivation of this argument, however, is still very controversial. There is ongoing debate concerning what *exactly* the principle of natural selection is — or even whether there is one at all (for this skeptical position, see, e.g., Shimony 1989). I will enter this minefield by bringing up some earlier attempts at reconstruction. Alfred Russel Wallace (1969, 166) had already argued that the structure of natural selection could be derived from some proved facts (e.g., how organisms can multiply rapidly while the total population size stays the same) and their "necessary consequences." Similarly, Julian Huxley (1942, 14) writes in his *Evolution* that "Darwin based his theory of natural selection on three observable facts and two deductions from them." Huxley's reconstruction of the process of deriving the theory of evolution has been refined by Flew (1989); let us look more closely at this account.

According to Flew, the first deductive step is that there is the famous "struggle for existence" (SE) among biological entities once two conditions are fulfilled: first, that there are biological entities with a geometrical rate of increase (GRI),[8] and second, that the resources these entities need are limited (LR). From these premises it follows that

$$GRI + LR \rightarrow SE$$

We must be aware that the increase is only geometrical in general, since the ratio of offspring to parents may vary considerably.[9] What is important is that in all cases there is a tendency or potentiality not merely to a progressive or additive increase but to a multiplicative one. Yet, as it stands, Flew's reconstruction is too vague. We can spell out a third premise (that Flew seems to include in LR), namely that these biological entities depend upon other things and cannot live without them. (In the language of systems theory [Bertalanffy 1968], we could say that organisms are homeostatic open systems that depend on an exchange or interaction with the external world.)[10] This dependency has two aspects, the resources and their availability to the individual entity. Since we cannot limit resources to food but must also include space, air, et cetera, we might use the more general term *environment* to signify the set of all external factors with which the biological entity interacts or that influence its interactions. We can thus make a third premise explicit:

> *The persistence or nonpersistence of a biological entity in time results from its capacity to perform self-sustaining interactions with its environment.*

As it stands, the dependency is expressed for an individual entity only. But it yields the further judgment that the *total quantity* of biological entities that can exist will be a function of the total quantity of possible interactions with its environment. We can, therefore, modify the third premise slightly and come to a "rule of dependency" (RD), as we might call it:

> *The persistence or nonpersistence of a biological entity in time, and thus the maximum quantity of biological entities of this type that can exist, results from its capacity to perform self-sustaining interactions with its environment.*

From these three premises, it follows that it will not be possible for all individuals of some filial generation to persist in time—that is, there will be a "struggle for existence":

$$GRI + LR + RD \rightarrow SE^{11}$$

The second step of Darwin's argument allows for a similar reconstruction. Here the struggle of existence serves as a first premise and must be connected with the premise that there are "heritable variations" (V), as Darwin put it (or "mutations," as we would say today). Not all beings of a particular type of entity are identical; they differ to some degree with regard to their properties or attributes. For this second step, it is important to see the "attributes" of an entity as its capacity to interact with its environment in the broadest sense, extending from consuming resources up to producing other entities. To come to the conclusion of natural selection, we need to be more precise about the premises and spell out a third one (which, again, Flew seems to include in the other ones, this time in V). It is what we might call a "rule of property dependency" (PD):

> *The possible interactions of a biological entity with its environment depend upon the kind of environment and upon the properties of this entity.*

From this it follows that not all individual biological entities have the same capacity to interact with a given environment. This leads to the idea of "natural selection," since some entities (those whose properties are better suited for using the resources) will be able to perform more interactions, others less. Since we know from SE that not all can survive, those with a greater capacity for interaction will be positively selected, the others negatively selected:

$$SE + V + PD \rightarrow NS$$

To summarize our present conclusion, the natural selection of biological entities is a process constituted by two dependency rules, which can operate under certain conditions. The first rule says that existence depends upon interaction and the second that interaction depends upon properties. We can combine the two and state:

> *The persistence or nonpersistence of a biological entity in time, and thus the maximum quantity of entities of this type that can persist, results from the kind of environment and from the properties of this entity.*

In what follows, I will call this the "core principle"' (CP) of Darwinism because it is, if some antecedent conditions are given, necessary for natural selection to happen. Combining the first two steps, we can say:

$$GRI + LR + V + CP \rightarrow NS$$

Let us compare the present result with the current standard definition of natural selection. According to Mayr (1997, 309), natural selection is "the nonrandom survival and reproductive success of a small percentage of the individuals of a population owing to their possession, at that moment, of characters that enhance their ability to survive and reproduce." It is important to be aware of the exact role of the terms in this definition. "Nonrandom survival" (in other formulations "differential reproduction of heritable variations") is a description of *what* is going on in evolution; "owing to their possession . . ." states *why* this happens. Or, as we might say, differential reproduction is the particular *mechanism* of biological evolution, and natural selection (and CP as its essential moment) is the *explanation* of this mechanism. It is important to note that differential reproduction is not itself an explanation. As Brandon (1996, 47) rightly remarks: "Unfortunately, most theorists have confused this mechanism with its Darwinian explanation. The mechanism is compatible with several distinct causal explanations. For example differential reproduction of heritable variation can be due to random drift" (in a similar spirit, see also Williams 1973, 89). Darwin's insight is that natural selection is the *vera causa* of the evolution of species (i.e., for nonrandom survival or differential reproduction)— and this explanation is ultimately referring to the properties of an entity.

The third step of Darwin's argument begins with natural selection as a first premise and adds another one, namely that we are looking at long periods of time during which the environment remains sufficiently stable (T). Then the natural selection of slightly different variations can accumulate and lead to biological improvement (BI) and evolutionary progress:

$$NS + T \rightarrow BI$$

A fourth step is necessary to explain speciation, since the third clarifies evolution of a better form or new species from an old one but not the branching into several new ones. This needs two further premises: first, that a species inhabits a nonhomogenous environment with different selective pressures, so that divergent varieties are favored in different places, and second, that some reproductive barrier between these positively selected varieties must stop them from mixing with the remaining representatives of this type. Otherwise, there will be a swamping-out effect, making branching impossible. This is an important premise that Darwin did not consider sufficiently, although Lord Jenkins and Moritz Wagner brought it to his attention (cf., e.g., Darwin 1988b, 85).

Since natural selection is the fundamental insight behind Darwin's explanation of evolution, I will neglect the third and fourth steps in the subsequent discussion. They are not as central and, at least in the form stated, are only valid for sexually reproducing organisms.

The Explanatory Force of the Core Principle

For an explanatory principle to be fundamental in the strictest sense, it seems necessary to fulfil several criteria. In general, such a principle must be consistent but also simple and not derived from any proposition within the theory. It must have stability and must not itself be subject to possible changes or alterations (see Beatty 1981). Natural selection (and CP at its heart) seems to have exactly this kind of simplicity, priority, and timelessness: it is not itself part of the evolutionary change. (In this context it is noteworthy that the reach of natural selection as a fundamental theory goes beyond biological phenomena. It can be applied *whenever* there are entities with varying properties that can reproduce at a geometrical rate and that depend on limited resources.)[12]

Obviously, to be a fundamental principle of a theory that is supposed to *explain* evolutionary change, natural selection must also have some explanatory content. Yet one might wonder whether the explanation it provides goes very far. In this spirit, some authors have argued that in evolutionary biology all the work is done by other principles, such as those of variation and heredity, while natural selection itself appears to be a "null theory" (Shimony 1989, 256). According to this view, there is no first principle with explicatory value: "Whether the probabilities of being preserved can be ascribed to the varieties of an organism, and with what definiteness and what qualifications, are local and contingent matters" (263).

It is surely right that *this kind* of specific explanation cannot be achieved by Darwin's theory. The insight that natural selection happens — and, in more detail, that natural selection results from several premises — will not by itself answer more concrete questions. They can be settled only by further empirical research: to find out what property of a biological entity allows it to persist in time will require many facts about a situation and the involved entities and will include many general laws of physics, chemistry, heredity, and variation. Consequently natural selection (or CP) does not function as a specific causal law among others (like the law of gravity). Rather, all concrete laws that we need to account for an evolutionary process are additional information about the ways in which the persistence of entities tends to depend upon their properties. (See, e.g., the Hardy-Weinberg law, which requires mathematical knowledge, such as the laws of stochastics, and knowledge of the specific conditions, such as sexual reproduction and random choice of partners, to be applicable.)

But all of this does not show that natural selection or CP contains no genuine explanatory value. Darwin's principle provides the *type* of explanation that evolutionary processes can find. This we might call a metalevel explanatory role: it states that entities' causal interactions with their environment determine their fate (and thus sum up to long-term changes). Thus in the world as it is (and has been), the properties of entities are, *ceteris paribus,* causally responsible for their persistence and proliferation. Consequently natural selection and CP serve as a framework within which other, more concrete explanations can be developed by adding further information. Following Brandon (1996, 51), we might call natural selection a "schematic law" that shows the way knowledge about particular developments can be gained.

It has often been argued that Darwin's theory does not explain anything because it is incapable of making predictions. But for three reasons this objection does not hold. First, the theory allows at least predictions of a hypothetical type: for example, that if the environment changes in a certain way (such as the air becoming polluted), then entities with certain properties (such as bright white wings) are likely to be replaced by entities with more beneficial properties (dark wings)— of course, only if they happen to turn up. Second, the principle of natural selection enables us to make predictions *retrospectively* about events that must have taken place. Obvious examples are the links between different branches in the kingdom of plants and animals. That is why Darwin saw his theory as supported by the discovery of *Archaeopteryx,* a former form of life that must have had existed according to his explanation of evolution. Third and more generally, even if the predictive power of a theory is restricted, this does not automatically nullify its explanatory power. Given our epistemological limitations, explanation and prediction do not have to go together. A theory can fail to permit predictions (e.g., if there are too many future events unknown to us), but it can still explain why things have happened once these events have taken place. In this sense Scriven (1959, 481) rightly remarks on natural selection that "its great commitment and its profound illumination are to be found in its application to the lengthening past, not to the distant future."

However, all of this does not imply that every development can be explained through natural selection. In certain situations, individual biological entities perish or persist due to factors wholly unconnected to their properties. Some moths, for example, though blessed with superior properties, may still be caught in a spider web before proliferating. CP does not imply the Panglossian view, to use the famous phrase of Gould and Lewontin (1979), that all features of everything in the living world are adaptations—that is, necessary interactive elements. The theory allows for neutral properties, possibly even for impeding ones, if supportive ones outweigh them. All of this amounts to saying that even a fundamental principle will be applicable—and thus explain developments—only if several conditions are satisfied. The actual developments that we observe in nature will always be mixedly due to natural selection and to other causes. That is why the dependency of persistence upon properties can be expressed only as a statement of tendency rather than as a strict necessity. (This restriction is mirrored by the explicit or implicit employment of probability that is common in recent formulations of the principle of natural selection; see, e.g., Brandon 1978; Hodge 1987.)[13] The principle gives a *ceteris paribus* explanation for situations in which certain conditions are fulfilled (like entities that can reproduce), but not all situations are of this type.

Why Natural Selection Does Not Contain a Tautological Element.

One might wonder, however, whether the proposed core principle of natural selection fully escapes the famous allegation of being tautological. After all, it seems so obviously true that the persistence of entities in a certain environment is dependent upon their properties that it is questionable whether it has any real content—that is, whether it provides synthetic knowledge.

Let me argue that it does: it is not necessarily true (in the way analytic judgments are) because we can think of counterexamples. Malebrancheanism, for example, is a logically possible theory that makes the persistence of entities dependent not upon their features but directly upon God. We can imagine a voluntarist form of such creationism, according to which God made individual organisms not because of their properties but simply because he wanted to do so. (This is, of course, not the version of special creationism that was current at Darwin's time. According to the tradition of William Paley and the *Bridgewater Treatises,* God had carefully adjusted each species to its environment to make it survive in the best way possible.) One might object that even for the voluntarist, God's craftsmanship would be constrained by the features of his creatures: if some organisms were unsuited for the particular environment, they would perish as soon as they were created (imagine, e.g., a moth that had no organs of respiration). So God too could create only entities with features that would allow them to interact with their environment successfully. But this objection is not cogent. It is logically possible that a Malebranchean God might re-create even the most unsuited biological entities again and again. They might perish in a moment—but he would be free to remove their corpses and replace them permanently by new creatures. Hence things *could* be otherwise than stated in CP.[14]

Another counterexample would be a world of permanent changes. As Scriven rightly observes (1959, 478), "In a world where accidents were extremely frequent and mobility very low, Darwin would never have supported this claim: there would not be enough correlation between the possession of observably useful characteristics and survival to make it plausible." This thought experiment shows that CP is no analytic truth in the sense of being valid for all possible worlds that we might imagine. We can deny CP without logical self-contradiction, since the concept of "properties" does not imply by itself the persistence or nonpersistence of the subjects of which they are predicated. Thus CP is nontautological and can make a claim to providing a genuine explanation.

The accusation of tautology, however, is rightly directed against a certain (explanatory) use of the Spencer's term *survival of the fittest*. Darwin used the term in later editions of the *Origin of Species*.[15] It has notorious difficulties, especially if one is not precise about its exact function. The correct intuition underlying the phrase is that of several competing entities; which ones survive and proliferate is not arbitrary or determined by chance but depends on the properties of the entities. Yet the expression "survival of the fittest" appears to contain *further* substantial information about what these properties are—namely those that "make the entity survive in the struggle for existence." This is no real addition to our knowledge. If persistence in time depends on properties, then we can always add that x's feature "to persist better in time than y" will, *ceteris paribus,* make x persist better in time than y. But this is tautological—it is like adding to the insight that "the color of things depends on their ability to reflect light of a certain kind" the further (pseudo-) information or specification that "green things are those that reflect green light." In this sense it has often been argued, for example by the early Popper (1963), that "survival of the fittest" is tautological without explanatory content, since there is no way to define fitness other than in terms of survival.

This problem is not solved by simply replacing the term by similar ones. We still face the same difficulty when we say that entities are "adaptively superior" or "adaptively more complex" (Maynard-Smith 1969), that they possess the "characters that enhance their ability to survive and reproduce" (Mayr 1997, 309), or that there is a "greater ability of phenotypes to obtain representation in the next generation" (Wilson and Bossert 1971). In all of these cases, the terms used do not really explain anything (at least beyond CP).

To escape from this allegation of tautological emptiness, there have been many suggestions for replacing *fitness* with different characteristics that explain *concretely* why one organism wins in the struggle for existence.[16] And indeed, if this could be done, the charge of tautology would be avoided and we would have gained some substantial knowledge. Yet the problem with these attempts is that nature is not sufficiently uniform to allow for this replacement. Any specification of beneficial properties will remain relative to the environment. It holds in certain situations and fails to hold in others, and many contingent facts about particular situations determine whether it holds or not. And even if there are broad and well-defined classes of situations in which some general claims about supportive properties can be made, it is, again, a matter of contingencies whether an environment belongs to one of these classes. Further, any attempt to substantiate the properties will depend on the investigative framework. Which entity is fitter, one that proliferates enormously, dominates a habitat, and perishes after a short time or one that has never been very successful in terms of numbers but has representatives over the ages, such as the so-called living fossils? Do we look for fitness of genes or of phenotypes? There is no single, general, quantitative definition of fitness or the unity of evolution; every definition remains to a certain extent relative to our question. S. C. Stearns rightly observes that we must ask: "What questions do you want to answer and with which assumptions are you satisfied?" (qtd. in Weber 1998, 202). It does not help to make only comparative judgments about some feature making its possessor "fitter" than another—it is still necessary to make any such judgment relative to the environment and investigative framework.

At this point one might suspect that the expression *fittest* (or *fitter*) contains one item of substantial information that the suggested reconstruction (and CP) neglects, namely that the persistence of entities will always be relative to the persistence of other entities. But this information is also captured by CP. First, the crucial role that competitors play for the chances to survive is an essential part of what has been called the "environment" of a biological entity. Entities are part of each other's environment. Their activities will determine which interactions are possible for a certain entity, given the properties it has and given the properties of all other entities. Second, it is, of course, always possible to infer a comparative judgment on the basis of judgments about different entities. In general terms, we might conclude that more representatives of a biological entity A (e.g., a moth with darker wings) will persist in time than those of an entity B (e.g., a moth with lighter wings) when A possesses features that are more efficient or better suited for some interaction with the environment (e.g., based upon the different camouflaging effects of different wing colors). But any comparative judgment that "A is fitter than B" will be logically derived: it is the general result of the comparison of two noncomparative judgments about the quantities of individual entities A and B that can exist in a certain environment.

We can thus conclude that if the term *fitness* is defined generally enough, then it is of no explanatory value beyond CP. If, on the other hand, it is defined in a more substantial way, it is insufficiently general to explain all biological evolution. It will remain relative, either to the specific situation or to the framework of investigation. In contrast to this, CP has the advantage of being an explanatory principle that is sufficiently general to be true for all investigative frameworks, even if no further quantitative specification is made. And it does not pretend to give information about what the entity properties are.[17]

THE STATUS OF DARWIN'S INSIGHT

Natural Selection and Darwin's Standards for a Satisfying Explanation

Natural selection was the first theory of evolution acceptable in the light of the powerful explanatory ideals for science that Darwin himself had used to criticize creationism. Most importantly, by referring to natural selection, Darwin could offer a *vera causa* for evolution (something he was aware of—see, e.g., the postscript of his letter to George Bentham on May 22, 1863).[18] Both demands for a *vera causa* are met. First, the principle of sufficient reason is satisfied because evolution is explained through causes in the sense of ateleological efficient causes (i.e., the ones we know from physics) and because these causes seem to be sufficient to explain the grand changes in the long run (as Darwin tried to show through his discussion of "variation under domestication"). To be sure, natural selection as a causal mechanism is peculiar and different from other causal explanations in that it allows for many different causal interactions to contribute to the developmental process. CP does not spell out a cause *sui generis* but gives an explanation of how the myriad of causal interactions work together. Still, all of these interactions are within the well-established framework of a lawlike causal order. Darwin's explanation no longer employed directed powers (such as those claimed by Lamarckian theories of an inner goal-directedness). That is also the reason why the second demand for a *vera causa* is satisfied, namely independent evidence for the existence of the causes at work. Especially Darwin's long discussion of variation and selection under domestication was supposed to provide evidence for their reality. He showed that there are variations of the properties among plants and animals and that these properties allow different types of interaction with the environment. The same is true for CP: we find independent evidence for a dependency between properties of entities and their persistence in time in certain situations. We can look, for example, at Darwin's experiments with seeds in saltwater where he wanted to show how long they could survive in an inimical environment. (Obviously, *Darwin* never argued for the independent reality of CP. I have introduced this principle to make the implicit structure of natural selection explicit, but it was not part of Darwin's own understanding of his theory.)

Natural selection is also acceptable by the two other fundamental principles mentioned above. First, the mechanism Darwin saw at work requires no divine miracles through which entirely new structures suddenly come into existence—according to natural selection, nature does not take a jump. Every instance of evolutionary progress happens gradually, and all biological entities are in continuity with their predecessors. In addition, Darwin's explanation stated that the world as a whole changes according to the principle of continuity. We have seen above that natural selection (and CP) would not explain evolution and development in a world of catastrophes where "accidents were extremely frequent." Second, natural selection is a highly economical explanation in that one (simple) mechanism is supposed to account for a multitude of phenomena in areas as different as paleontology, embryology, and taxonomy.[19]

A Priori and Empirical Elements in Natural Selection

Following Julian Huxley's and Flew's suggestion, I have analyzed Darwin's explanation of evolution as being centered around a deductive core. The "long argument" is not inductive and is not based on circumstantial evidence. And since it results from deductive reasoning, its conclusions have a strict necessity but only on the basis of the given premises. Inferences can never be stronger than the premises on which they rest. As Flew (1959, 73) remarks on this point, "Though the argument itself proceeds *a priori,* because the premises are empirical it can yield conclusions which are also empirical."[20] This is surely true for the premises GRI, LR, and V, the three conditions that are necessary for natural selection to take place. But what about CP, which carries the explanatory weight of Darwin's insight? Is anything about this principle knowable a priori even if it is not regarded as tautological?

Let me argue that there is, since CP follows from the principle of sufficient reason if certain conditions hold and can thus be understood by "unprejudiced minds of sufficient grasp" without relying on empirical research. The main condition is that some resources are limited and that this is decisive for the existence of the entities under discussion. If we have excluded that God creates species directly and that there is any internal vitalist force at work, given that both are not satisfying *verae causae*,[21] then the existence or nonexistence of some entity will depend on causal interactions with its environment. This is simply what follows from the principle of sufficient reason. To be sure, this still might take place in either of two ways.

First, it might be what we might call a random event. This is what we might expect in Scriven's world of permanent change (see above) or, similarly, when Cuvier's catastrophes happen. Even then, there would still be a causal explanation for why something perished or persisted in time, but this cause might simply have been effective because it was at a certain point at a certain time. In this case the cause would have nothing to do with the individual properties of the entity.

Or we might have what we might call a "Darwinian situation," where a limited resource was decisive for the existence or nonexistence of some entities. This situation can also be expressed as follows: entities' interaction with this limited resource—how intensely they use the resources—will determine the entities' fate. Interactions, however, are possible only if we presuppose the capacity of entities to interact. These capacities (or dispositions) can also be called the "properties" of an entity. Thus in this situation the properties of the entities will play a pivotal role; as CP states, in this situation the persistence of an entity in time will depend on its properties. Accordingly, having different properties (being a variation) means that different interactions are possible. If two entities compete in a Darwinian situation—one that allows only interactions with the environment to explain the existence or nonexistence of the entities—then the different properties will decide the fate of the entities. All of this is exactly what CP claims. In other words, CP does not result from inductive reasoning but is already *presupposed* by the exact account of the Darwinian situation and the type of causes that are relevant in it. That is why the explanation given by natural selection can be discovered in an a priori way: it is given with the exact account of a certain situation in which interactions with an environment (and limited resources) are seen as decisive.[22] It is no objection against this reconstruction of the a priori element in natural selection that people have not known it all the time. Even a self-evident insight needs to be discovered—and its discovery is often much more difficult than its communication (for the difficulty of arriving at the a priori truth of natural selection, see Hartmann 1950, 646 ff.).

This result is not negated by the fact that the possible interactions of an entity can be expressed only as a statement of tendency. Certainly, there are many contingent facts that explain why some interaction does not take place and some entity perishes independently from the properties it may have (e.g., because it is at the wrong place at a certain time). And similarly, a fortunate spatiotemporal location may allow some interaction though the entity's properties are not well suited. All of this does not show that CP has become invalid. It is rather a situation that is not purely "Darwinian" and will therefore include different, non-Darwinian explanations (and, as stated above, many situations are of this mixed type).

Let me emphasize again that this does not mean that natural selection employs no empirical premises. Without experience, we cannot conclude that the principle will be instantiated in our world, since we cannot know a priori whether there are entities of the required type or Darwinian situations. Science does not presuppose these entities for natural selection to be possible—and there seem to be no other compelling reasons for their existence.[23] It is, consequently, a result from empirical research that biological evolution takes place and also that the evolution of species is explained by natural selection (given other entities, other explanations are imaginable).

The Peculiar Relation between Natural Selection and Its Own Premises

At this point we should mention one peculiar feature of natural selection: natural selection is capable of shaping the empirical world in a way that makes it satisfy the premises that it needs. Natural selection seems to promote (further) instantiations of natural selection. It does so by positively selecting varying entities that reproduce at a geometrical rate. If there are entities with different rates of reproduction, then the entity that reproduces at a higher or even geometrical rate will, *ceteris paribus,* be positively selected. It has a greater probability of being present over time at a greater quantity than an entity that reproduces at a lower rate. (Of course, the exact rate of reproduction that is most successful in a certain environment cannot be predicted. It depends upon all sorts of details, contingent facts, and constraints, which will determine whether a low or high reproduction rate will be more beneficial for the numbers of an entity.) In a similar spirit, we might even argue that natural selection supports reproductive entities over nonreproductive ones in general: given the way the world is, an entity with a reproductive capacity has, *ceteris paribus,* a greater probability of being present in higher numbers than an otherwise identical entity without this capacity. All of this does not mean that natural selection produces the empirical entities that it needs as premises, but it means that their presence is causally supported by natural selection once they have turned up. Furthermore, we might argue for the positive selection of entities that *vary,* another element that we have seen as essential for biological evolution. This argument requires an additional but rather weak assumption, namely that we live in a dynamic, continuously changing world. In this world any environment will alter over time, and this implies that the interactions that are decisive for the persistence of an entity or its offspring may no longer be possible after a certain time. Again, we can conclude that a type of entity that reproduces with some variations will have a higher likelihood that some of its members will persist in time than the members of a type that reproduces without variation. If the entity has slight variations in the next generation, there will be a greater chance that one of them will be sufficiently well equipped for a changing environment. This is, of course, to be taken with a grain of salt, but in the long run the strategy of variation will pay off as long as the world is in permanent change. Of course, the exact degree of variation that is beneficial for existence will again depend upon many empirical aspects of the environment. It also follows that natural selection will be favored if these beneficial variations are heritable. Entities that lack this ability will have to start from scratch each generation, as we might put it.[24] (It is interesting to add that this kind of reasoning also yields the conclusion that there should be a positive selection of Lamarckism when it turns up. If there were an entity that could transmit acquired capacities, then this would seem to have some advantages relative to an entity that had to rely upon the random process of trial and error—its adaptation would be much faster. And, indeed, we do find a Lamarckian pattern of inheritance when it is possible, as in the case of cultural evolution.)[25] This "self-affirming ability," as we might say, of natural selection, that it supports the realization of its own premises, is again something that we can discover without or before looking at the empirical world. We know it because it is implied by the principle of natural selection. Nonetheless, it should find some confirmation in the case of Darwinian evolution in the empirical world. After all, we expect empirical events to be in accordance with correct a priori conclusions. And indeed, we can refer to Manfred Eigen's hypercycle theory (Eigen and Winkler 1975), which is based upon this very idea. One central point behind his principle of self-organization is that, given a certain environment, and given that reproductive structures have turned up, they are positively selected over nonreproductive ones. In a similar spirit, we find observations within evolutionary theory that biological evolvability itself is *likely* to evolve (see Dennett 1995, 220 ff.). Nature seems to be in favor of life. Thus natural selection shows a kind of self-affirming structure: it is a process that causally affects (and supports) its own applicability. This is a noteworthy peculiarity of natural selection that I can only briefly mention here.

To summarize, Flew is right in claiming that we know only empirically that an instance of Darwinian evolution has taken place. This is something the theory of natural selection has in common with other scientific explanations—once we have discovered a physical law, we still need empirical evidence to find instantiations of it. But it is important to stress the special status of Darwin's central insight regarding natural selection and CP. CP is not, like other natural laws, entirely the result of inductive reasoning; instead, it provides an explanatory framework that can be discovered to a certain extent a priori (in the sense outlined above). This is not the case with most, possibly all other scientific theories.

NOTES

1. Quoted in the historical sketch that Darwin included in the *Origin* from the third edition onwards.

2. It is important to stress at the very beginning that I consider Darwin's explanation and neo-Darwinism to be based essentially upon the same reasoning. There are, of course, many differences, most famously Darwin's belief in the inheritance of acquired characteristics. But although our biological knowledge about details has grown immensely and has led to the dismissal of many of Darwin's assumptions, his central idea is still valid.

3. Thus Darwin's usage follows the Kantian one, according to which every natural event, such as the appearance of a new species, must be the result of a natural cause—rather than that of, say, Leibniz or Spinoza, both of whom use the principle with regard to both natural causes *and* rational reasons.

4. There is a far-reaching debate about whether we are justified to do so and why we should regard the principle of sufficient reason as a priori. Is the principle of sufficient reason analytically true? Kant has argued that the proposition "Every effect has a cause" is analytically true while "Every event has a cause" is not (thereby agreeing with Hume): the idea of an event does not contain the idea that something is its reason. Is the principle justified? Kant tried to provide a justification by what he called "transcendental deductions," while Schopenhauer, who argued that his entire philosophy stemmed from this a priori principle of sufficient reason, remarked that it could not possibly be proved (*Über die vierfache Wurzel des Satzes vom zureichenden Grunde*, § 14). We can set these questions aside; for our purpose, it is sufficient to state its a priori character in the sense that it is presupposed by science.

5. The principle of continuity has also been regarded by Leibniz as basic for physics and the scientific understanding of the world. The importance of Lyell for Darwin can hardly be overestimated. Darwin took the first, just published volume of the *Principles of Geology* (1830–33) with him on the *Beagle*, and he was always very conscious of his debt to

Lyell—as he wrote to L. Horner (Aug. 29, 1844), "I always feel as if my books came half out of Lyell's brain" (Darwin 1996, 83).

6. It should be added that, besides these principles, Darwin also gave profound *theological* reasons to dismiss creationism. First, God would be a great betrayer if he created, for example, fossil shells to make us believe that some mussels lived a long time ago (see Darwin 1988a, 120). Darwin's second theological objection was the problem of theodicy. If God created all species individually, then he must be held responsible for all individual acts the animals perform according to their species nature. Darwin mentioned the cat, which "plays" with the mice when it kills them, and the cuckoo, which lays its eggs in the nest of other birds so that the cuckoo's brood can live while the foster brothers must die. It would be much more satisfying, Darwin argued, if God had not created these species directly but were only responsible for the general laws of evolution (Darwin 1988a, 174 ff.). This would not solve the problem of theodicy entirely, but it would make it less oppressive. Third, he argued in a Leibnizian sense that it is in some way more appropriate for God to act through laws than through individual acts of creation: "How far grander than idea from cramped imagination that God created . . . the Rhinoceros of Java & Sumatra. . . . How beneath the dignity of him." (See Darwin 1987, Notebook D, 36/37, p. 343.)

7. It should be noted that this is neither a historical or psychological inquiry into what motivated Darwin nor an account of how he himself understood his achievement. Rather, I will try to uncover the deeper logic of Darwin's insight.

8. I have changed Flew's wording to a certain extent and introduced the term *biological entity,* which enables me to leave the question of the unity of evolution open.

9. *Offspring* must not be understood in too limited a sense. Depending on what sort of biological entities we talk about, whether genes, chromosomes, organisms, kin groups, populations, or species, we will have to change the content of *offspring* appropriately.

10. Expressed in this way, this is true only for organisms as evolutionary units. If we take other biological entities, such as genes, the dependency has to be expressed in different terms.

11. Since this third premise is a regular, general dependency, we can regard it as a law and thus see the first step of Darwin's argument as following the Hempel-Oppenheim scheme.

12. For example, it can be adjusted successfully to cultural phenomena. Dawkins (1976) coined the term *meme* for those minute elements of culture, like an idea or a tradition, that are similarly subject to natural selection. Stephen Toulmin has argued powerfully that natural selection and evolutionary processes apply to theories (including the theory of evolution itself). Others, like Wheeler (1974), go so far to explore the idea of a Darwinian evolution of different cosmic systems. Darwin's explanation has, as Dennett (1995, 50) put it, "substrate neutrality."

13. It should be noted that CP is compatible with a deterministic and an indeterministic view of the world. Darwin himself was a determinist, as we can infer from his

statement that "chance" is merely a disguise of our ignorance of the causes (Darwin 1988a, 95). Many neo-Darwinians, on the other hand, would rather locate us in a world of propensities. (On this subject, see chapter 14 of this book.)

14. The only logically necessary constraint for God is that he can create no impossible (i.e., inconsistent) entities. But this is substantially less than CP claims.

15. Darwin followed the recommendation of A. R. Wallace in order to evade any associations with a selecting agent that the term *natural selection* might suggest.

16. See Sober's (1993, 69–73) introduction to the concept of "fitness" and the discussion by Weber (1998, 178–202).

17. There is, however, one possible way to avoid the dilemma of tautology or limited validity with regard to the expression *survival of the fittest:* if we understand it as a summary formulation of an exhaustive description of all situations, containing all particular facts about the environment and the entities, all causal interactions and all laws that must be applied (e.g., those of physics, chemistry, geology, or biological variation), and all possible investigative frameworks. In this case, the account of *fittest* would provide a fully satisfying explanation of differential reproduction (and thus be an adequate analysis of "natural selection"). But this would require an endless list that was very complex and detailed—more or less a complete record of nature. A job for a Laplacian demon, no doubt. If we took *fittest* as shorthand for all these details, it would indeed give the cause of every single entity and its fate. But given the knowledge that we mortals have, this is not likely to happen. Until then, *survival of the fittest* is not an adequate reconstruction of the *explanation* that natural selection provides.

18. Even if all of this does not show that natural selection is the right explanation of evolution, this being independent from the question of whether it is a *vera causa* explanation. There can be rivalry in *vera causa* explanations, as the recent debate between Dennett and Gould has shown.

19. It is worth noting that even the theological reservations that Darwin harbored are given better responses. Paleontological relicts, rudimentary organs, and the like are precisely what they seem to be. They reveal a story of natural history that involves heredity, development, and migration—rather than some "mockery" created by a maliciously deceptive God. Also, the problem of theodicy becomes less pressing if God acts only indirectly through secondary causes. In a rather Leibnizian spirit, one can argue that the pains resulting from the evolutionary process were the unfortunate but possibly inevitable side effects of a simple, causal law of creation that has made the world evolve. Darwin (1988a, 174 ff.) follows this traditional line of reasoning when he writes that he finds more satisfying the idea that God did not create cats, the cuckoo, or other vicious animals directly.

20. Yet he adds, after looking at the premises, "[A]ll these propositions are nonetheless contingent and empirical for being manifestly and incontestably true" (174).

21. And since no entity can be its own cause. Neither Darwin nor I have argued this, but it follows for the same reasons as those for why God cannot be a proper cause. To

claim that things could "cause" themselves would be a use of *cause* entirely unrelated to the way science uses the term.

22. It must be added, however, that all of this does not show that CP is justified, so long as we have not justified the principle of sufficient reason (cf. n. 4). Since my aim is to analyze Darwin's explanation and not to justify the principle of sufficient reason, I do not want to discuss the problems of its proof in any detail.

23. To be sure, science must transcendentally presuppose the existence of rational beings for its own possibility, but this is a different matter. For example, it does not presuppose that these beings reproduce themselves.

24. Against this, one might want to argue with George C. Williams's (1966) well-known a priori argument against the possibility of the positive selection of a high mutation rate. He states that the mutational rate has "only one possible direction, that of reducing the frequency of mutation to zero" (139). His central argument is that the ability to have high variation among one's offspring cannot be positively selected, since *this ability itself* would then always be altered. But this is not necessarily the case—it is quite possible that a biological entity might possess the ability to have a slightly varying set of produced entities but that this ability *itself* might be generally excluded from being altered. And we can conclude that, at least in some environments, there should be a positive selection for this kind of self-protecting variability.

25. With cultural evolution, it even seems possible that entirely separate cultural traits amalgamate to new and successful forms—another strategy that is stochastically privileged.

REFERENCES

Beatty, J. 1981. What's wrong with the received view of evolutionary theory. In *Proceedings of the Biennial Meeting of the Philosophy of Science Association,* vol. 2, ed. P. Asquith and R. Giere, 397–439. East Lansing, Mich.: Philosophy of Science Association.

Bertalanffy, L. V. 1968. *General system theory.* New York: Braziller.

Brandon, R. 1978. Adaptation and evolutionary theory. *Studies in the History and Philosophy of Science* 9:181–206.

———. 1996. *Concepts and methods in evolutionary biology.* Cambridge: Cambridge University Press.

Breidert, W. 1976. Kontinuitätsgesetz. In *Historisches Wörterbuch der Philosophie,* vol. 4, ed. J. Ritter and K. Gründer. Basel: Schwabe.

Darwin, Charles. 1862. *On the various contrivances by which British and foreign orchids are fertilised by insects and on the good effects of intercrossing.* London: Murray.

———. 1987. *Charles Darwin's notebooks, 1836–1844.* Ed. P. H. Barrett et al. Cambridge: Cambridge University Press.

———. 1988a. *The origin of species, 1859.* Vol. 15 of *The Works of Charles Darwin,* ed. Paul H. Barrett and R. B. Freeman. Washington Square, N.Y.: New York University Press.

———. 1988b. *The origin of species, 1876.* Vol. 16 of *The Works of Charles Darwin,* ed. Paul H. Barrett and R. B. Freeman. Washington Square, N.Y.: New York University Press.

———. 1993. *The autobiography of Charles Darwin.* Ed. Nora Barlow. New York: Norton.

———. 1996. *Charles Darwin's letters: A selection.* Ed. Frederick Burkhardt. Cambridge: Cambridge University Press.

Dawkins, R. 1976. *The selfish gene.* Oxford: Oxford University Press.

Dennett, D. 1995. *Darwin's dangerous idea.* New York: Penguin Books.

Eigen, M., and R. Winkler. 1975. *Das Spiel.* Munich: Piper.

Flew, A. G. 1959. The structure of Darwinism. *New Biology* 28:25–44.

———. 1989. The structure of Darwinism. In Ruse, 1989, 70–84.

Gould, S. J., and R. Lewontin. 1979. The spandrels of San Marco and the Panglossian paradigm: A critique of the adaptationist programme. *Proceedings of the Royal Society* B205:581–98.

Hartmann, N. 1950. *Philosophie der Natur.* Berlin: De Gruyter.

Hodge, M. J. S. 1977. The structure and strategy of Darwin's long argument. *British Journal for the History of Science* 10:237–46.

———. 1987. Natural selection as a causal, empirical, and probabilistic theory. In *The probabilistic revolution,* ed. L. Krueger, G. Gigerenzer, and M. S. Morgan, 233–70. Cambridge, Mass.: MIT Press.

Huxley, J. 1942. *Evolution: The modern synthesis.* London: Allen and Unwin.

Mayr, E. 1997. *This is biology.* Cambridge, Mass.: Belknap Press.

Maynard-Smith, J. 1969. The status of neo-Darwinism. In *Towards a theoretical biology,* ed. H. C. Waddington, 82–93. Chicago: Aldine.

Oldroyd, D. 1980. *Darwinian impacts.* Milton Keynes: Open University Press.

Popper, K. R. 1963. Science: Problems, aims, responsibilities. *Federation Proceedings* 22:961–72.

Ruse, M., ed. 1989. *Philosophy of biology.* New York: Macmillan.

Scriven, M. 1959. Explanation and prediction in evolutionary theory. *Science* 130 (3347): 477–82.

Shimony, A. 1989. The nonexistence of a principle of natural selection. *Biology and Philosophy* 4:255–73.

Sober, E. 1993. *Philosophy of biology.* Dimensions of Philosophy Series. Oxford: Oxford University Press.

Wallace, A. R. 1969. *Natural selection and tropical nature: Essays on descriptive and theoretical biology.* Farnborough: Gregg. (Orig. pub. 1891.)

Warnock, G. J. 1953. Every event has a cause. In *Logic and language II,* ed. A. Flew, 95–111. London: Blackwell.

Weber, M. 1998. *Die Architektur der Synthese.* New York: De Gruyter.

Wheeler, J. A. 1974. Beyond the end of time. In *Black holes, gravitational waves and cosmology: An introduction to current research,* ed. M. Rees, R. Ruffini, and J. A. Wheeler. New York: Gordon and Breach.

Williams, George C. 1966. *Adaptation and natural selection.* Princeton, N.J.: Princeton University Press.

Williams, M. B. 1973. The logical status of the theory of natural selection and other evolutionary controversies. In *The methodological unity of science,* ed. M. Bunge, 84–102. Dordrecht: Reidel.

Wilson, E. O., and W. H. Bossert. 1971. *A primer of population biology.* Sunderland, Mass.: Sinnauer.

CHAPTER 4

Darwinism and Naturalism
Identical Twins or Just Good Friends?

Michael Ruse

L et us start off like good philosophers by asking about the meanings of terms. What is meant by *Darwinism?* What is meant by *naturalism?*

By *Darwinism* I understand a commitment to evolution—that all organisms, living and dead, have slowly emerged from just a few primitive forms, probably ultimately from inorganic matter. I understand also a commitment to natural selection, that the chief mechanism of change is differential reproduction brought on by the struggle for existence, with the major consequence being adaptation. I do not think that Darwinism necessarily implies that selection is the only mechanism or that every last detail of organic life is adapted, but selection as cause and adaptation as effect are the overwhelming factors (Ayala 1985; Ruse 1982, 1984).

By *naturalism* I understand a commitment to unbroken law, excluding miraculous interventions, that leads to explanation and prediction (Ruse 1998). That is to say, speaking sociologically, I understand the kinds of things that go on today in physics and chemistry, the most successful of the natural sciences. Like most people, I would distinguish between *methodological naturalism* and *metaphysical naturalism*. The former is an attitude and practice one has in everyday life in the laboratory and field. The latter is an epistemological and ontological commitment to the nature of knowledge and of the real world (Johnson 1995; Ruse 2001). The metaphysical naturalist denies that there is anything beyond scientific knowledge. Such a person I take to be an atheist (not just an agnostic) and a materialist (inasmuch as materialism makes sense today).

What is the relationship between Darwinism and naturalism? Start with methodological naturalism. I take it that the Darwinian (meaning the person accepting Darwinism) is bound to be a methodological naturalist. Darwinism is a natural science, and methodological naturalism is what natural science is all about. In a way, this seems to me to be an analytic statement like "All bachelors are unmarried." I am not sure that a Darwinian necessarily is going to be a theory reductionist, meaning that ultimately all science is going to be absorbed by physics and chemistry; but, in any case, I do not see that reductionism of this ilk is a necessary component of (a consequence of) methodological naturalism.[1]

Is the methodological naturalist going to be a Darwinian? Even if they all are, this is not an analytic connection. If the world were eternal in an Aristotelian sort of way, the methodological naturalist would not be an evolutionist and hence would be no Darwinian. Of course, the real world as we know it is not eternal. Hence, as a matter of fact, all methodological naturalists could be Darwinians. On this matter, my own feeling is as follows. Given what we do know today, all methodological naturalists will be (should be) evolutionists. Given what we know today, Darwinism is the most sensible position for the evolutionist to take (Ruse 1982). However, I recognize that serious scientists today—scientists whose opinion deserves respect—differ about how far Darwinism should extend (Ruse 2000). Some, and I am one, argue that it should extend a very long way (Dawkins 1976; Dennett 1995; Ruse 1982). Richard Dawkins (1983, 1986), indeed, argues that adaptive complexity can be explained only by selection and that such complexity is ubiquitous. I agree. However, while none would deny such complexity or the significant role of selection in its causal origin, there are those would deny that complexity is ubiquitous. They would point to factors other than selection, like genetic drift and developmental constraints, to explain the nonadaptive (Gould 1980; Lewontin 1974; Kimura 1983; Depew and Weber 1994). These people are methodological naturalists without being full-blooded Darwinians. I think they are wrong, but they are not stupid or scientifically irresponsible.

Turn now to metaphysical naturalism. Everything I said about the relationship between methodological naturalism and Darwinism holds for metaphysical naturalism as well. The metaphysical naturalist today is going to be a Darwinian, although how much and how far is an open question. The interesting question is about the relationship the other way. Is the Darwinian necessarily committed to being a metaphysical naturalist? There are many who think so, from ardent Darwinians like the biologist Richard Dawkins (1986, 1995, 1996) and the philosopher Daniel Dennett (1995) to the evolution-rejecting lawyer Phillip Johnson (1991, 1995) and the Notre Dame philosopher Alvin Plantinga (1991, 1995, 1997). All of these people think that if you are a Darwinian you are well on the way to materialism and atheism. I disagree. I will consider three areas where Darwinism may seem to lead to atheistic materialism, suggesting that in all three cases one can reject the conclusion even though one continues to accept Darwinism. For the purposes of this discussion, I will take the alternative to atheistic materialism to be Christianity of a fairly conservative kind. In the context of a discussion about Darwinism, there are good historical reasons for accepting this dichotomy (Ruse 1996, 1999a, 1999b). But if one wanted to extend the discussion to other religions and faiths (or simply to skeptical agnosticism), I am sure the same or closely related conclusions would obtain. I will assume that the conservative Christian is a traditional Christian and that hence—in the tradition of Augustine and Aquinas, as well as of the great Reformers—an absolutely literal reading of the Bible is neither obligatory nor necessarily warranted. For this reason, here I will ignore the Genesis issues (Ruse 2001, 2003).

First, let us take up the issue of *miracles,* meaning such things or events as the turning of water into wine at Cana, or the raising of Lazarus, or the resurrection of Jesus (Swinburne 1970). The metaphysical naturalist would deny that these events were miraculous—deny them completely, probably—and inasmuch as the Darwinian is a methodological naturalist, it would seem that he or she would go the way of the metaphysical naturalist on this matter. Surely miracles are by definition events that break natural law—water does not spontaneously turn into wine—and hence the Darwinian must turn from them?

There are two (legitimate and respectable) ways you can wriggle out of this problem. First, you can deny that miracles necessarily involve a breaking of nature's laws. You could, for instance, take the water-into-wine episode as symbolic, and more a measure of the happiness that Jesus brought to the wedding feast, perhaps making the father of the bride bring out his really good wine that he was keeping for himself. The Resurrection likewise could be regarded as a resurrection of the spirit—on the third day, the hitherto-dispirited disciples felt Jesus living in their hearts. Beside this fact, the rotting body of Jesus is irrelevant. Second, you can invoke the traditional distinction between the "order of nature" and the "order of grace," arguing that the special events of the Atonement and the like were laid over and on the normal course of nature (McMullin 1993; Ruse 2001). After all, they would hardly have been that miraculous were they common events! Of course, at

some level you are stepping out of science here—out of Darwinism and methodological naturalism—but you are hardly denying Darwinism. Jesus did not make
new species. The point is that you argue that Darwinism only goes so far, and then
the miraculous takes over. Methodological naturalism is being constrained, not denied. Metaphysical naturalism is being denied.

Second, there is the *God* question, something answered negatively by the metaphysical naturalist. Here we need to distinguish reason (natural religion) and faith
(revealed religion). Start with natural religion. Darwinism explains adaptation,
the key premise of the teleological argument for God's existence (Ruse 1999a,
2003). Does this mean that God's existence is denied? Obviously, not in itself.
Darwin himself probably believed in God (in a Deistic fashion) right through the
time he published the *Origin of Species* (Ruse 2000). One simply argues that God designed through the medium of unbroken law. The teleological argument is gone
but not the conclusion. Dawkins, however, argues that the case is stronger than
this. Selection depends crucially on pain and suffering (the struggle for existence)
and leads to horrible adaptations—parasites and so forth. This denies the Christian God (Dawkins 1995, 1997). But one can surely respond in a Leibnizian fashion
that if God created through law, then the pain and suffering were unavoidable side
effects (Reichenbach 1976, 1982). God's powers and love are not denied by that
which he could not do. Indeed, if Dawkins is right, and adaptive complexity demands selection, one has additional reason to argue that the pain and suffering are
not God's fault!

Turn now to revealed religion. Darwinism dispels mystery and mysticism. Is
this not incompatible with the mystery that Christians always locate in and insist
is a central part of their religion? Dan Dennett (1995) argues this. But can one not
counter that animals such as we humans, with adaptations for coming out of the
trees and going onto the plains as scavengers, might not be expected to be able to
peer into the ultimate mysteries of the universe? We have the adaptations needed
for going into the garbage and offal business. Not for looking at God. The important evolutionist J. B. S. Haldane (1927, 208–9) put it well:

> Our only hope of understanding the universe is to look at it from as many
> different points of view as is possible. This is one of the reasons why the data
> of the mystical consciousness can usefully supplement those of the mind in its
> normal state. Now, my own suspicion is that the universe is not only queerer
> than we suppose, but queerer than we *can* suppose. I have read and heard many
> attempts at a systematic account of it, from materialism and theosophy to the

Christian system or that of Kant, and I have always felt that they were much too simple. I suspect that there are more things in heaven and earth than are dreamed of or can be dreamed of, in any philosophy. That is the reason why I have no philosophy myself, and must be my excuse for dreaming.

Finally, *humankind*. Can a Darwinian give humans the special status that is demanded by the Christian? Take the matter of souls. I think that ultimately one has to bite the bullet here but that there is a bullet—or rather there are two bullets—waiting to be bitten. One could argue that souls evolved, up from the first primitive forms (Ruse 2001). One would be inclined to argue this if one tied souls tightly to consciousness and rationality. And if one did this, I see no reason why Darwinism would in any way deny or downplay the significance of souls/minds.[2] Or alternatively (and I think this is the position that really conservative Christians like the pope would take), one could simply argue that souls are a nonempirical concept (John Paul II 1997). At some point in the past, God dropped souls on the human species. Before "time x," no souls; after "time x," souls. It is as simple as that. I confess that this second option seems to me to go against the spirit of Darwinian gradualism, but (given that souls are nonempirical) it does not go against the letter (McMullin 2000; Ruse 2001).

What about the necessity of humans appearing? I doubt that God's plan insists that we have five fingers rather than six, but the very appearance of intelligent life of some form—life that can reason and act morally—is not something that could be left to chance. God is totally free, so our appearance is not logically necessary; but, within his scheme of things, our appearance is not contingent. Oysters are not made in God's image, neither are oak trees. Humans are God-like, even if we are nevertheless limited. Does not Darwinism, being essentially nonprogressive—there is no guarantee that the course of life will be from the monad to the man—go against this and in favor of atheism? Such would certainly seem to be the stand of Stephen Jay Gould (1989, 318): "Since dinosaurs were not moving toward markedly larger brains, and since such a prospect may lie outside the capabilities of reptilian design (Jerison, 1973; Hopson, 1977), we must assume that consciousness would not have evolved on our planet if a cosmic catastrophe had not claimed the dinosaurs as victims. In an entirely literal sense, we owe our existence, as large and reasoning mammals, to our lucky stars."

However, against this I would point out that the really fanatical Darwinians, from Charles Darwin (1859) himself to Edward O Wilson and Richard Dawkins, have always inclined to the view that evolution is progressive and that Darwinism is the cause.[3] First Wilson on progress: "[T]he overall average across the history of life has moved from the simple and few to the more complex and numerous. During the past billion years, animals as a whole evolved upward in body size, feeding and defensive techniques, brain and behavioural complexity, social organization, and precision of environmental control—in each case farther from the nonliving state than their simpler antecedents did" (Wilson 1992, 187). He concludes: "Progress, then, is a property of the evolution of life as a whole by almost any conceivable intuitive standard, including the acquisition of goals and intentions in the behaviour of animals."

Then there is Dawkins, who (like Darwin) argues that there are arms races that pit prey against predator and lead eventually to the emergence of intelligence and moral ability. As the prey gets faster, so also the predator gets faster; as the shell gets thicker, so also the teeth and jaws get stronger. Overall, it is thought that this kind of comparative progress leads to a kind of absolute progress, which is to be expressed in terms of brains and intelligence (Ruse 1996). Dawkins especially draws attention to the way in which military arms races have evolved, from a focus on such things as more efficient armor and weapons of destruction to a focus on the use of computers and similar electronic hardware (Dawkins and Krebs 1979; Dawkins 1986). Similarly, in the animal world, we have the evolution of organisms with ever greater and more powerful on-board computers. Humans or humanlike creatures may not have been absolutely necessary end consequences of evolution—presumably somebody might have won all of the arms races earlier on—but their appearance is far from a matter of brute coincidence. They are just the sorts of things one might reasonably have expected.

I am not sure how much I myself subscribe to these kinds of views.[4] But if you do not like this kind of science, you can always take the Augustinian position of someone like Ernan McMullin, who points out that God supposedly stands outside time and that thought, act, and completed product are as one with him. For God, there is little need of fancy mechanisms of progress. He knew that humans would emerge from life as we have it now here on earth, and that is enough. "It makes no difference, therefore, whether the appearance of *Homo sapiens* is the inevitable result of a steady process of complexification stretching over billions of years, or whether on the contrary it comes about through a series of coincidences that would have made it entirely unpredictable from the (causal) human standpoint. Either way, the outcome is of God's making, and from the biblical standpoint could properly be said to be part of God's plan" (McMullin 1996, 156–57).

God is not simply forecasting on the basis of what will happen. There is an act of creation that unfurls through time for us but that is outside time for God and hence for which beginning, middle, and end are all as one. "Terms like 'plan' obviously shift meaning when the element of time is absent. For God to plan is for the outcome to occur. There is no interval between the decision and completion. Thus the character of the process which, from *our* perspective, separates initiation and accomplishment is of no relevance to whether or not a plan or purpose on the part of the Creator is involved" (157). Hence "the contingency or otherwise of the evolutionary sequence does not bear on whether the created universe embodies purpose or not. Asserting the reality of cosmic purpose in this context takes for granted that the universe depends for its existence on an omniscient Creator" (157).

I rest my case. In all of these areas—miracles, God, and humankind—I argue that although the Darwinian is a methodological naturalist, there is no need for him or her to be forced into metaphysical naturalism, where this is understood as materialistic atheism. One can continue to accept some traditional beliefs about the existence and nature of God and of his intentions for and actions here on earth. So my conclusion is that Darwinism and naturalism are certainly not identical twins and that although Darwinism and methodological naturalism are indeed very good friends, when it comes to metaphysical naturalism the relationship becomes much more distant, if indeed existing at all.[5]

NOTES

1. Ernest Nagel's *The Structure of Science* (1961) is the *locus classicus* for discussions of theory reduction. In my *Can a Darwinian Be a Christian? The Relationship between Science and Religion* (Ruse 2001) I look at other questions to do with reduction in the science/religion discussion. For what it is worth, from my *Philosophy of Biology* (1973) right up to my recent *Darwin and Design: Does Evolution Have a Purpose?* (2003), I have argued that the teleology of evolutionary biology precludes a full theory reduction of biology to physics and chemistry.

2. I take it that this would be a form of Traducianism. See Hodge (1872).

3. The third edition of the *Origin* is the crucial version, for here Darwin starts to insert progressivist sentiments in a big way. See Darwin (1959); also Ruse (1996), Ospovat (1981), and Richards (1987).

4. Actually, it would be disingenuous not to admit that, on both scientific and theological grounds, I am uncomfortable. I am not a lover of notions of biological progress (Ruse 1998), and I always worry about theology being too closely linked to science. Science has a nasty habit of being overturned, and then you are left with no science and bad theology.

5. If I were inclined to write more, then I would go on to consider the question of morality and ethical dictates. Christianity has many such dictates—love your neighbor, for instance—and it is often felt that a nonreligious position, Darwinism in particular,

cannot embrace these moral imperatives. Supposedly, one is driven as a naturalist to ruthless, amoral competition and a lack of any ethical sensitivity. I disagree entirely and have argued the case at length in my *Can a Darwinian Be a Christian?*

REFERENCES

Ayala, F. J. 1985. The theory of evolution: Recent successes and challenges. In *Evolution and Creation*, ed. E. McMullin, 59–90. Notre Dame: University of Notre Dame Press.

Darwin, C. 1859. *On the origin of species*. London: John Murray.

———. 1959. *The origin of species by Charles Darwin: A variorum text*. Ed. M Peckham. Philadelphia: University of Pennsylvania Press.

Dawkins, R. 1976. *The selfish gene*. Oxford: Oxford University Press.

———. 1983. Universal Darwinism. In *Molecules to men*, ed. D. S. Bendall. Cambridge: University of Cambridge Press.

———. 1986. *The blind watchmaker*. New York: Norton.

———. 1995. *A river out of Eden*. New York: Basic Books.

———. 1996. *Climbing Mount Improbable*. New York: Norton.

———. 1997. Human chauvinism. *Evolution* 51 (3): 1015–20.

Dawkins, R., and J. R. Krebs. 1979. Arms races between and within species. *Proceedings of the Royal Society of London, B* 205:489–511.

Dennett, D. C. 1995. *Darwin's dangerous idea*. New York: Simon and Schuster.

Depew, D., and B. Weber. 1994. *Darwinism evolving*. Cambridge, Mass.: MIT Press.

Gould, S. J. 1980. Is a new and general theory of evolution emerging? *Paleobiology* 6:119–30.

———. 1989. *Wonderful life: The Burgess Shale and the nature of history*. New York: Norton.

Haldane, J. B. S. 1927. *Possible worlds, and other papers*. London: Chatto and Windus.

Hodge, C. 1872. *Systematic theology*. London: Nelson.

Hopson, J. A. 1977. Relative brain size and behavior in archosaurian reptiles. *Annual Review of Ecology and Systematics* 8:429–48.

Jerison, H. 1973. *Evolution of the brain and intelligence*. New York: Academic Press.

John Paul II. 1997. The pope's message on evolution. *Quarterly Review of Biology* 72:377–83.

Johnson, P. E. 1991. *Darwin on trial*. Washington, D.C.: Regnery Gateway.

———. 1995. *Reason in the balance: The case against naturalism in science, law and education*. Downers Grove, Ill: InterVarsity Press.

Kimura, M. 1983. *Neutral theory of molecular evolution*. Cambridge: Cambridge University Press.

Lewontin, R. C. 1974. *The genetic basis of evolutionary change*. Columbia Biology Series 25. New York: Columbia University Press.

McMullin, E. 1993. Evolution and special creation. *Zygon* 28:299–335.

————. 1996. Evolutionary contingency and cosmic purpose. In *Finding God in all things,* ed. M. Himes and S. Pope, 140–61. New York: Crossroad.

————. 2000. Biology and the theology of the human. In *Controlling our destinies: Historical, philosophical, ethical, and theological perspectives on the Human Genome Project,* ed. P. R. Sloan, 367–93. Notre Dame: University of Notre Dame Press.

Nagel, E. 1961. *The structure of science: Problems in the logic of scientific explanation.* New York: Harcourt, Brace and World.

Ospovat, D. 1981. *The development of Darwin's theory: Natural history, natural theology, and natural selection, 1838–1859.* Cambridge: Cambridge University Press.

Plantinga, A. 1991. An evolutionary argument against naturalism. *Logos* 12:27–49.

————. 1995. Methodological naturalism. In *Facets of faith and science,* ed. J. M. Van der Meer. Lanham, Md.: University Press of America.

————. 1997. Methodological naturalism. *Perspectives on Science and Christian Faith* 49 (3): 143–54.

Reichenbach, B. R. 1976. Natural evils and natural laws: A theodicy for natural evil. *International Philosophical Quarterly* 16:179–96.

————. 1982. *Evil and a good God.* New York: Fordham University Press.

Richards, R. J. 1987. *Darwin and the emergence of evolutionary theories of mind and behavior.* Chicago: University of Chicago Press.

Ruse, M. 1982. *Darwinism defended: A guide to the evolution controversies.* Reading, Mass.: Benjamin/Cummings.

————. 1984. Is there a limit to our knowledge of evolution? *BioScience* 34 (2): 100–104.

————. 1996. *Monad to man: The concept of progress in evolutionary biology.* Cambridge, Mass.: Harvard University Press.

————. 1998. *Taking Darwin seriously: A naturalistic approach to philosophy.* 2nd ed. Buffalo, N.Y.: Prometheus.

————. 1999a. *The Darwinian revolution: Science red in tooth and claw.* 2nd ed. Chicago: University of Chicago Press.

————. 1999b. *Mystery of mysteries: Is evolution a social construction?* Cambridge, Mass.: Harvard University Press.

————. 2000. *The evolution wars: A guide to the controversies.* Santa Barbara, Calif.: ABC-CLIO.

————. 2001. *Can a Darwinian be a Christian? The relationship between science and religion.* Cambridge: Cambridge University Press.

————. 2003. *Darwin and design: Does evolution have a purpose?* Cambridge, Mass.: Harvard University Press.

Swinburne, R. G. 1970. *The concept of miracle.* New York: Macmillan.

Wilson, E. O. 1992. *The diversity of life.* Cambridge, Mass.: Harvard University Press.

Darwinism's Multiple Ontologies

David Depew

ICONIC DARWINISM AND ITS DISCONTENTS

In recent decades, George Gaylord Simpson's 1959 complaint that "one hundred years without Darwin are enough" has found a measure of resonance. In the wake of the recession of the two other great nineteenth-century systems of thought, Marxism and Freudianism, any number of discourse communities have been "taking Darwinism seriously," in Michael Ruse's phrase.[1] We have Darwinian accounts of emotion, behavior, cognition, language, economics, morality—all topics that for most of the twentieth century were protected from what Daniel Dennett calls "the universal acid" of the theory of natural selection.[2]

This sea change was provoked, I imagine, by discoveries in the mathematics of kin selection. The case of the Hymenoptera, with their lopsided haplodiploid genetic system, has become as paradigmatic of Darwinian thinking as the beaks of Darwin's finches or the differently colored pepper moths of industrial Britain.[3] It has encouraged Darwinians to extend the theory of kin selection to diploid organisms by a variety of additional instruments—game theory, for example, which assures us that in any evolutionary stable population there will be as adequate a supply of doves as of hawks.[4] The result is a proliferation of selectionist explanations of behavior and cognition that collectively have eroded a standing objection that had kept Darwinism safely away from the human scene: that, even in a world conceded to be generally selfish, natural selection seemed incapable of explaining the amount of cooperative and altruistic behavior that we do in fact find among animal and human communities.

The erosion of this barrier has in turn had the effect of destroying the truce that the makers of the modern synthesis had encouraged between physical and cultural anthropology, according to which the former were to be Darwinian and the latter roughly Lamarckian.[5] More generally, it has disturbed the boundary markers of what Bruno Latour calls "the modern constitution," which keeps the natural sciences separated from the social sciences and the social sciences separated from the value-preserving, aesthetics-centered, but decidedly nonscientific humanities.[6] These disturbances have been intensified recently by the now widespread perception that molecular genetics has at last made biology a technoscience, capable for the first time in its history of manipulating objects that were hitherto largely theoretical. In the concomitant recirculation of old topoi about how humans can now direct their own evolution, we cannot help recognizing the recrudescence in biology and psychology of themes from the radical Enlightenment. Such themes were a prominent feature of the physical and chemical sciences in the first flush of discovery but have largely been domesticated in them. In contemporary evolutionary and molecular biology, on the other hand, we are currently hearing a good deal of crowing about how genetic Darwinism "blocks the exits"—Dennett's phrase again[7]—that afford metaphysical solace and religious sanctuary only at the cost of childlike illusion.

These developments have provoked a significant shift in the iconic meaning of the term *Darwinism*. By *iconic meaning* I mean the sort of imagery that crosses from one discipline to another, in part by circulating through the general public.[8] The latest iconic image of Darwinism springs from the perception that the logic of Darwinism requires that *something* must be selfish in the sense of being favored by selection's winnowing process. If cooperating and even self-sacrificing organisms do not fill the bill, then perhaps genes do. Accordingly, today's iconic Darwinism is tilted toward versions of natural selection in which genes, viewed as coding pieces of DNA, are taken to be the primary units over which selection ranges and, in their proliferation, to be the beneficiaries of the selection process. The bits of morphological structure, including neural architecture, that favor successfully proliferating genes are construed as modularized units of function. Variant versions of these can be recombined and recycled in different populations, species, and lineages to form selection-driven adaptations. Among these functions are behaviors, emotions, and various cognitive skills, including perhaps language use.

I do not deny that many eminent genetic Darwinians and philosophers of evolutionary biology reject this picture or that many of those who support some of it may be repelled by the rest. It is, after all, only an icon. What I do maintain, however, is that these days, when strong inferences are made from the very idea of natural selection—"the best idea anyone has ever had," as Dennett puts it[9]—to conclusions that purport to offer definitive solutions to philosophy's questions—metaphysical, epistemological, and normative—by pulling them into the presumably empirical orbit of Darwinism, one or another aspect of the iconic image of Darwinism generally does the persuasive work.

My aim here is to unsettle such inferences. There are two aspects to my argument. The first has to do with the identity, stability, and referential range of the term *Darwinism*. Inferences from the very idea of Darwinism to matters hitherto considered safe from its grasp clearly depend on what Darwinism is taken to be. The term is, I think, essentially contested enough so that we cannot prevent Darwinian rhetors from at least attempting to corner the market on the idea of natural selection. Nice try, we might say, once again borrowing sneers from Dennett.[10] Nonetheless, such attempts anachronistically rely on treating natural selection as an idea that has only recently come to clarity about itself. In the current iconic view, for example, Darwin's frustration with the apparent inability of natural selection at the organismic level to explain altruistic behavior—he was forced to invoke unstable combinations of sexual selection, group selection, and the inheritance of acquired characteristics—could not be relieved until the discovery of genes and the development of kin selection theory gave us something closer to the very *nature* of Darwinism. In a similar vein, Dennett thinks that the "universal acid" that makes Darwinism "reductionism incarnate"[11] had to wait for the emergence of the technological notions of feedback and computational programming to reveal its own identity.

Such "retrospective coronations"—Dennett again, though turned in an unwelcome way on his own thinking[12]—depend, I assert, on a picture in which what I take to be a historical fact is obscured: that the identity of the Darwinian tradition has been knitted and reknitted together in response to a series of contingencies; that at any number of points it might have perished; that this knitting up has been achieved by way of often agonistic conversations, punctuated by temporary peace treaties, in which different bits of empirical data have been assimilated to the core idea of natural selection by way of different, often conflicting conceptual frameworks; and that in consequence the ever-widening array of empirical facts that evolutionary theory has incorporated and will presumably continue to incorporate as best it can—it is, after all, still a progressive research tradition—are best represented as a totality when they are distributed over a plurality of interpretive frameworks, some of which do well on one set of data while others do well on another. The teleological logic of iconic Darwinism may rely tacitly on a retrospectively crowned, and factually inadequate, image of the meaning of the term *Darwinism*. But if I am right about the pluralistic conditions for the historical development of the Darwinian tradition, it might also follow that if the iconic image were to further consolidate its growing hegemony over the Darwinian research tradition it might obstruct the further empirical articulation of the idea of natural selection. Not content merely to stand on the shoulders of giants, genocentric Darwinians might press so heavily on the shoulders of Darwinism's giants that the latter would be pounded down into the earth, where they would be rendered invisible and silent.

The second aspect of my argument contends that appeals to natural selection to "block the exits" to philosophy's traditional monopoly on questions about mind and knowledge (and to the religiously transcendent) depend more on the conceptual schemes adopted by their authors than on any empirical evidence encoded by means of those schemes. Interpretive frameworks, at least the ones I have in mind, have generally to do with the nature of the relationship between the parts (including genes) of organisms and the organisms that contain these parts; between organisms and populations; and between populations and species. Since issues about these relationships bear upon the ontology proper to describing and measuring living things and the processes in which they are imbricated—they are the very stuff philosophy of evolutionary biology is made of—conceptual schemes that figure these relationships can be considered ontologies. These schemes vary. In their variation they provoke intense theoretical quarrels and large extrapolations from known facts. Taken apart from the empirical information and explanations they encode, however, such frameworks are prone to generate transcendental illusions in Kant's sense; when proponents of each framework are left to their own devices, they spin their wheels without gaining sufficient traction in matters of fact to justify the large inferences that the preferred conceptual vocabulary and system of tropes seem to license. This tendency is particularly marked whenever we talk about the implications of Darwinism for matters of behavior, mind, and language.

The antidote is reasonably obvious. The impulse that commends the deployment of a plurality of conceptual schemes in order to advance the empirical adequacy, or at least respectability, of Darwinism is the same impulse that, if followed, will block strong inferences from the alleged meaning of Darwinism to philosophical or religious conclusions. The form of my argument is simple. Such inferences are always the work of preferred interpretive schemes. The very effort to silence other schemes, however, obscures some of the empirical data that must be weighed before any inferences about the implications of Darwinism can be ventured. In consequence, a discursive space must be left open in which are weighed the advantages and disadvantages of *various* schemes for interpreting Darwinism and the larger implications they *might* sustain. Versions of naturalism that collapse this space by demanding too much continuity between sciences and reflective discourses about it are as dangerous as versions of philosophy that insulate metaphysical, epistemological, and normative propositions from the deep insights of evolutionary science.

DARWINISM DEMARCATED

Inferences about Darwinism's alleged ability or inability to dispose of philosophy's traditional problems or religion's traditional aspirations depend on what one means by Darwinism. I have already implied that Darwinism is co-extensive with the idea of natural selection; natural selection, I have said, provides the thread that has tied the Darwinian research tradition together through its many vicissitudes. There is an important distinction to be drawn, however, between the bare notion of natural selection and a potentially open-ended array of different and contesting interpretations by means of which that bare notion has been and presumably will be articulated under the pressure of new discoveries and problems. Thus I want to defend the centrality of natural selection in demarcating Darwinism both from wider interpretations, which tend to identify it with transmutation plus monophyletic descent with modification no matter by what means, whether natural selection or something else, and from narrower interpretations, which presume that if and when it becomes *strenge Wissenschaft* evolution by natural selection can, should be, and even will be articulated in only one way.

There was a moment in Darwinism's career when one might reasonably have circumscribed it solely as commitment to transmutation combined with monophyletic descent of all taxa from a common ancestor. The negative reception that at first greeted *On the Origin of Species* was reversed within a decade by recourse to just such a broad definition. Once it took hold, this broad definition gave rise to a wild assortment of interpretations of Darwinism-as-evolutionism promoted by a no less wild profusion of metaphysical frameworks, from purely idealist to grossly materialist. These interpretations were metaphysical, in the casually Kantian way that I am stipulatively using the term, because the interpretive frameworks tended to drag the empirical evidence around rather than the reverse. After all, there wasn't a lot of empirical evidence available at the time. Soon enough, however, only the materialist interpretations seemed to count as Darwinian; those holding idealist interpretations began to think of themselves as neo-Lamarckians or as neo-Geoffroyeans. It was under a materialist banner, for example, that Clemence Royer's translation of *On the Origin of Species* exported Darwinism to France, that Ernst Haeckel's *Darwinismus* was diffused throughout Germany, and that Chernesevsky and other political radicals promoted "Darwinism without Malthus" in Russia.[13] At home in Britain, and in the United States, where one could not as easily plead misunderstanding due to different cultural contexts, Darwinism was more or less co-opted by Spencerism.

Strikingly, none of these interpretations had much truck with natural selection, and many of them actively opposed it. In reflecting on this well-known fact, we might take a tolerant, permissive view of the identity of Darwinism by agreeing with David Hull, who argues that these self-proclaimed Darwinisms without natural selection should be allowed to count as Darwinian, at least by Darwinians, because by its very nature Darwinism cannot have an essence, only a history;[14] and we might correspondingly disagree with Peter Bowler, who speaks of this diffuse period as "the non-Darwinian revolution."[15] Bowler's claim is somewhat excessive: he neglects the many selectionists who were at work during this period, whose biostatistical inquiries led to the reknitting by the new synthesis of the Darwinian tradition in the twentieth century.[16] I am more ready to concede to Hull, then, that evolutionism of all sorts once *was* Darwinism.[17] At the same time, it is clear that monophyletic descent alone has not been the primary means by which the unity and projectibility of the Darwinian tradition have been carried over some very difficult moments into the present. Most versions of nineteenth-century Darwinism-as-evolutionism failed to generate empirical research programs that had much staying power or were able to give birth, through closely engaged controversy, to successor programs—programs that, with a loss here and there, tended to consolidate the gains made and to move on new problems, such as speciation or the articulation of variation in terms of mechanisms of inheritance. All that was, and still is, the work of the idea of natural selection, and indeed of natural selection encoded within a fairly large range of competing interpretive frameworks.[18]

There is, however, a still more telling reason for demarcating the Darwinian tradition by means of the idea of natural selection and for trying to evade Hull's accusation that this identity involves essentialism by invoking, as I have been doing, Darwinism's conceptual pluralism. The shift away from thinking of Darwinism as monophyletic evolutionism *simpliciter* was not a matter of mere conceptual drift. Nor was its recentering on natural selection when it rose from its rumored death-bed a reprehensible suppression of Darwinism's original pluralism with respect to evolutionary forces. Late-nineteenth-century evolutionism, even when it traveled under Darwin's name, was intrinsically directional: the heated question of the day turned on whether evolution was actually directed or tended by some inner impulse to move from monad to man. Natural selection, by contrast, is at best only accidentally directional and quite possibly not directional at all. The reason is that natural selection ties the very idea of an organism to available variation in a population and to selection pressures operative in a given environment. The more this idea takes hold, the more difficult it is to see any sort of cumulativeness regarding what is more successful, more adapted, more complex, more progressive, or more anything else as one moves from selective context to selective context.

At an extreme, the "cumulativeness problem" can suggest that a strongly directional Darwinism is inconsistent with the very idea of natural selection. If you want your Darwinism to include natural selection, let alone make it primary, you had best give up on direction except in a weak ex post facto sense. This is exactly the inference that tended to take hold as the modern synthesis hardened, although, as Ruse has shown, it was not the private view of most of its pioneers.[19] The reason for this neo-Lyellian dogma, as it might be called, is not hard to find. The modern synthesis gave new life to natural selection by linking it to genetics through a conceptual framework in which organisms were construed as members of statistically identifiable gene-exchanging populations and in which the notion of a species itself was adjusted to "population thinking." The setting in which even such a basic concept of species made any sense at all was biogeographical.[20] The modern synthesis's defense of natural selection was predicated on privatizing, as it were, evolutionary directionality: you might see it from some points of view, but from other points of view you might see no progress at all.

As Ruse and others have recently pointed out, however, a shift away from this view and toward a sort of objective-directionality-pushed-from-behind has become increasingly respectable among Darwinians.[21] Game theory and mathematical population ecology have given us a fairly well-developed notion of evolutionary arms races and in so doing have refreshed Darwin's own intuition that what looks like evolutionary progress is an unavoidable, and hence unintended, consequence of the fact that as occupied niches close, selection is forced, willy-nilly, to produce more complex morphological and psychological entities to exploit available resources.

It is just here that we can vividly see the workings of rival conceptual frameworks *within* contemporary Darwinian discourses. The fact that genes are highly conserved across taxa—arguably one of the greatest discoveries of recent gene-sequencing programs—fits rather nicely into an interpretive framework in which units of genetic-morphological-behavioral functioning are conceived of as modules that can be shuffled and reassembled as natural selection, under the pressure of evolutionary arms races, explores what Dennett calls "design space." It is relatively easy to see on this view how species might be more or less "advanced," in the sense that weaponry is more or less advanced.[22] This idea is rhetorically powerful in part because it preserves the commonsense conviction that there is in fact some sort of evolutionary direction, thereby appealing to a general public that, as Stephen Jay Gould lamented, has always been loath to give up this intuition and has held it against modern Darwinism that it does.

Predictably, this interpretation is less attractive to those who, like Gould, sponsor versions of the modern evolutionary synthesis that stress many-to-many relationships between genes and traits, the relativity of genetic effects to particular contexts, the great breadth of norms of reaction, and statistically based population thinking rather than Lego-block constructionism. Gould's own stress on contingency and irreversibility was simply an extreme case of this protest. At the same time, it must be acknowledged that in the dawning age of genetic biotechnology the burden is increasingly being shifted to those who think that the new genetics does *not* affect the stability or adequacy of the modern synthesis. While those who offer new articulations of genetic Darwinism—Richard Dawkins, for instance, or Dennett—often proclaim that all they are doing is articulating the modern synthesis in a slightly different but equivalent way, their ways of articulating it tend to suggest that in its canonical organism-population-centered formulations, the synthesis of the 1940s is not quite up to the challenges posed by molecular genetics. Clearly, one's opinion on this question depends greatly on one's preferred conceptual framework for encoding the results of genetic Darwinism. This fact calls for some reflection.

DISPUTED ISSUES IN CONTEMPORARY DARWINISM

The synthetic theory, at its most theoretical, was supposed to be sublimely indifferent to what kinds of genetic variation might turn up and what mechanisms might produce it. As for distributing variation once produced, it was assumed that Mendel's laws would not in any significant way be violated. Many advocates of the modern evolutionary synthesis do not think that the ascendancy of molecular genetics and biotechnology has done anything to affect these assumptions. For Darwinians of this persuasion, the reference of the term *gene* was fixed prior to the rise of molecular genetics. It meant "allele." Others, however, such as the "modularists" I have been mentioning, think that genetic Darwinism should be reformulated in terms more accommodating to the new genetics. To be sure, molecularists like to cash in on modern Darwinism's "every snowflake is different" concept in order to tell us that our individual identities are written in our genes—that is, our alleles. On the whole, however, the molecular gene refers to a structure that, while it certainly varies, can also be seen to be "the same" across many species and other taxa. Molecular geneticists refer to just such structures when they now tell us that we have only thirty thousand or so genes rather than the expected eighty to one hundred thousand.[23] There need be no contradiction here; one need only speak of alleles *of* genes. Nonetheless, what is foregrounded about genes by the premolecular modern synthesis—and has of late been defended in terms of an explicit conception of genes as *difference* makers and *difference* explainers rather than constructive elements[24]—is precisely what is put into the background by the molecular account of the gene. Such perspectival shifts bear rather closely on interpretations of particular facts, as well as on opinions about the extent to which Darwinism forces us to revise our picture of the world.

This issue exercises contemporary Darwinians and Darwinism-oriented philosophers mightily. Iconic Darwinians such as Dawkins work hard at devising hybrids between the Mendelian gene concept and the molecular gene concept in order to recast the synthetic theory in a different, presumably more precise language that describes genes as immortal bits of DNA.[25] They think that everything that can be stated in an organocentric-populationist version of Darwinism can be stated in a genocentric framework but that more can be added. Adherents of traditional, organism-population-centered versions of the modern synthesis and their philosophical allies tend to doubt this; perhaps the genocentric view subtracts something important. Developmental systems theorists, meanwhile—at least those who will allow themselves to be called Darwinian—suspect that the solid insights of the modern synthesis about the context dependency of adaptedness, snowflake-like individuality, and the primacy of the organism as an agent in its own world can be preserved in the age of biomechanical reproduction only by reversing the casual complicity of the makers of the modern synthesis in what Evelyn Fox Keller calls the discourse of "gene action."[26]

The developmental systems approach demands a bit of explication. In the eyes of Paul Griffiths, Susan Oyama, Lenny Moss, and others, the notion of a genetic program has become an albatross from which the Darwinian tradition can free itself only by thinking of genes as merely one among a whole raft of developmental resources that interactively come together in the development of an organism.[27] In Lakatosian terms, the genetic program idea is not part of the identity-constituting core of modern synthesis but a contingent, dispensable analogy that is now a drag on the progressive problem-solving prowess of the research program. Ernst Mayr commended the notion of a "genetic program" at a time when that phrase was attached rather innocently to the notion of cybernetic feedback. It helped him resolve the problem of biological teleology. In the era of iconic Darwinism, however, when the hardware-software distinction has become a key trope, the same phrase connotes the idea that organisms are not coupled to their environments and to their own development by massive feedback but are instead printout from a digital-like program.[28] This image seems to compromise the agency of organisms by treating them as passive effects of their own genes. For developmental systems theorists, then, only by jettisoning the excess baggage of "genetic programs" can contemporary Darwinism restore the primacy of the organism and make peace with the developmental biology that the modern synthesis strategically but unfortunately marginalized in midcentury.

I now reach the key point: each of these ways of adjusting empirical innovation to Darwinian tradition clearly uses different ontologies about the relationships between genes and organisms, organisms and populations, and populations and species. Dennett's preference for an aggregative, decompositional, module-centered conceptual framework, for example, undergirds his conception of natural selection as an algorithm and his related defense of the claim that Darwinism is "reductionism incarnate." To block the exits to the philosophically a priori and to the religiously transcendent, all empirical data are to be fitted into a scheme that depends heavily on an analogy to our current technology. Darwinism of this sort is a Darwinism of building blocks. This framework explicitly construes evolutionary adaptation, moreover, as an effect of design, albeit by the "generate-and-test" methods of natural selection, rather than as a process that explains what at best only *looks like* design and is often mere bricolage, which was the dominant view in the modern synthesis. The use of highly decompositional conceptual frameworks then licenses progress-oriented inferences from the starting point when naked RNA began to catalyze itself through the successive ratcheting up of ever more complex organisms. Rhetorically, this view makes it easier to debate creationists; both sides presuppose design and argue head on about who or what the designer might be.

Unfortunately, however, this set of inferences reaches its goal only by subtly converting a framework for processing empirical knowledge into a materialist metaphysics by promising that all the details will eventually be filled in. It thereby converts a useful ontology into a bad metaphysics by appealing to the Enlightenment rhetoric that one can find in Dawkins, Dennett, and, most recently, explicitly, and confidently, E. O. Wilson's *Consilience* to make up the gap between the real and the promised. Wilson virtually confesses that this is so. "Human beings," he writes: "[a]re obsessed with building blocks, forever pulling them apart and putting them back together again. . . . The cutting edge of science is reductionism, the breaking apart of nature into its natural constituents. . . . Given enough components assembled the right way, chemists may someday produce a passable living cell."[29]

On the basis of our presumed assent to this building-block ontology and its past triumphs, we are then called on to give our assent to the following:

> Because human genetics is still in its infancy, there is a near absence of direct links between particular genes and behaviors underlying universal culture traits. . . . As a result, the exact nature of gene-culture coevolution can in most cases only be guessed. . . . These shortcomings are conceptual, technical, and deep. But they are ultimately solvable. Unless new evidence commands otherwise, trust is wisely placed in the natural consilience of the disciplines now addressing the connection between heredity and culture, even if support for it is accumulating slowly and in bits and pieces.[30]

It seems to me that the prudent inference is to suspend judgment, even about conceptual issues. Wilson's faith in "the consilience of disciplines *now* addressing the gene-culture relationship" depends almost entirely on fixing a given conceptual scheme by embalming it in the ambient fluid of Enlightenment ideology. To me it is more likely that by the time all the details *are* filled in they will not fit within one interpretive framework and that even if they do it will not be *this* framework—which, if the littered trail of past scientific speculations is any guide, will have gone the way of the inevitably outdated technological models that form its tropological core.

It is not just, or even primarily, because of its metaphysical excesses, however, that this way of talking about evolution by natural selection is vulnerable. It is more importantly vulnerable because it tends to screen out statistical, populational, and biogeographical considerations that, in other versions of genetic Darwinism, support many-to-many relationships between genes and traits. These considerations throw cold water on the very possibility of considering the effect of genes taken out of a particular selectionist context and counsel Darwinians to shift the ground of the debate rather than to debate design with creationists. Either things can be designed or they can evolve.

The Darwinian research program first developed by Theodosius Dobzhansky within the broader context of the modern synthesis has been prominent in defense of this alternative view. This idea involves the following: first, finding ever new sources of variation—from meiotic recombination to protein polymorphism and, most recently, gene reduplication; second, discovering the clever ways in which natural selection shapes, out of this stupendous array of heritable variation, adaptations that enable populations not only to meet the challenges that directly confront them but to retain variation so that it can be used to meet the challenges of future environments; and, third, reconstructing selection's subtle solutions to the particular problems of particular populations of particular species in particular environments at particular times. Taken together, these considerations tend to interrupt straightforward inferences about progress from selective context to selective context. Typically, solutions to concrete adaptive problems are accessible only to careful, ex post facto narratological reconstructions and exhibit the spirit of a satisfying trade-off rather than sleek optimization. The relationship between sickle cell anemia and malaria in West Africa exemplifies both these characteristics.

One might plausibly view developmental systems theory as fueling this approach by adding still more sources of variation and still more ways of evolving adaptations. Developmental systems theorists regard genes as merely one sort of "developmental resource." Only in combination with a wide array of other such resources do genetic variations enter into the reliable reconstruction of a breeding population across generations. Among these resources might be parts of the environment that are themselves reliably reconstructed by the agency of organisms themselves, which ensures higher heritability in the next reproductive cycle. Developmental systems theorists thus preserve themes that are prominent in Richard Lewontin's version of Dobzhansky's program by adopting a more process-oriented, self-organizing ontology than Lewontin himself might commend.

NATURAL SELECTION AND THE QUESTION OF AGENCY

Nowhere is the rivalry between these broad interpretations of natural selection more intense, more directly relevant to questions about the evolution of cognitive abilities, or more germane to whether issues previously left to the tender mercies of philosophers will be or should be handed over to scientists than on the question

I have just mentioned: the agency of organisms. Are they actors in their own worlds, even makers of these worlds, as Lewontin is sometimes prone to say,[31] or passive pawns of their genes and of environments that force highly constrained behavioral moves on them? It is possible that conflicting views about the evolution of mind and language sort themselves out primarily on the basis of this issue.

The issue of evolution and mind is not a new one, particularly among American Darwinians. If you stood on any corner in Cambridge, Massachusetts, in, say, 1894, and inquired at the top of your lungs whether anyone wanted to talk about Darwinian theories of mental capacities and functions, about as many people might pour out into the street then as in 1994. We owe it to the researches of Robert Richards[32] and, in the case of Dewey, Peter Godfrey Smith[33] that we are now aware of a discourse in which, as early as 1875, William James could write, "Taking a purely naturalistic view of the matter it seems reasonable to suppose that, unless consciousness served some useful purpose, it would not have been superadded to life."[34]

The problem that exercised James and Dewey was social Darwinism. Social Darwinians relied on Spencer's belief that adaptation is the result of the environment's ability to work directly on the developing organism to argue that a strenuously laissez-faire economic environment was the best way to keep humans adaptively tuned to their environments. Spencer viewed organisms as generally passive, embattled, hostages to their surroundings—although he hoped that in the long run better adaptation would make people happier. Obsessed from his youth with free will, James took an exactly opposite view. For him, and even more for Dewey, the problem was to let organisms be problem-solving agents in their own worlds without regressing to vitalism or panpsychism.[35]

James and Dewey were certainly not alone in opposing social Darwinism, the iconic Darwinism of its day, which Dewey called "the ordinary biological theory of society."[36] Unlike religious populists such as William Jennings Bryan, however, James and Dewey did not reject evolutionary theory altogether as a way of protesting social Darwinism. Nor did they resubscribe, as much of the clerical intelligentsia was inclined to do, to vague, ultimately untenable neo-Lamarckian theories of cosmic evolutionary progress. Instead, they opposed social Darwinism by denying that it was good Darwinism in the first place. James and Dewey viewed habits and mental representations as adaptations that enable humans to respond to changing, complex environments by anticipating how to respond to those environments; and they thought of consciousness, and even self-consciousness, as having evolved by natural selection as a way of accessing just these kinds of selective advantage. Dewey's early critique of the reflex arc concept convinced him that the agency of organisms is enhanced, not diminished, by behavioral and cognitive adaptive mechanisms of this sort. For fitness, Dewey wrote, includes "the ability which enables organisms to adjust themselves without too much loss to sudden and unexpected changes in the environment."[37] Natural selection, accordingly, which *ex hypothesi* enhances fitness, will select for the problem-solving capacities Dewey called "intelligence," which he defined as the ability of organisms to anticipate the consequences of their behavior and to guide that behavior toward useful ends by amending their environments. What Dewey called "intelligence" was not a push-pull mechanism, as it was in the empiricist tradition that Dewey rejected in both his non-naturalistic and naturalistic phases. Indeed, Dewey felt that empiricism undermined Spencer from the outset and even peripherally tainted the work of James.[38]

One indication of the authenticity of Dewey's claim to have become a Darwinian, rather than some other sort of evolutionist, is his idea that the most important "influence of Darwinism on philosophy" is its stress on particular responses to particular environments. "If organic adaptations are due simply to constant variation and the elimination of those variations which are harmful in the struggle for existence," he wrote in his famous 1910 essay of the same name, "interest shifts from the wholesale essence back of special changes to the question of how special changes serve and defeat concrete purposes."[39] Readers have often dismissed this claim as blandly uninformative. It is true that Dewey never gave up the notion of higher and lower organisms; capacities that enable organisms to deal with their environments, and even to construct those environments, become incrementally heritable traits of progressively more adapted, because adaptable, lineages and species. At the same time, Dewey's sensitivity to the fact that the reference of ideas to world is constrained by their tie to selective environments was strong enough for him to reject intrinsically directional evolution, which in an uncharacteristically witty remark he called "design on the installment plan," and to develop pragmatism. Ideas, he argued, which necessarily circulate in social space rather than in the isolated, semi-Cartesian minds of individuals, can secure no reference beyond their ability to reconstruct (or retard the reconstruction of) the environment. Thus for good reason pragmatism, until its physics-oriented, behaviorist-leaning rearticulation at midcentury, was known to friend and foe alike as "the biological theory of mind"[40] and was opposed by the philosophical establishment as a threat to its very existence.

James's and Dewey's approach did not fare well after the full force of Weismann's denial of the heritability of acquired characteristics began to sink in. But neither did any other pre-Weismannian theory of evolution. Social Darwinism, which relied heavily on the inheritance of acquired characteristics, began to be displaced by eugenics, and orthogenetic Christian evolutionism drifted ever further toward the fundamentalism that eventually consumed Bryan. To be sure, pragmatist Darwinisms made a brave stand by having recourse to the hypothesis known later as "the Baldwin effect," after the American child psychologist James Mark Baldwin. That notion, advanced simultaneously by three desperate authors at the same time Dewey became a Darwinian, 1895–96, was an attempt to find a middle way between what Dewey called "extreme Weismannism" and Lamarckism. It suggested that if a particularly useful bit of learning could be maintained across generations in a population with some mimetic ability, that bit of learning might stabilize the cultural and physical environment in which the learning took place in such a way that whatever genetic variation subsequently arose would be forced to shift in the direction of learned behavior. If this was so, Baldwin argued in 1896, "the direction at each stage of a species' development *must* be in the direction ratified by intelligence"—intelligence in Dewey's sense—for the simple reason that the fitness of organisms not carrying the trait, whether phenotypically or genotypically, would be lowered.[41]

A clever idea, which has in recent years had its innings again.[42] With the coming of genetics, however, the Baldwin effect, and pragmatic Darwinism as a whole, became a degenerating research program. The founders of the modern evolutionary synthesis were generally skeptical about it. Any novel behavior had to already be licensed by a preexisting, if only occasionally activated, gene complex; and even if a novel behavior was conceded *ex hypothesi* to precede genetic change in a particular circumstance, it was difficult to see why, if the behavior was so effective at the phenotypic level, it would be replaced with genes, which would tend to reduce the very behavioral plasticity they were supposed to foster. It was even more difficult to see *how* this might happen.[43] By then, however, Dewey was in no position to respond to these difficulties. Even though, ironically enough, his own Columbia University was during this entire period the site where Dobzhansky's action-oriented, problem-solving version of the modern synthesis was being articulated, he ceased to follow developments in evolutionary biology after he had made up his mind in the 1890s how it worked; and in the course of having to defend himself against realist philosophers he naturally took to arguing for his account of mental adaptation by increasingly philosophical and decreasingly Darwinian means.[44]

What may have gone overlooked in this story, however, is that, while Dewey did not seem to have been affected by Dobzhansky, Dobzhansky may well have been at least peripherally influenced by Dewey's notion of problem solving. The very conception of organic flexibility that Dewey defended at the phenotypical level was defended by Dobzhansky at the genetic level. Dobzhansky saw adaptive evolution as a problem-solving process. "The method by which a living species responds to environmental challenges is natural selection," he wrote.[45] Literally, remarks like these refer to genetic variation as exploring genotype space for solutions to the adaptive problems set by particular, and particularly capricious, environments. Nonetheless, Dobzhansky clearly meant also to imply that over time natural selection would favor phenotypic adaptations, many of them behavioral and cognitively anticipatory, that would enable organisms, and not just gene pools, to deal with environmental complexity. Indeed, the variation-maintaining principles postulated by Dobzhansky, when combined with recognition that natural selection is a two-step process that begins in each generation with how the organism performs at the phenotypic level, virtually imply, in the words of Robert Brandon, that "[a] range of conditions favoring the evolution of phenotypic plasticity also favors the development of cultural transmission. One would expect a species subjected to such conditions to become highly plastic and to develop culture. . . . A high degree of genetic determination of behavioral traits would be a rarity, not the rule. . . . The behavior of a cultural species could not be predicted simply on the basis of the central tenet of sociobiology."[46]

I think it likely that this view flourished in the pragmatic environment at Columbia in which Dobzhansky worked—an environment permeated with the anti-eugenic democratic thinking that Dewey and the totalitarianism-hating refugee Dobzhansky shared.[47] The same agent-centered vision of the adaptive process was defended by Lewontin, who was schooled by Dobzhansky at Columbia and who differed from his mentor only in ascribing even *more* agency to organisms. "It is not that organisms find environments and either adapt themselves to them or die," Lewontin wrote. "They actually construct their environments out of bits and pieces," both by internalizing the information that enables them to deal with their world and by changing their world to meet their needs and desires.[48] Lewontin, by his own account, saw in Marxist dialectics an antidote to Dobzhansky's overly deterministic-sounding notion that environments posed problems to which organisms had to adjust.[49] He may have heard echoes in this wording of the pop-psychological "adjustment theory," flat behaviorism, and complaisant preference for consensus over conflict into which Dewey's pragmatism had degenerated by the 1950s. In turn, developmental systems theory can be seen as outdoing Lewontin in promoting agency by about as much as Lewontin outdid Dobzhansky, or Dewey outdid James. They were all on the same side, but they saw traces of their opponents' thinking in that of their mentors.

I confess my own attraction to the action-oriented view of organisms. Nonetheless, a fair-minded judge must concede that it is based on an ontology of the person as a deliberator and maker that *precedes* the versions of Darwinism it seeks to commend. It projects this model onto other species. It gets its metaphorical suggestiveness, its normative appeal, and its reassuring quality by intimating that very little that we prize in pre-Darwinian ontologies of the person—suitably amended Aristotelian ontologies, for example—needs to be given up, even after mental and behavioral functions have been naturalized by an encounter with Darwinism. In this vein, Dobzhansky writes, "It has become a commonplace that Darwin's discovery of biological evolution completed the downgrading and estrangement of man begun by Copernicus and Galileo. I can scarcely imagine a judgment more mistaken."[50]

My own view is that the action-oriented account expresses facts about organisms—particularly ethological, biogeographical, and populational facts—that are obscured by building-block materialism. It is very difficult to get back from Dennett's or Dawkins's descriptions of the evolutionary process, for example, to what organisms and populations are actually *doing* in the wild. For built into the building-block framework is always the lurking suggestion that Lewontin identifies and rejects: that the environment and the genes are conspiring to make us do something by producing in us an adaptation that merely leads us to think that *we* are doing it. In Dawkins and Wilson, our genes fool us into cooperation and altruism in a way that is favorable to their interests but only incidentally to ours; and even in Dennett, who has an admirable penchant for thinking of organisms as agents in their worlds—in reviving consideration of the Baldwin effect, for example[51]—there is always an implicit suggestion that in building modular assemblies like us natural selection solves the genes' problems by tricking us into adopting an intentional stance that merely leads us to interpret our environment as a scene of purposive activity.[52] It is far more flattering to say that we really *do* understand our environments and that when we act in these environments it is truly *we,* as integrated organisms, that do the acting. Nonetheless, I admit that in projecting onto all organisms an ontology of the person that bears traces of a prenaturalist past, what seems a reassuring framework for Darwinism can potentially become as question-begging and metaphysical in an invidious sense as the building-block approach of its opponents, as when Lewontin uses Marxian materialism to express it.

In the heat of debate, every conceptual framework probably tends to become excessive, to make up for the knowledge it cannot (yet) encode by calling on the framework itself to make up the difference. Sometimes philosophers or their philosophies are enlisted to freeze-frame the intensively discursive, open-ended discussions in and through which knowledge of the evolutionary process is acquired by putting their a priori blessing on one interpretive framework or another. In science, however, as in human affairs, you cannot get wholesale what must be purchased retail. Evolutionary knowledge, no less than political or economic wisdom, depends on well-informed judgment; and acute judges alone will know how to separate the chaff of conceptual framing from the wheat of knowledge.

I have illustrated Darwinism's reliance on conceptual frameworks by considering two concepts: progress and agency. These are big themes—perhaps too big. One might object to the role I have assigned to multiple interpretive frameworks in articulating Darwinism's key notion, natural selection, by asserting that evolutionary scientists, as opposed to those who package the science for public consumption, do not depend on conceptual framing at all. They simply do the work of observation and experiment, and the work adds up.[53] I do not think, however, that this is so. Even the most technical evolutionary work takes place within research programs whose identities are held together by conceptual frames and obligate metaphors. Even if this were not so, moreover, the question of conceptual framing would necessarily arise whenever even the least hint is given that evolutionary science might have larger implications. But such hints are everywhere and have an influence on the formulation of Darwinian research programs.

NATURALISMS WORTH HAVING AND NOT HAVING

Reasserting my main claim about ontological pluralism leads me to ask what, if any, larger implications of the relationship between Darwinism and metaphysics flow from the pluralism I have postulated as a condition of Darwinism's growth. Surely, one implication must be that Darwinians should reject the view that one conceptual frame or another is uniquely consistent with all possible objects of experience and knowledge. If pluralism is a necessary condition for the pursuit of knowledge about evolution, and indeed for a realistic interpretation of the particulars that constitute that knowledge, Darwinians should oppose those who would epistemologically and metaphysically police the first-order sciences. Darwinian naturalism actually implies Darwinian pluralism. For naturalism has historically asked philosophy to be responsive to the changing play of the natural sciences and to remove the heavy hand of aprioristic metaphysical and epistemological doctrines from the scene of scientific inquiry. Naturalism has thus helped Darwinism secure for itself a space in which it could grow robust enough to deal, as it is now beginning to do, with cognitive functions. One might well ask, then, what sort of naturalism can block philosophical apriorism without at the same time undercutting the conceptual pluralism that, in evolutionary biology at least, is a beneficent midwife to knowledge.

Surely there can be no objection to a naturalistic approach to human behavioral, cognitive, and ethical traits in the sense that Darwin was a naturalist when he looked at worms, orchids, and finches, or Wilson at ants, or Mayr at birds, or Dobzhansky at flies. Natural selection, articulated in one way or another, serves as both a heuristic stimulus to and a suitable background warrant for increasingly good explanations of behaviors, affections, and cognitive capacities considered as natural phenomena exhibited by natural beings in a natural world. The work of a naturalist in this "muddy-boots" sense easily becomes, as a point of professional pride, natural*ism,* in the weakest possible sense, when it embodies what to a practicing naturalist is a reasonable determination never to cry uncle and give in to a supernatural explanation, no matter how tempting it may be. Thus Darwinism has a built-in tendency toward naturalistic, and away from supernatural, explanations.

Since naturalism in this sense has its roots in and gets its warrants from the experience of naturalists, Darwinians of the mind should be expected to make *substantive,* rather than hand-waving, appeals to actual Darwinian research and reasoning. We should not be treated to spectacles in which self-described naturalists swear up and down that they are Darwinians without telling us, or even thinking they *can* tell us, anything much about how such phenomena as consciousness or language *actually* evolved. It is just here, however, that a stronger, perhaps objectionable sense of naturalism sometimes appears to compensate for this missing but always promised information—a scientistic naturalism that asserts that questions once exclusively philosophical will eventually be answered by science and that consequently holds that philosophy will lose both its unique subject matter and its unique discursive forum, if it has not already lost it. The most objectionable aspect of scientistic naturalism is that it is so readily converted into full-flown metaphysical materialism of the sort that, oddly enough, flies in the face of naturalism's chief virtue—its opposition to philosophical apriorism. Scientistic naturalism also conflicts with the conceptually pluralistic conditions under which evolutionary knowledge has been and probably will continue to be acquired. For no matter how sensitive to and reliant on the results of up-to-date science they may be, those who take this view sooner or later will find themselves opting for one interpretive scheme or another on grounds that exceed the work that the scheme does in organizing and explaining particular bits of knowledge and that screen out the bits of knowledge preserved by other frameworks.

Darwinians not only can, therefore, but should reject the collapse of a naturalistic stance into scientistic and metaphysical naturalism. For this view diminishes, or even destroys, the discursive space that must be kept open if discussions about the adequacies and inadequacies of various conceptual frameworks for capturing, analyzing, and diffusing knowledge of the evolutionary process, and for estimating its current and prospective effect on human affairs, are to be carried on. Certainly, discoveries coming from evolutionary psychology and ethology can and should be allowed to throw cold water on, or even rule out, some revered philosophical theories, and may suggest others. But even if one were limited to a purely naturalistic account of all human psychological functions, one's naturalism, since it depends on discussions about the uses and limits of this or that ontology for this or that issue, would still call for a kind of discussion — the kind of discussion to which this essay purports to be a contribution — that would never fully coincide with the work of science itself. If Dobzhansky's and Lewontin's example is apposite, moreover, such discussions will leave room for a far wider range of conceptual frameworks than scientistic naturalists are typically comfortable with. A good case can be made that this forum is philosophical, whether a scientist or a philosopher avails him- or herself of the opportunity to speak in it.[54] This apparently was the position of Dobzhansky himself. Far from subscribing to an approach in which philosophy served as an "underlaborer" to science, he relied on evolutionary scientists to serve "as purveyors of raw materials with which philosophers operate when they formulate and try to solve their problems."[55] Something very close to this view has also been consistently maintained by Dobzhansky's former colleague Marjorie Grene, who has done much to shape philosophy of biology into such a forum.[56] It is also the view that I have advocated in this essay.

NOTES

1. M. Ruse, *Taking Darwin Seriously* (London: Basil Blackwell, 1986).

2. D. Dennett, *Darwin's Dangerous Idea* (New York: Simon and Schuster, 1995), 521.

3. W. D. Hamilton, "The Genetical Evolution of Social Behavior, I and II," *Journal of Theoretical Biology* 7 (1964): 1–52.

4. J. Maynard Smith, *Evolution and the Theory of Games* (Cambridge: Cambridge University Press, 1983); R. Axelrod, *The Evolution of Cooperation* (New York: Basic Books, 1984).

5. The treaty, as I call it, was sealed at the University of Chicago's 1959 centennial commemoration of the publication of *On the Origin of Species,* led by the anthropologist Sol Tax. It was generally agreed by all participants that the genes we have made us cultural beings. In Tax's view, nineteenth-century progressive evolutionism had given biological accounts of cultural evolution a bad name; cultural anthropologists generally avoided them. The modern synthesis, Tax wrote, "brought Darwin and evolution back into anthropology, not by resurrecting analogies, but by distinguishing man as a still evolving species, characterized by the possession of cultures which change and grow nongenetically." Sol Tax, quoted in Vassiliki Betty Smokovitis, "The 1959 Darwin Centennial Celebration in America," *Osiris,* 2nd ser., 14 (1999): 291. Eugenic evolutionists, notably R. S. Fisher and H. Muller, were absent from this celebration and perhaps were consciously excluded from participating in it. A scandal ensued when Julian Huxley came close to some sort of eugenic rhetoric in his keynote address. See Smokovitis, "1959 Darwin Centennial Celebration."

6. B. Latour, *We Have Never Been Modern* (Cambridge, Mass.: Harvard University Press, 1993).

7. Dennett, *Darwin's Dangerous Idea,* 419.

8. I have developed the notion of "iconic Darwinism" from the notion of the gene as "public icon" in P. Beurton, R. Falk, and H.-J. Rheinburger, eds., *The Concept of the Gene in Development and Evolution* (Cambridge: Cambridge University Press, 2000), ix. I believe I am entitled to illustrate the notion of iconic Darwinism by reference to Dennett's quips because Dennett assumes in *Darwin's Dangerous Idea* the voice not of a scientist, or even a philosopher, but a rhetorician: "This book is largely about sciences, but is not itself a work of science. . . . Scientists do, however, quite properly persist in holding forth, in popular and not-so-popular books and essays. When I quote them, rhetoric and all, I am doing what they are doing: engaging in persuasion." *Darwin's Dangerous Idea,* 11. In responding, my voice in this essay is that of a *rhetorical critic.* This means that I have placed what people have said about the assumptions and implications of Darwinism not in a space of context-independent reason giving, but in terms of what it seemed persuasive to someone to say in one particular context or another in order to secure an assumed audience's consent to his or her own views about how best to collect and explain a variety of facts and to draw this audience away from rival interpretations toward one's own.

9. Dennett, *Darwin's Dangerous Idea,* 21.

10. Ibid., 393.

11. Ibid., 82.

12. Ibid., 96.

13. On the reception of Darwinism in France and Germany, see P. Corsi and P. J. Windling, "Darwinism in Germany, France, and Italy," in *The Darwinian Heritage,* ed. D. Kohn (Princeton, N.J.: Princeton University Press, 1985), 683–729. On the Russian reception, see D. Todes, *Darwin without Malthus* (New York: Oxford University Press, 1989).

14. D. Hull, "Darwinism as an Historical Entity: A Historiographic Proposal," in Kohn, *The Darwinian Heritage, 773–812.*

15. P. Bowler, *The Non-Darwinian Revolution* (Baltimore: Johns Hopkins University Press, 1988).

16. See especially J. Gayon, *Darwinism's Struggle for Survival* (Cambridge: Cambridge University Press, 1998).

17. Darwinism as evolution generally is the kind of Darwinism that some creationists, ever anxious to reiteratively reperform nineteenth-century debates, still hope it to be—for at least two reasons: by conflating Darwinism and evolutionism they can, in their public debates, evade the subsequently acquired explanatory power of natural selection; and, in the absent space thus created, they can all the more easily point, as evidence of their belief that Darwinism is a pseudoscience or a religion, to the outbursts of fealty to metaphysical materialism to which Darwinian rhetors are, whether through frustration or triumphalism, occasionally prone. They could not have been more delighted, for example, when Richard Lewontin, more frustrated, I think, than triumphal, acknowledged in a now famous piece in the *New York Review of Books* that commitment to materialism precedes and guides Darwinism. R. Lewontin, "Billions and Billions of Demons," *New York Review of Books,* January 9, 1997, 28–32. In making this rhetorical misstep, Lewontin failed to pause long enough to make clear that his brand of materialism was supposed to be agent centered and agent empowering—active materialism in Marx's sense—and not the reductionistic, atomistic, and Hobbesian passive materialism of his genocentric opponents. Differences between rival views about how selection bears on the agency of organisms gain whatever relevance they have only when evolution generally, and indeed evolution by natural selection, is already presupposed to be a fact. Standard-issue creationists, however, could not be expected to grasp Lewontin's meaning, let alone evaluate it, precisely because for them the transmutation of species was not conceded to be a fact in the first place. To them Lewontin was just another materialist.

18. Darwinism's only rival for sheer historical continuity of this sort is, I think, the structuralist morphological approach to descent from a common ancestor that goes back to Geoffroy, a tradition that is still alive in Brian Goodwin and others. Proponents of structuralist approaches to evolution certainly rely on preferred ontological frameworks: Goodwin, for example, is quite insistent that species are stable natural kinds rather than historical individuals à la Hull. But at least some of its ontological frameworks are sufficiently tied to experimental and observational results that they don't become mere metaphysical surrogates for science, as many late-nineteenth-century versions of evolution certainly did—Henri Bergson's, for example, which was inherited by Pierre Teilhard de Chardin.

19. M. Ruse, *Monad to Man: The Concept of Progress in Evolutionary Theory* (Cambridge, Mass.: Harvard University Press, 1996), 450–55.

20. E. Mayr, *Animal Species and Evolution* (Cambridge, Mass.: Harvard University Press, 1963).

21. Ruse, *Monad to Man,* 483–84.

22. See Dawkins, *The Selfish Gene,* 2nd ed. (Oxford: Oxford University Press, 1989); Dennett, *Darwin's Dangerous Idea.*

23. This announcement gave rise to some amusing reflections in the popular press. Some people worried that thirty thousand genes were not enough to generate the uniqueness of all human individuals—as if eighty to a hundred thousand were much better. Others concluded that identity must not be encoded in the genes after all. Perhaps it was a matter of development, or even of acculturation. The entire discussion suffers from the confusion introduced by biorhetoricians who sought funding for the Human Genome Project in the 1980s by promising that it was the key to human identity. See especially the essays by Walter Gilbert and James Watson in *The Code of Codes,* ed. D. Kevles and L. Hood (Cambridge: Cambridge University Press, 1992).

24. K. Waters, "Why the Anti-Reductionist Consensus Will Not Survive." *PSA* 1 (1990): 125–40.

25. Dawkins, *The Selfish Gene.*

26. E. F. Keller, *Refiguring Life: Metaphors of Twentieth Century Biology* (New York: Columbia University Press, 1995). According to Keller, molecular biology, if left to its own devices, would have arrived at a far less reductionistic view of gene-cell-organism dynamics than today's iconic Darwinians want to impose on it. See E. Keller, "Is There an Organism in This Text?" in *Controlling Our Destinies,* ed. P. Sloan (Notre Dame, Ind.: University of Notre Dame Press, 1997).

27. See P. Griffiths and S. Oyama, eds., *Cycles of Contingency* (Cambridge, Mass.: MIT Press, 2001).

28. See D. Depew, "From Heat Engines to Digital Printouts: A Tropology of the Organism from the Victorian Era to the Human Genome Project," in *Memory Bytes: History, Technology, and Digital Culture,* ed. L. Rabinovitz and A. Geil (Durham, N.C.: Duke University Press, 2003), 50–51.

29. E. O. Wilson, *Consilience* (New York: Alfred A. Knopf, 1998), 50, 54.

30. Ibid., 172–73.

31. R. Lewontin, "Gene, Organism, and Environment," in *Evolution from Molecules to Men,* ed. D. S. Bendall (New York: Cambridge University Press, 1983), 273–85.

32. R. Richards, *Darwin and the Emergence of Evolutionary Theories of Mind and Behavior* (Chicago: University of Chicago Press, 1989).

33. P. Godfrey Smith, *Complexity and the Function of Mind in Nature* (Cambridge: Cambridge University Press, 1996).

34. William James, "*Grundzüge der physiologischen Psychologie* by Wilhelm Wundt," *North American Review* 121 (1975): 201, quoted in Richards, *Darwinism,* 433. Looking back in an autobiographical sketch to the 1880s and 1890s from the vantage point of 1930, Dewey wrote, "We have not yet begun to realize all that is due to William James for . . . the return to a . . . biological conception of the *psyche* [that has been given] new force and value

due to the immense progress made by biology since the time of Aristotle." He meant Darwinism. J. Dewey, "From Absolutism to Experimentalism," in *Contemporary American Philosophers*, ed. G. P. Adams and W. P. Montague (1930; reprint, New York: Russell and Russell, 1963), 2:24.

35. This is Richard Rorty's way of putting it in "Dewey between Hegel and Darwin," in *Consequences of Pragmatism* (Minneapolis: University of Minnesota Press, 1982), 54–68.

36. J. Dewey, review of Lester F. Ward, *The Psychic Factors of Civilization* [1894], in *The Early Works, 1882–1898,* 5 vols. (Carbondale: Southern Illinois University Press, 1967–72), 4:208.

37. J. Dewey, "Evolution and Ethics" [1896], in *Early Works,* 5:41.

38. Ibid., 55:41. In his first flush of enthusiasm for Darwinian psychology, Dewey blames just about everyone for regressing to empiricism—James Mark Baldwin, for example, and Lester Frank Ward, both of whom were, in the larger scheme of things, as much his allies as that other regressor, William James.

39. J. Dewey, "The Influence of Darwinism on Philosophy" [1910], in *Middle Works, 1899–1924,* 15 vols., ed. Jo Ann Boydston (Carbondale: Southern Illinois University Press, 1976–83), 6:31–41.

40. Arthur O. Lovejoy, "The Thirteen Pragmatisms," *Journal of Philosophy* 5 (1908): 36–39.

41. J. M. Baldwin, "A New Factor in Evolution," *American Naturalist* 30 (1996): 441–51, 536–53. See also C. Lloyd Morgan, *Habit and Instinct* (London: Arnold, 1896), and "Of Modification and Variation," *Science* 4 (1896): 99: 733–39; H. F. Osborn, "A Mode of Evolution Requiring Neither Natural Selection nor the Inheritance of Acquired Characteristics," *Transactions of the New York Academy of Science* 15 (1896): 141–48.

42. See D. Depew, "Baldwin and His Many Effects," in *Evolution and Learning: The Baldwin Effect Reconsidered,* ed. Bruce H. Weber and David Depew (Cambridge, Mass.: MIT Press/Bradford Books, 2003), 3–31.

43. G. G. Simpson, "The Baldwin Effect," *Evolution* 7 (1953): 110–17.

44. On this point, I follow R. Rorty, "Dewey's Metaphysics," in *Consequences of Pragmatism.*

45. T. Dobzhansky, *The Biology of Ultimate Concern* (New York: New American Library, 1967), 41.

46. R. Brandon, "Levels of Selection: A Hierarchy of Interactors," in *The Role of Behavior in Evolution,* ed. H. C. Plotkin (Cambridge, Mass.: MIT Press, 1988), 82; reprinted in R. Brandon, *Concepts and Methods in Evolutionary Biology* (Cambridge: Cambridge University Press, 1996).

47. It is true that Dobzhansky brought much of his conception of organisms as agents with him from Russia and that he was influenced on this point especially by I. I. Schmalhausen. It is also true, as scholars have pointed out, that Dobzhansky got the challenge-response model from reading Toynbee's account of history and meeting with him. See

C. Krimbas, "The Evolutionary Worldview of Theodosius Dobzhansky," in *The Evolution of Theodosius Dobzhansky,* ed. M. Adams (Princeton, N.J.: Princeton University Press, 1994), 186. Nonetheless, Dobzhansky was talking this way long before his personal and literary encounter with Toynbee.

48. R. Lewontin, *Biology as Ideology* (New York: Harper Perennial, 1993), 109.

49. R. Levins and R. Lewontin, *The Dialectical Biologist* (Cambridge, Mass.: Harvard University Press, 1985).

50. T. Dobzhansky, *The Biology of Ultimate Concern* (New York: New American Library, 1967), 7.

51. Dennett relies on computer modeling to reaffirm the Baldwin effect. See G. E. Hinton and S. J. Nowlan, "How Learning Can Guide Evolution," *Complex Systems* 1 (1987): 495–502, reprinted in *Adaptive Individuals in Evolving Populations,* ed. R. K. Below and M. Mitchell (Reading, Mass.: Addison-Wesley, 1996), 447–53. Dennett gives the impression that he can recruit the Baldwin effect to his cause because it is, in his view, an uncontroversial component of the modern synthesis. This is incorrect. I cannot help thinking that Dennett's own conceptual scheme has blinded him to how problematic the Baldwin effect actually is in standard versions of the modern synthesis.

52. D. Dennett, *Consciousness Explained* (Boston: Little, Brown, 1991).

53. I am grateful to Michael Ruse for raising this objection.

54. This essay has been an exercise in rhetorical criticism insofar as it attends not only to questions of fact *(questio factis)* and questions of law *(questio juris)* but to the third classical *stasis* as well: questions of proper jurisdiction. My claim that a forum must exist in which the interpretation of Darwinism is discussed in a way that slightly transcends scientific work itself is about proper jurisdiction and who has the standing to speak to the question.

55. Dobzhansky, *Biology of Ultimate Concern,* 10.

56. M. Grene, "Puzzled Notes on a Puzzling Profession," paper presented at the 61st Annual Pacific Division Meeting, American Philosophical Association, San Francisco, March 27, 1987. On Marjorie Grene's naturalism, see David Depew, "Philosophical Naturalism without Naturalized Philosophy: Aristotelian and Darwinian Themes in Marjorie Grene's Philosophy of Biology," in *The Philosophy of Marjorie Grene,* ed. Randall E. Auxier and Lewis Edwin Hahn (Chicago: Open Court, 2002), 284–309.

PART II

Is a Non-Naturalistic Interpretation
of Darwinism Possible?

Within evolutionism, the great alternative to Darwinism was Lamarckism (in fact, Darwin himself accepted the idea that acquired modifications could be inherited and tended to grant it more and more importance in the course of his life; only August Weismann freed Darwin's theory from all Lamarckian remnants). It seemed therefore essential to begin this section with the essay of a highly respected biologist such as Rupert Riedl, who perseveres in asking questions that according to him cannot be answered by orthodox Darwinism alone. Can an approach based on the systems theory of evolution shed light on phenomena that Darwinism tends to ignore or take for granted? Are there ontological assumptions of the classical philosophy of nature that have to be given a new interpretation on the basis of Darwinism but that still retain part of their validity? Can Lamarckism be modeled within Darwinism? In fact it is quite obvious that the mechanism of natural selection as such is compatible with a host of possible evolutions; to explain why the evolution of life on our planet—which includes the astonishing phenomenon of convergence, best known from the Metatheria and the Eutheria—took the course it took, we need more than the principle of natural selection, namely other biological "laws." How can these be formulated, perhaps explained? And does the phenomenon of retroactive causality, so important in the organic world, oblige us to abandon a more linear way of thinking even if the latter, itself a result of evolution, is more natural to us?

A greater openness to a non-naturalistic interpretation of Darwinian biology may be achieved if one analyzes the concept of nature Darwin himself had. The questions the essays by Phillip Sloan and Robert Richards address are of the following type. What are the philosophical sources of Darwin's great discovery? How does his scientific theory differ from his philosophical interpretation of his theory? How did his own philosophical interpretation of his theory evolve? What presuppositions did he make? Which consequences did he draw from it? To go more into the details: What is Darwin's conception of nature? Is it a value-free concept, or does he ascribe to nature a particular value? Is his attitude of reverence toward nature determined by a pantheistic view of it? If this is the case, how is it linked to German *Naturphilosophie?* Strands of influence of Romantic *Naturphilosophie* can clearly be found in his writings, not based on a direct knowledge but mediated by one of the heroes of his youth, Alexander von Humboldt. Are the struggle for life and the suffering it causes compatible with the recognition of harmony in nature? Does the confutation of the physico-theological proof in Paley's version, which Darwin had absorbed as a young student in Cambridge, entail the impossibility of seeing any design in nature as a whole? How is the attempt of a uniform interpretation of nature linked to the Unitarian faith both his grandfathers shared? How clearly did Darwin understand that the rejection of a literal belief in the Bible—which in his case, as he writes in his autobiography, was based on skepticism with regard to the miracles, on the arguments of the contemporary historical analysis of the gospel, and on a criticism of some of the moral ideas to be found in the Bible—entails little with regard to the truth of theism?

Finally, of course, what is the place of mind in nature? What solution of the mind-body problem did Darwin endorse? What solutions are compatible with the Darwinian theory? (Alfred Russel Wallace, after all, embraced a very different one than Charles Darwin himself, even if he certainly had grasped the principle of natural selection.) For there is little doubt that Darwin is the founder of evolutionary psychology—mental capacities are, according to him, as much subjected to evolution as morphological and physiological traits, and in their formation natural and sexual selection play a fundamental role.

After these two historical essays dedicated to Darwin himself, three essays address the issue of the compatibility of Darwinism with philosophical traditions that are explicitly non-naturalistic. Certainly, showing that certain tenets are compatible with a possible interpretation of Darwinism does not justify these tenets, but at least it demonstrates that one easy way of "confuting" them—appealing to Darwinism—is not acceptable. There may be good arguments against Kant's ethics or Hegel's philosophy of nature, but the possibility of an evolutionary explanation of biological species or of human behavior is not such an argument. Recognizing such a compatibility is philosophically important, even if the positive justification of the approaches demonstrated to be compatible with Darwinism can hardly be offered in the essays of this volume.

Darwinism has philosophical consequences both for metaphysics proper and for philosophical anthropology. In which of the many senses of the term *metaphysics* can Darwinism be of any use for this philosophical discipline? Is metaphysics simply to be intended as the clarification of the basic concepts of the sciences? Jean Gayon follows Kant when he distinguishes four further concepts of metaphysics — transcendental metaphysics (or epistemology), special metaphysics, metaphysics of nature, and metaphysics of morals. The questions he and others pursue are: Can the purity of logical knowledge be affected by the evolutionary origin of our cognitive processes? Did Darwin really add anything relevant to the criticism of the physico-theological proof achieved in the eighteenth century by Hume and Kant? Does the historicity of the evolutionary process imply that it cannot have any relevance for a metaphysics of nature? (Obviously, this is related to the problem of whether metaphysics is a doctrine valid for all possible worlds, as Russell assumes; whether it is an inductive discipline based on the analysis of our world, as Spencer and Bergson believe; or whether metaphysics wants to explain fundamental traits of our world by referring to their value, as Leibniz and Hegel try to do.) And how does an assumed relevance of Darwinism for ethics avoid the naturalistic fallacy?

One of the most disturbing results of Darwinism, at least the result that most clearly seems to lead to a rejection of all forms of idealism, is Darwinism's denying that there could be a goal of evolution. Of course, nobody wants to state that the adaptation enforced by natural selection implies any direction of evolution — the most primitive organisms are as well adapted to their environment as the Eutheria. Still, increasing levels of complexity are visible in natural history (besides the continuous survival of more primitive life forms), and it is not only parochial anthropocentrism to see in the human spirit something of unique importance, for it alone could develop Darwinism and a general theory of life. What are the causes for this evolution toward complexity, which must be at least compatible with Darwinism? Are there factors that render it even likely, as the search for new environmental niches? How can one argue that animals are a more complex form of life than plants? What philosophical criteria could be adduced to see in the human mind a final outbidding of nature — doubtless a product of nature, but with a capacity of transcending nature, as the capacity of developing general theories about nature seems to suggest? What can be said about those natural laws that an evolution culminating in the human mind necessarily presupposes? Dieter Wandschneider undertakes the endeavor of exploring whether a teleological interpretation of natural developments in the Hegelian tradition could possibly be harmonized with natural selection as the prime motor of evolution.

Vittorio Hösle's starting point is the recognition that the Platonic theory of ideas has been severely damaged by the Darwinian revolution of the species concept. But even if the ontological center of gravity in Darwinism has moved from a timeless idea of species toward concrete populations, does this really entail that *all* ideas—including those of the true, the good, and the beautiful, the so-called transcendentals—must be naturalized? There are strong philosophical arguments against naturalizing them, one being that such an operation might endanger the truth claim of Darwinism itself. Still, after Darwin the defense of the non-natural status of the transcendentals has to be rendered compatible with the results of evolutionary epistemology, ethics, and aesthetics. One way to make them compatible is to ask the following questions: Do not organisms strive after knowledge, after goodness, after beauty—for is not adaptation a form of correspondence, life a fascinating mixture of egoism and altruism (and genetic egoism probably the only way altruism can get a foothold in the empirical world), and the irreducibility of sexual to natural selection at least compatible with an idealistic theory of beauty? In any case, Darwin himself never shared the ideas that something is good because it is successful in evolution or that something is beautiful because it is sexually attractive—on the contrary, Darwin's insistence on the impossibility of subsuming sexual selection under natural selection clearly demonstrates that he had a nonreductionistic concept of beauty. An interesting point of Hösle's essay is that it might be easier to reconcile a determinist than a nondeterminist interpretation of Darwinism with a teleological and religious worldview.

CHAPTER 6

A Systems Theory of Evolution

Rupert Riedl

WHAT DO WE NEED FROM SYSTEMS THEORY?

When I was a student of Bertalanffy in the late 1940s, systems theory was under-stood as a part of theoretical biology. The same position was taken by Paul Weiss, and we felt that what was new in systems theory was the application to biology of "recursive" or "recurrent" causality: that is, the idea that every biological effect in living systems, in some way, feeds back to its own cause. An interwoven causality was the new perspective, in opposition to linear causality thinking, as supported by positivism. The latter in turn was superseded by "pragmatic reductionism," which proposed to explain complex systems sufficiently from their constituents, much as we attempt to understand our everyday lives and the business world.

In the last fifty years, theoretical biology followed different directions at differ-ent universities. Systems theory became commonplace, except that the necessity of adopting recursive causality did not attract much attention.

A comparison of the two approaches of linear causality and recursive causality reveals their metaphysical underpinnings. If, by linear causality, the "big bang" made everything, including the earth, and the "blueprint," which was nothing but genetic instruction, made humans, then humanity either has been planned by the Creator via his direction of evolutionary pathways, as Teilhard de Chardin theorized, or, as Jacques Monod suggested, is a senseless product of accidents that derives from complex functions put together by chance effects. On the other hand, if recursive causality is adopted, evolution, put concisely, seems to be neither a planned pre-stabilized harmony nor a chaotic process without harmony. Evolutionary principles of self-organization allowed a poststabilized harmony to develop, producing sense and purpose within creatures and allowing even God to be revealed or sensed as a necessary hope.

REASONS TO ADOPT THE SYSTEMS THEORY OF EVOLUTION

It is obvious that we are in many respects genetically well prepared to deal with our daily lives, but it is less obvious that in other respects we are not—as when we are confronted with large numbers and dimensions, emergent processes, and complex systems. In problem-solving strategies, the decision-making tree leads us to expect and select regular, deterministic, indication-dependent, functional, and linear processes, but in complex systems many processes are irregular, undeterministic, and independent from their indications, and almost all of them are nonlinear (Brehmer 1980). Yet our tendency to simplify problem solving is itself biologically justified (Riedl 1992) because in many cases, particularly for our ancestors, quick decisions were more important than disentangling complexities.

The environmental problem is caused by a linear causality concept of profit maximization that is too simple. Now that atmosphere, sea, forests, and soil have suffered so much from humans, the biosphere is rebelling with ozone holes, increasing temperatures, rising sea levels, and deteriorated landscapes. The way we see ourselves in the universe depends a great deal on how we understand our genesis. The systems theory of evolution can provide a new basis for this understanding. Further, the synthetic theory of neo-Darwinism, as it is still providing the conceptual grounds for evolutionary theory as discussed in textbooks, is not wrong but merely incomplete, as I will elaborate below.

WHAT IS MISSING FROM CURRENT EVOLUTIONARY PERSPECTIVES

Many writers of modern texts avoid discussing questions that still have not found an answer. They seem to be writing with the purpose of furnishing facts for exams. This obscures the current views in science and does not encourage critical perspectives.

Evolution is taken to be linear. Central to this process is the DNA "blueprint," which suffers accidental mutative changes (true physical mistakes). By processes of transcription and translation that include factors of dominance and inter-change of alleles, gene pool size, and gene flow, the strands produce proteins and, in a way, shape organisms into new species by allopatric speciation and give rise to the entire diverse animal kingdom by environmental and intraspecific selection. This straightforward concept has been somewhat complicated by the metaphor of an "epigenetic landscape," as Waddington (1957) called it, and the growing knowl-edge of hierarchical dependencies in the action of regulatory genes. It is uncertain, however, whether epigenetics still fits into the linear conception just described. Deviations from the established pattern have appeared, as first elaborated by Gould (1977), Alberch (1983), and Raff and Kaufman (1983), but they have not changed the dominant conceptions.

Below, I consider two sets of open questions and phenomena that cannot be ex-plained by synthetic theory but that my systems theory can explain: (1) the con-tinuing ability of complex systems to be adaptable and (2) phenomena of "old pat-terns," macroevolution, morphology, and systematics.

THE CONTINUING ABILITY OF COMPLEX SYSTEMS TO BE ADAPTABLE

What allows complex systems to continue to be adaptable? May we expect an evolution of evolutionary processes, and if so, what may be the consequences?

General Thoughts about Complexity

The genome of a garden snail possesses about 108 base pairs in three 3×10^7 triplets. They become translated into twenty-one amino acids. These resemble the twenty-six letters of our alphabet. Such a quantity of letters, for example, is contained in the twenty volumes of the *Encyclopedia Britannica* (two 5×10^7 triplets). Assume that a point mutation, which changes one triplet, would improve the fitness. Should we then expect that the correction of a single letter in the *Encyclopedia* could im-prove its success on the market?

In addition, to find this single base pair or letter and change it into the right one by blind trial and error would require approximately 10^7 (ten million) new editions. Admittedly, no comparison is fully satisfactory. But let us go one step further. Few words alter their meaning by changing one letter only. For example, no single letter change in *and* (such as *fnd, aod,* or *anr*) will change the word to *for*. Certainly all three changes necessary to accomplish the overall word change could coincide (with regard to genome changes, Simpson had already made us aware of this in 1955). But in our case this would require 1021 attempts—an impossibility for both the snail and the *Encyclopedia*.

Adaptability of complex systems needs larger units; we assemble letters most functionally to words and sentences. General thoughts about handling complexity do not indicate what form these larger units may have. In principle they could have any form. A hierarchical pattern would be reasonable because deposition and searching for deposited information would proceed by the logarithm of 2.

Although the existence of larger gene units to handle complexity is not a possibility but a necessity, the mere existence of this complexity still does not explain enough. The first crucial point is to come.

Evidence for Larger Gene Units beyond Neo-Darwinian Explanation

Gene regulation, from Lac-operon to homeobox genes, underlines the existence of larger gene units. But morphological concepts allow us to predict the fundamental principle. I put the evidence in seven categories. None of the topics are addressed by the synthetic theory.

1. *Synorganization* and *coadaptation* make us expect an intermodality of changing organs. If a mutation of a longer neck in giraffes is selected for, it cannot be limited to the backbone—the spinal cord and all the rest must elongate too.

2. *Heteromophoses* demonstrate that even in the case of a somatic mutation in a regeneration bud, a supergene finds all the somatic genes that are necessary to form a widely complete, complex organ at a wrong place, such as an antenna instead of an eye. This phenomenon is one among a group of facts that led Darwin (1875) in his theory to speculate on the existence of inner mechanisms that are explained neither by Lamarck's theory of active adaptation nor by his own theory of selection.

3. *Homeotic or system mutations* support the same hypothesis of a supergene controlling a complex of other genes with regard to gene mutants. We see, for example, that the legs of a fly can be produced in different degrees of completeness and in several places but only with leg characters and only on such places where appendages (extremities) have been expected: antennae, aristae, mouthparts. The supergene must be switched on to develop more or less of the pertinent structures, but only those that belong to a leg, and the relation to former appendages must be linked to the phylogeny of the fly.

4. *Cartesian transformation* (Thompson 1942) makes evident that gradients of alteration must direct neighboring parts to change: for example, bones of a skull must be able to flatten, to elongate, or to shorten harmoniously.

5. *Regeneration,* such as rebuilding a limb in a salamander or a tail in a lizard, or even restoring the layers of a wounded part of the skin, indicates that the necessary instructions are activated only in pertinent positions.

6. *Phenocopies,* ontogenetic alterations by disturbances, tell us the time and sequence in which master switches act, namely in the sequence of phylogenetic innovations.

7. The term *homeosis* covers most phenomena listed in 1–6, and we take it as an indication of an inner order, based primarily on the construction and design of the organism and its history.

In summary, we may predict specific structures in the development of the epigenetic system; their form will correspond to functional units of the phene system. I expect a mechanism (1–3) that makes, in hierarchic patterns, such genes interdependent; the genes code for functionally interdependent phenes, forming (4) gradients over (5) the sequence of the phylogenetic steps of innovation of the organism. These depend primarily (6–7) on the design of the history of the organism itself, of its past, and less on its present fitness conditions.

A Model for Linking Genes That Code for Functionally Linked Phenes

Almost all features of an organism—that is, all that are functionally adjoining—are interdependent. Interdependence with new features develops by their approaching each other to form a new function. A simple example can be given by the development of a joint.

Take Devonian fishes, such as *Eusthenopteron* or *Laugia* (figure 6.1). The stylopodia of the front fins are very different in shape—compact in *Eusthenopteron,* slender and elongated in *Laugia*—and they do not really form movable joints with the zeugopodia with special muscles and ligaments, as we know from recent *Sarcopterygii.* They simply approach each other, making the fin both stiff and flexible.

In the upper Devonian and lower Carboniferous, the joint is formed, as in an *Ichthiostega* and a *Cacops,* and the usual theory (figure 6.2) lets us expect that the new close contact between stylo- and zeugopodium corresponds to the very precisely operating joint of our elbow.

In *Eusthenopteron* and *Laugia* there is still much freedom in the form of the stylo- and zeugopodium. Probably length and breadth of the approaching parts

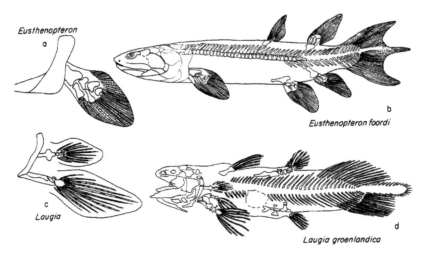

Eusthenopteron
a

b
Eusthenopteron foordi

c
Laugia

d
Laugia groenlandica

FIGURE 6.1. Early crossopterygian fish (*Sarcopterygii,* lobe-finned fish). (a–b) Order Osteolepiformes, Lower to Middle Devonian. (c–d) Order Coelacanthiformes, Middle Devonian to recent. Note (a–c) the different form of the stylopodium of front fins (in *Laugia* the back fins have moved close to the front fins). *Source:* Riedl (1978). All figures by permission of author.

were independently coded by different genes. There is at least no need and no in-dication for their being under a common command.

The Chances or Speed of Adaptability

Since we will soon consider very small numbers, let us be generous, taking a high mutation rate for each feature ($10-5$) and a high probability ($10-1$) of changing a phene in the right direction, like making a feature longer, shorter, more com-pact, or more slender. Already a population of 106 individuals could expect one favorable change in each generation.

Now, what will happen if two front parts of bones approach each other, form-ing the function of a demanding, precise-working joint, as in the case of our elbow? If adaptation to new behavior requires broadening the joint, this will work only if the socket and head of the joint change at the same time in the same way. We had this problem before. It is possible that we will have it again, but we will have to wait remarkably long for its next appearance. Under the given assumption the proba-bility would decrease to about $10-12$: in other words, our population of 106 speci-mens would have to wait 106 generations to expect this.

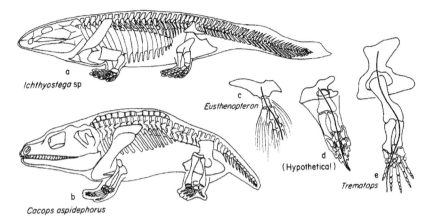

FIGURE 6.2. Early and primitive amphibians (Labyrinthodonts) and the theory of the evolution of limbs. (a) A representative of the Ichthyostegalia from the Upper Devonian. (b) A temnospondyl (Lower Carboniferous to Upper Triassic). (c–e) Theory of the phylogenetic transition from osteolepiform pelvic fin to the posterior pentadactyl limb from a temnospondyl. The probably homologous bones of the axis are connected with a line. *Source:* Riedl (1978).

Today we know that many genes are related: for example, put under the domain of an operator gene. The operator locks and opens by regulating molecules. We also know that the number of such master-gene systems has increased remarkably in the course of evolution, much more than the number of structural genes. No general model seems to be at hand today to explain how such master genes develop and how they assemble the structural genes that they direct. But no doubt they have developed, even in sub- and superimposed hierarchical orders, though again under the blind condition of trial and error.

In my mind, however, how they develop is a matter of secondary importance. The crucial question is: Which of them will be counteracted by the production of a subvital or lethal mutant, and which will be successful and spread in the population? Those mutants that hinder necessary adaptation will slowly become extinct in a population; those that speed up adaptation will be preserved and spread out within the population.

In our example of forming a joint, remember that all the processes are meant to act by pure chance. If genes that coded for length and breadth of one of the two bones became linked, this development might be favorable for the bone. It could shrink or strengthen the bone proportionally. But it would weaken the joint. If, however, genes that coded for the shape of socket and head of the joint became linked, the success would be remarkable. If we admit that the chance that this regulator gene will make the right change corresponds to that of other genes, then our population could expect the adaptation of the joint in each generation. It would not have to wait a million generations.

In an evolution that also involved competition for the speed of adaptation, the success rate would be increased from $10-12$ to $10-6$, a millionfold. But even the probability that such a master gene would develop might be a hundred thousand times less likely than the development of a new structural gene. The process sketched above would still be advantageous, selected for, and fostered in the population.

Feedback Causality

Some of my fellow biologists used to follow me up until this point in the story. They might have found this game of numbers not really necessary or enlightening, but they appreciated the collection of morphological facts (Riedl 1978). A hundred reviewers were mainly friendly, some surprised, others even excited. My fellow evolutionists, later on, did not know what to do with it. Why?

A personal story may throw light on their objections. Ernst Mayr, at that time almost a fatherly friend, wrote me a flattering letter stating, in summary, that no further evolutionist would ever find it possible to bypass the contribution made by my book. In the years to come, after he may have read it more closely, he himself never cited it. During a personal conversation about feedback causality ten years later, Ernst became furious, and I had to learn from him that "there is the blueprint! And the blueprint makes everything!" In a way, we come back to metaphysics, and whether we can expect to explain even complex systems with a straightforward linear causality.

What happens when genes coding for phenes of opposite or interdependent functions are randomly linked? The first group will produce impediments, the other much success. What happened to the genome? Something will be introduced into and preserved in its structure that has something to do with functional interdependencies of its products.

In our example of a joint, a structure is put into the epigenetic system that corresponds or stands vicariously or representatively for a specific phene, a joint. No epigenetic system coding for a joint was previously in the genome because there was also no joint in the phenes. Shall we say that the genome has "learned" what a joint is? This is, perhaps, saying too much. Nevertheless, if the model is accepted, we can expect that presumably all or at least most of the interdependencies of the phenes have left their traces or marks in the epigenetic system.

How can we now label such a transfer of information? The transformation from the phenes is obviously almost entirely chemically coded, from translation up to most inductive processes. The feedback is obviously not chemically coded. I have used the phrase "by stochastic processes," but this sounds rather vague. However, whatever label one may prefer, I have no doubt that this feedback of information transfer exists because it comprehensively explains the puzzling *and* seemingly different phenomena I listed above.

Lamarckism Coming through the Back Door?

Fellow biologists who disliked this idea (Riedl 1977, 1978) took it as a hidden or camouflaged Lamarckism, sneaking environmental conditions into the genome. This criticism was never published, but it was laboratory gossip, good enough to set the complicated matter aside.

Of course, the last word about fitness is always spoken by the environment. But which feature of the organization of an organism is free for adaptation to improve fitness is quite another question. For example, as soon as an exoskeleton is optimized, there is no way to change it into an endoskeleton, although fitness, in many cases, would improve remarkably by such a change. And such patterns of fixations go deep into the whole organization. All mammals, with only two exceptions, have seven cervical vertebrae, although giraffes would do better with more and dolphins with less; in the first case they are maximally stretched, in the latter case extremely compressed and widely fused.

According to Lamarck's conception, the genome actively "learns" from the environment, presumably keeping adaptability extremely open and dissolving patterns of order. In contrast, according to my conception, the genome "learns" only from its own products, keeping adaptability in complex systems for new characters but reducing adaptability for old characters dramatically. It produces order in living organisms.

What Feedback Causality Can Explain

Consider a hierarchy of *epigenetic units,* corresponding widely to the hierarchic patterns of the phene system. If they are coding for large functional interdependencies of the organism, this explains (1) synorganization, coding hierarchically for a hierarchy of organs and organ parts, acting from somatic and germ cells as well; (2) heteromorphoses and (3) homeotic mutations, coding for subunits being balanced by superimposed units; (4) Cartesian transformation; (5) regeneration being switched on in a sequence of developmental steps; (6) phenocopies; and, if one takes all these actions together as forming "inner order" in development, the overall principle of (7) homeosis in development.

Summing up: we expect all the pertinent functional principles of the phylogenetically transmitted fitness condition to be copied by feedback causality. The epigenetic system of each stem of organisms, Waddington's "epigenotype," is an "imitatory epigenotype" (Riedl 1977).

Wagner's Corridor Model

A concept analogous to mine has been developed by G. P. Wagner. I started from morphological features, deriving structures of the epigenetic system; Wagner started from gene structures, namely from pleiotropies. If you take a corridor, along which the genome of a population is forced to climb a hill of fitness, much of the success will depend on the pleiotropic genes.

If the two or more phenes for which such a gene is coding point in the upward direction of the corridor, then this gene will bring advantage to the increase of fitness and spread out within the population. If the contrary is the case, one phene being changed in the right way and the other not, this gene will not have success in the population.

The result is similar to mine. Those multifunctional genes that code for phenes that need to change into the same direction to increase fitness will be selected (see, e.g., Wagner 1983, 1988).

Confirmation for our concepts is now coming mainly from molecular geneticists and developmental biologists, but this is not the topic of this chapter.

UNEXPLAINED PHENOMENA OF "OLD PATTERNS," MACROEVOLUTION, MORPHOLOGY, AND SYSTEMATICS

My theory had to develop a model to explain how complex systems can still be adaptable. It also accounts for unexplained phenomena, or those that do not have a common or overall explanation, that are the consequence of complex systems' retaining their adaptability. These phenomena can be divided into four large groups—phenomena of "old patterns," macroevolution, morphology, and systematics—concerned mainly with directedness, order, and predictability in evolution. To make clear why the second set of phenomena is a consequence of the first, I will provide a metaphor.

Two gamblers, "Black" and "White," play dice in front of the king. Each has at the start two dice; one is red and the other yellow. They are allowed to roll the dice, but here is the catch: they are both blind. The king's role is to reward the "double-6," and he will tell the winner when this occurs. The game stands for competition, each gambler for a gene pool, the dice's eyes for mutation, the king for environmental selection, and the profit for fitness (figure 6.3).

Both players throw their dice, knowing that in the long run they will have success within about thirty-six throws. As soon as White learns of his success, he glues the dice together. He will win from now on at least every sixth time. Black will quickly fall behind.

Now the king alters his role, honoring red-6 yellow-2. Black (the unspecialized genome) will be content with his slower success. White will not win any more trials as long he does not get his two dice apart.

Assume that the game gets more complex, with four and eight dice, and that what is honored is the "quadruple 6." If White manages to get the four dice glued together, then he will still win every sixth time, while Black falls back from a success of $1/36$ to chances of $1/1,296$ to $1/1,679,616$. But if the king changes the game again, then White again will be in trouble.

With growing complexity, Black will again fall back in terms of his rate of adaptability, and under each condition of the environment White has advantages as long as the conditions are not changed. Black will slowly organize his dice. White will have to keep what he has and search for a better environment. One cannot cheat probability unless one pays for it by reducing further possibilities of adaptability. This is the solution to the forthcoming problems of organized genomes.

Findings of Old Patterns

Five groups of phenomena—the survival of atavisms, the spontaneous revival of atavisms, induction, rudimentation, and Haeckel's phylogenetic law—demonstrate constraints in the form of old patterns within the developmental process. And if we adopt the concept of larger genetic units, we find them now at all levels that were of importance for fitness in the past phylogeny of the organism.

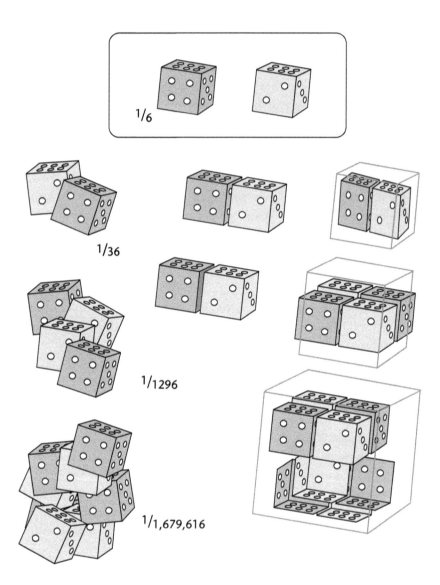

FIGURE 6.3. The game of the two blind gamblers. In the first game both players have the same chance of throwing a 6. In the second and third game the chances of throwing a double or quadruple 6 are reduced (unorganized genome), except that White maintains his chances if he glues the dice together (organized genome) as soon as he is informed by the king that the dice are in the right position. In the fifth game the chances for Black become nearly nonexistent. Nevertheless, White will get into trouble as well if the rules of the game change (see text).

Survival of Atavisms. The survival of atavisms, such as the appendix in humans, is puzzling. Without surgical interventions the appendix is open to a remarkable pressure for elimination. Why, then, does it still exist? It is assumed to have important functions for immunization in early development and to be a remnant from times when it was a large organ.

Spontaneous Revival of Atavisms. The spontaneous revival of atavisms was one of the phenomena that inspired Darwin in his "pangenesis theory" to assume "inner mechanisms." Referring to Etienne Geoffroy Saint Hilaire, who already knew such cases, Darwin stated (1875, 368), "What can be more wonderful than that characters, which have disappeared during scores, or hundred, or even thousands of generations, should suddenly reappear perfectly developed?" He had pigeons and fowls in mind. Today we know of numerous cases, even in humans: tailed children, surpluses of nipples and alveoli, faces covered with hair, the so-called "dog-man," even cervical fistulae that are remnants of gill slits. For the last 150 to 200 years that phenomenon has puzzled us.

To make sure one appreciates the strangeness of such phenomena, imagine a contemporary horse appearing with three toes, as was the case with the ancestors of horses. This is as astounding as if an automobile with Bronze Age wheels emerged from the factory. We would think that this modern factory still had a Bronze Age and, presumably, a medieval department alive and well and that it produced all wheels first in the Bronze Age department but forgot to roll some of the wheels into the Middle Ages and further into the last department, which would have changed the Bronze Age products into modern ones.

If we build a home, we do not start with a bower and rebuild it into a wooden cabin, then finally remake it as a brick home. We do not repeat our history materially, but ontogeny does. We must assume that the old units of instructions cannot be skipped: once they were indispensable for fitness, but now they remain indispensable to carry on instructions for further changes. Biological systems have to stay with this assumption, and this doubtless restricts further possibilities of adaptation.

Induction. Induction, in developmental biology "the transfer of instruction," signifies, as I see it, the transition from a phylogenetically older structure to a potential for a newer organ that is building on it. Using my automobile analogy, it corresponds with how the departments communicate with each other. Among a hundred cases, let us take the dorsal cord as an example.

All chordates, consequently all embryos of vertebrates, start with a dorsal nerve cord. This tells the dorsal muscle plate to divide in segments, the segments define where the spinal column should put the vertebrae, the vertebrae define where the spinal ganglia have to emerge, and these, in turn, organize the whole nervous system of the body. For example, if one removes the dorsal cord of a frog embryo and puts it under the ventral skin, the ventral muscle plates start to divide in segments, as they would otherwise never do. If one takes the notochord from a primitive fish and puts it under a chicken's skin, the chicken embryo still understands the message.

This remarkable finding tells us that even the instruction to build up functional units is preserved, corresponding to fitness conditions over the whole time of the species' phylogenetic development: 450 million years. Since even sender, "language," and receiver are preserved over the whole phylogenetic time, one rightly speaks of *homodynamy,* homologous messages. And despite all the mutative bombardment occurring over that whole time, changes are causing lethal damage.

Rudimentation. In rudimentation, or the process of devolution to rudiments of complex organs, as in the case of the eyes of cave-dwelling fish and amphibians, the order of disintegration is opposite to that of building up by induction. The rudimentation of eyes, for example, involves the reduction first of the vitreous body, then of the bulbous and lens, and then of the longest remaining traces of the optic nerve, which, as the "eyestalk," was first represented in embryonic development. It seems as if even in rudimentation the chain must be opened from the end link if a great disturbance is not to occur.

Haeckel's Phylogenetic Law. Haeckel's law, commonly described in textbooks, that ontogeny recapitulates phylogeny, has become almost a commonplace, but until now there has actually been no explanation for why such a law must be operative. From the evidence we have gathered in this chapter, the explanation emerges: the old structures and their information transfer turn out to be indispensable for all further development.

It has been debated whether Haeckel's phylogenetic law is actually a law or simply a rule. It is, as we now see, definitely a law if one distinguishes between cenogenetic and palingenetic characters, the old versus the new. The first are adaptations to larval or embryonic life, such as the floating devices of starfish larvae or the umbilical cord of mammalian embryos. This is because every ontogeny has its phylogeny. The palingenetic ones are certainly recapitulations.

Summary. Thus the larger genetic units discussed earlier in this chapter, corresponding to the hierarchy of pertinent functional units, reappear as "old patterns" in the ontogenetic process. They cannot be superseded because they are being preserved as principles for further construction, and clearly they narrow further alternatives for adaptive radiation. This was in principle predicted by Karl Ernst von Baer (1828).

Findings of Macroevolution or Cladogenesis

Most of the findings of highest significance for macroevolutionary changes stem from paleontology. Important discoveries go back to the nineteenth century. The field has been mostly ignored in the synthetic theory of evolution, with its concentration on "microevolution" that ends where species turn into genera.

Parallel Evolution. Parallel evolution has remained a puzzling phenomenon for the new synthesis. The classical examples are the wolf and the Tasmanian devil, a marsupial. The skeletons, skulls, and rows of teeth in these two forms are much more similar than the similar behaviors could explain. And given that true mammals separated from marsupials about a hundred million years ago, the phenomenon is surprising. Perhaps the ground plan of their ancestors was so rigid that whatever carnivore developed from it had to take exactly the same path. Admittedly, this is speculation, but the puzzle remains a problem of "directed" evolution.

Orthogenesis. Trend and orthogenesis are more serious problems. Hundreds of sequences of fossils, from Foraminifera to snails to horses, to name just a few, are documented. And plotting the findings on a vertical axis of time and a horizontal axis of "morphological distance," a measure of acquired structural difference, shows that the upward trend straightens more and more with passing time. It was long discussed how "ortho" (straight) an orthogenesis had to be in order to be a true orthogenesis. This was too academic. It is surprising enough that one finds trends everywhere doing almost no meandering despite enormous spans of time and changes in environmental conditions. We must therefore assume that keeping the functional units is often more vital than changing them by adaptation. But why do trends straighten as time passes?

Concave Curves. The term *concave curves* is not found in every biological dictionary but is often used by systematists. The term may not have been felicitously coined, but it stands for a remarkable phenomenon. If trends have the tendency to straighten with time, this indicates that structural changes are reduced. The time-structure curve changes from a type-producing (typogenetic) to a type-conserving (typostatic) phase. This greatly influences the patterns of the phylogenetic trees, in details as well as in the whole picture. All the branches tend to become straighter the higher one goes in the tree. We are accustomed to seeing such pictures, but it remains a good question why the phenomenon occurs. It can be understood by an increase of optimization of functional and genetic units and their interactions, making the probability of success of any deviation less likely. I spoke (Riedl 1978) of functional burden. Using Wagner's approach (1983, 1988), one could speak of genetic burden. I have found that in comparative anatomy, microscopically as well as in ultrastructures, the process of fixation can be explained by the effects of four types of functional and genetic burdens—standard parts, hierarchical, interdependent, and traditive, including their subforms (Riedl 1978).

Typostrophe. The term *typostrophe* means that new types may repeatedly develop, like stanzas in songs. Many phylogenetic pathways demonstrate that, in the long run, typostatic phases allow for the offspring of a new typogenetic phase. This seems a contradiction of the former phenomenon, yet it actually clarifies it, as the following simple example shows.

Phylogeny in chordates at first showed considerable experimentation with the notochord. In larvae, it was found only in the tail, or running from the tail end to the tip of the nose, or reaching only from the tail to the head. It remained at the same places in adults. But as soon as it became the basis of the developing axial skeleton, no such changes occurred any longer (figure 6.4). The same sort of variation occurred at first in the number and then in the axis of paired limbs in vertebrates, but as soon as the pattern in front and hind limbs became optimized, all tetrapods kept it, regardless of whether they were humans, bats, horses, or dolphins. But the pattern also made limb girdles necessary, and as soon as the axial skeleton was required to carry pectoral and pelvic girdles, the regions of the vertebral column become differentiated and fixed.

Organ systems become increasingly burdened by new systems they have to carry. As a result, the chances of adaptive radiation are reduced dramatically. But on the end of functional chains, new freedom opens the chance of success of evolutionary inventiveness. A puzzle for former evolutionists is thus solved.

Additive Typogenesis. Additive typogenesis describes the same phenomenon from its other side. It raises the question of why in evolution some organ systems are

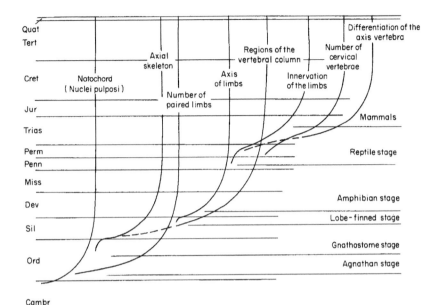

FIGURE 6.4. The building of typogenetic on typostatic phases. A sequence of examples is taken from notochord and axial skeletons to the differentiation of the vertebrae, and from primitive fish to mammals, by plotting geologic time versus morphological distance (the amount of changes). Note that the fixation of organ systems coincides with new systems building on them. *Source:* Riedl (1978).

free to change while others become fixed in what Wimsatt (1986) has labeled "entrenchment." This, for us, is no longer a problem. We now see clearly why some organ systems drift into fixation, and we can predict which others gained freedom in adaptive radiation.

Summary. Thus we find that optimized interdependency of organ systems leads to a reduction of adaptive freedom that can result in up to a hundred million years of canalization. The same occurs with functional burden and, if the systems are copied, with genetic burden, while new units at the ends of functional interdependencies are free for adaptive innovation and evolutionary novelty. We find also that the restrictions have nothing to do with a general lack of mutative impacts; instead, they concern elimination of mutants that do not fit the "interior requirements" of functional and genetic organization of our "White" gambler.

Justification of Morphology and Its Major Terms

Long before his investigation of the "Academy Controversy" between Cuvier and Geoffroy Saint Hilaire, Goethe (1795) dealt with problems of "morphology," and Richard Owen (1847) introduced the concept of "homology," separating it from "analogy," which before then had generally been used to describe similarity.

Morphology was originally thought to investigate the principles that lead us to compare structures and forms. Today, I understand it to be guided by inborn modes of gestalt perception, itself an interaction between perception and theory formation based on experience and expectation. It is based on *simul hoc*—expected coincidences of phenomena—that are comparable to David Hume's *propter hoc*—expected succession of phenomena, namely the so-called "cause," which he took as a "need of the soul" (Riedl 2000). This idea of Hume's inspired Kant to undertake his critical investigations.

Later, morphology as a way to discover principles that underlie structures and forms was confused with morphology as an attempt to explain those principles. And *morphology* as a term was taken to be, not the precondition for comparative anatomy, but comparative anatomy itself. This led to the characterization of morphology as an outcome of "German idealistic philosophy" (Ernst Mayr) and as a discipline with little interest in epistemology or the methodological problems of an assumed gestalt perception. Consequently, morphology was either distrusted or left to intuition. But intuition is not sufficient to make decisions in controversies.

Homology. *Homology* is the central term of morphology. Homologies are essential or substantial similarities having to do with the general organization of a group of organisms, in contrast to analogies, which are accidental similarities. But what is essential and what is accidental? Intuitionists see homologizing as an art, fostered by great experience, whereas rationalists see homologizing as avoidable or nonsense. Most biologists take it as an irritating or cumbersome puzzle.

By drawing on and synthesizing the homology criteria of Remane (1971), I have, however, shown that homologizing can be developed into a "probability theorem" (Riedl 1977, 1978, 1992, 2000). The probability of taking two features as homologues depends on the number of unrefuted confirmations and confirmations of predictions in further anatomical and systematical similarities. These homologues form hierarchies and pile up into hundreds and thousands of confirmations, making many homology expectations close to certain.

For the topic of this chapter, however, the explanation of the existence of homologues is simple: homologues are perceptible because of their partial resistance to adaptation, specifically with regard to either form or position, and in homo-dynamy also with regard to function. Functional burden and, if copied, genetic burden make it understandable why homologues must exist. They turn out to be the more conservative the more functions they carry. They correspond exactly to all present and past functional interdependencies for fitness conditions that were important in both the phylogeny and the ontogeny of the organism. We have no doubt that these structures have also been permanently pressured by mutative al-terations. But with the few exceptions of system mutations, these mutants must have been eliminated as subvital or lethal forms.

Hierarchic patterns of homologues consist of "frame homologues" that contain further subhomologues. Thus within the homologues of the mammalian skeleton we find those of the backbone, then those of the backbone of the neck region, then those of the atlas (first vertebra), then those of the atlas's arc. Further dissection—for example, to the plate of the front joint of the arc, the lowest or "minimum ho-mologue" of this specific series—would reveal still other homologies.

Homonomy. Homonomies are homologues in mass production, identical units spread over large parts of the body. In our example of dissecting a joint surface, we would find bone trabeculae; within them, bone cells; within the bone cells, their mi-tochondria; and within mitochondria, again homonome structures, namely specific biomolecules. Symmetries (as in anemones) or metamers (as in earthworms' articu-lation) are similar to homonomies; they are also identical, functional units.

Homonomy makes evident the existence of both genetically coded and func-tional units that can be reproduced in remarkably high numbers, such as the little gray cells in our brain. This process follows the principle of "cheap order," organiz-ing and linking much material with small information. We do the same with tiles or paving stones. This is all the more remarkable, given that, with the exception of the cilia, homonomes are all predisposed to be mass products, such as identical muscle cells, hairs, teeth in sharks, or legs of arthropods, such as millipedes. Only later in phylogeny may they gain individual characters in what we call differentiation.

Type and Ground Plan. These terms both refer to the overall design of a related group of organisms—what Goethe had in mind when he coined the term *mor-phology* as a new scientific field. If one puts the homonomes together and stacks on them the hierarchy of homologues, one ends up with a hierarchy of ground plans, such as the hierarchy in which the ground plan of butterflies is specialized on that of insects, that of insects on that of arthropods, and that of arthropods on that of articulates. As in the explanation of homologues and homonomies, we find now the complete set of conditions of functions and fixations of functional units. They repeat fitness conditions that originated in the times when, in chronological order, articulates, arthropods, insects, and butterflies developed. And it is the same chronological sequence in which we find new characters with much freedom for adaptive radiation that have additionally restricted and fixed the foundations on which they had to be built.

Summary. Thus we find the terms used in morphology justified because they describe phenomena that not only are indispensable for discovering comparable structures and forms but also correspond to principles of the reduction of adaptability, as in the example of the "White" gambler.

The Nature of Systematics and of the Natural System

Biologists, working with half a million systematic categories, have identified more than two million species, and it can be asked whether this large system has been discovered or invented. Are the systematic categories a product of nature or of our need to create order for purposes of description? This question was not pressing when neither the process of discovery nor the cause of fixations was really understood.

Systematic Categories. With regard to systematic categories, it has been asked what it actually means to select one special character to define a systematic unit. What would happen if one took another character? The grouping would look different. And who could decide which of the two similar categories corresponded to nature?

This uncertainty is known as the "weighting problem." It is based on a misunderstanding that itself originates from an ignorance of how gestalt perception works and from an ignorance of the principle of morphology that directs the process of discovery. We think in fields of similarities. That is, contemplating a gestalt immediately brings to mind all presumably comparable cases one has run across. In this process, several and sometimes many characters are always involved. And for concept formation along a path of optimization, one solution that produces the fewest contradictions or exceptions is finally selected.

As for concepts of systematic units, they describe both the highest probability of the stepwise bifurcation of species and the fixations of their characters along the paths of evolution.

The Nature of the Natural System. Rationalists among our fellow biologists have argued about the term *natural system.* They consider "system" to be a manmade conception and have claimed that if nature is in the mind it does not follow a system and if it is a system it is no longer natural.

This is misleading. The nature of the natural system is actually most natural, broadly designed by the laws of evolved evolutionary processes. There are the needs to adapt complex systems, including gambling under competition for the speed of adaptation. The natural system represents not only the hierarchy of types and archetypes of phenes but, correspondingly, a hierarchy of genotypes and archgenotypes.

Summary. The remarkable accomplishment of reconstructing the paths of development of millions of fossils and recent organisms demonstrates two things. First, our unconsciously working, inherited abilities of gestalt perception and thinking in similarity fields work surprisingly well, in fact nearly perfectly. Second, the uncovered patterns of similarities are the result of an evolving evolutionary mechanism developed to keep increasingly complex mechanisms adaptable.

The systems theory of evolution provides an explanation for many large groups of phenomena regarding old patterns, macroevolution and cladogenesis, systematics, and the natural system, which demonstrate restrictions of adaptive freedom. The underlying principle is that retaining the adaptability of complex systems requires the organization of the epigenetic system, corresponding to the hierarchy of functional interdependencies of the phenotypes imposed on probability. In the "accounts of probability," keeping the adaptability of new organization is paid for by a reduction in the adaptability of the basic organization.

SO WHAT?

As I mentioned at the beginning of this chapter, the use of linear causality seems to me the main obstacle to dealing adequately with complex systems. Industries and business have fostered it too long. It is, besides greed, the main cause of the environmental problem. It is not yet fully seen that complexity needs a more elaborated causality concept. Perhaps this systems theory contributes to the understanding that we are not only handling minor academic topics but opening up metaphysical perspectives that allow us to properly understand ourselves in our world.

REFERENCES

Alberch, P. 1983. Morphological variation in the neotropical salamander genus *Bolitoglossa*. *Evolution* 37:906–19.

Baer, C. E. von. 1828. *Über Entwicklungsgeschichte der Tiere*. Königsberg: Bornträger.

Brehmer, B. 1980. In one word: Not from experience. *Acta Psychologica* 45:223–41.

Darwin, C. 1875. *The variation of plants and animals under domestication*. 2 vols. London: John Murray.

Goethe, J. W. von. 1795. *Erster Entwurf einer allgemeinen Einleitung in die vergleichende Anatomie, ausgehend von Osteologie*. Vol. 2. Weimar: Böhlau.

Gould, S. 1977. *Ontogeny and phylogeny*. Cambridge, Mass.: Harvard University Press.

Owen, R. 1847. On the archetype and homologies of the vertebrate skeleton. *British Association Report* 1846:169–340.

Raff, R., and T. Kaufman. 1983. *Embryos, genes and evolution*. New York: Macmillan.

Remane, A. 1971. *Die Grundlagen des natürlichen Systems der vergleichenden Anatomie und der Phylogenetik*. 2nd ed. Königstein: Koeltz.

Riedl, R. 1977. A systems analytical approach to macro-evolutionary phenomena. *Quarterly Review of Biology* 52:351–70.

Riedl, R. 1978. *Order in living organisms: A systems analysis of evolution*. New York: John Wiley. Originally published as *Die Ordnung des Lebendigen: Systembedingungen der Evolution* (Berlin: Parey, 1975).

Riedl, R. 1992. *Wahrheit und Wahrscheinlichkeit: Biologische Grundlagen des Für-Wahr-Nehmens*. Berlin: Parey.

Riedl, R. 2000. *Strukturen der Komplexität: Eine Morphologie des Erkennens und Erklärens*. Heidelberg: Springer.

Thompson, D. 1942. *Growth and form*. Cambridge: Cambridge University Press

Waddington, C. 1957. *The strategy of the genes*. London: Allen and Unwin.

Wagner, G. P. 1983. On the necessity of a systems theory of evolution and its population biologic foundation: Comments on Dr. Regelmann's article. *Acta Biotheoretica* 32:223–66.

————. 1988. The influence of variation and of developmental constraints on the rate of multivariate phenotypic evolution. *Journal of Evolutionary Biology* 1:45–66.

Wimsatt, W. C. 1986. Developmental constraints, generative entrenchment, and the innate-required distinction. In *Integrating Scientific Disciplines*, ed. W. Bechtel. Dordrecht: Nijhoff.

CHAPTER 7

"It Might Be Called Reverence"

Phillip R. Sloan

> A very strong characteristic was his deep respect for authority of all kinds and
> for the laws of Nature. He could not endure the feeling of breaking any law of the
> most trivial kind, even the most harmless form of trespassing made him uncomfortable
> and he avoided it. . . . . As regards his respect for the laws of nature it might be
> called reverence if not a religious feeling. No man could feel more intensely the
> vastness or the inviolability of the laws of nature, and especially the helplessness
> of mankind except [*sic*] so far as the laws were obeyed.
>
> William Darwin to Francis Darwin, January 4, 1883[1]

When William Darwin penned these recollections of his father in the year after his father's death, he was trying to characterize an attitude toward nature that revealed something important about his father's thought. He refers in the comment to his father's attitude toward "laws of nature." He speaks of his father's "reverence" toward nature, and there is even a sense of a moral code dictated by nature that his father would not violate. It is my specific goal in this contribution to pursue these issues and use them to clarify Darwin's conception of nature and its relation to his conception of "natural law."

My focus is on Darwin's own metaphysical assumptions rather than on what may be called the metaphysical implications of his theory, and my approach will be primarily exegetical and descriptive. Clarifying these foundations nonetheless enables us to gain a clearer insight into the ways in which subsequent developments of Darwinism have overlain and even tended to reconstruct historic Darwinian formulations to suit other ends. Especially since the rise of the "new synthesis" in the 1920s and 1930s, in which natural selection theory and Mendelian genetics were brought into theoretical accord, a prominent historiographic tradition has imposed retrospectively on Darwin himself the language of "mechanism," "blind-ness," "chance," "contingency," and "randomness." As David Depew and others have shown, stochastic assumptions, imported from Ludwig Boltzmann's statistical me-chanics into the theoretical foundations of genetical and population-dynamical interpretations of Darwinian evolution by R. A. Fisher in the 1920s, made statis-tical randomness an integral part of neoselectionist Darwinian theory.[2] Coupled with the metaphor of natural selection as a "mechanism" that could be mathe-matically symbolized in the framework of theoretical population genetics by differ-ential survival ratios, Darwinian evolution has been recast in the modern period in very different terms than originally formulated. These assumptions have then been extended, either consciously or implicitly, to provide a warrant for a con-stitutive metaphysics of nature that is underpinned by the notion of contingency, to the extent that this is claimed to undermine all teleological philosophies of na-ture of the modern period articulated in different forms by Leibniz, Kant, Schel-ling, Hegel, Spencer, Teilhard de Chardin, and neo-Aristotelian and neo-Thomist thinkers.

I will begin from the other end of the story and explicate select aspects of Darwin's rich philosophy of nature that complicate this overly simplistic reading of Darwin. My thesis is that we can identify an original formative philosophy of nature in Darwin's writings, indebted in important ways to one strand of German *Naturphilosophie* that he encountered and deeply absorbed in the *Beagle* years through the writings of Alexander von Humboldt. This also supplied a context within which we can understand his initial discussions of "natural laws." My claim is that this early framework forms an enduring layer in his thought that was sub-sequently developed and modified as his scientific and theoretical vision elaborated, but that certain aspects of these foundations remained permanent features of Dar-win's intellectual framework. To this extent I shall endorse Robert Richards's re-cent argument that Darwin has fundamental historical connections to Romantic philosophies of nature.[3]

THE CONCEPT OF NATURE: THREE SCENARIOS

We begin by positioning Darwin's reflections on the concept of nature against the rich and complex discussion of this issue that had taken place in the scientific and natural-historical literature in the previous two centuries.[4] As a first sketch, we begin with the conception of nature that was fashioned by the main conceptual architects of the mechanical philosophy. In this conception, nature was deprived of internal teleological activity of the Aristotelian tradition. It also did not function as a designing world-soul in the Platonic-Stoic traditions. Descartes's comment in *Le monde* (1629–33) that "by Nature I do not here understand some Goddess, or some other kind of imaginary power, but I intend this word to signify Matter itself in so far as I consider it with all the qualities I have attributed to it,"[5] expressed succinctly a conception of nature as a passive mechanical framework on which a designing deity worked his will by imposing on it laws of motion. Robert Boyle's influential formulation of similar views in his *A Free Inquiry into the Vulgarly Received Notion of Nature* (1686) provided a more explicit and detailed development of this notion, and his discussions would reverberate through treatises on the concept of nature over the succeeding 150 years both within and outside the British tradition.[6] Boyle explicitly opposed his concept of nature to the classical Stoic-Platonic and medical usages of his contemporaries William Harvey, Francis Glisson, and Ralph Cudworth:[7]

> . . . [N]ature is the aggregate of the bodies that make up the world, framed as it is, considered as a principle by virtue whereof they act and suffer according to the laws of motion prescribed by the author of things. Which description may be thus paraphrased: that nature, in general, is the result of the universal matter or corporeal substance of the universe, considered as it is contrived into the present structure and constitution of the world, whereby all the bodies that compose it are enabled to act upon, and fitted to suffer from, one another, according to the settled laws of motion.[8]

I contrast this influential definition with that from a century later from the French natural historian Georges Louis LeClerc, comte de Buffon (1707–88), whose two substantial treatises on nature were published as the opening discourses to the twelfth and thirteenth volumes of his *Histoire naturelle, générale et particulière* in 1764 and 1765. In the first of these discourses, Buffon displayed both his continuity with and departure from the mechanistic tradition:

Nature is the system of laws established by the Creator, for the existence of things and for the succession of beings. Nature is not a thing, because this thing would be everything; Nature is not a being, because this being would be God. However, one can consider it as a living, immense power, which embraces all things, which animates all, and which subordinated to the first Being, commences action only by his order, and continues to act only by his concourse or consent. This power is from the divine power, its manifest part. It is at the same time the cause and the effect, the form [*mode*] and the substance, the design and the work. Very different from human art, whose productions are only dead works, Nature is itself a perpetually living work, an active worker without cessation, who knows how to employ all things, which working according with itself, and always on the same basis, far from exhausting itself, returns to it unexhausted. Time, space and matter are its means, the universe is its object, motion and life are its aim.[9]

The "phenomena of the world" are simply the "effects of this power." They are produced by "nature" through the action of "live forces" (*forces vives*). Nature also contains inherent ordering structures—the "internal molds"—that underlie organic organization and embryological formation. These molds also underlie the orderly succession of individuals in the generation of like by like within the historical lineage, forming the basis of Buffon's concept of biological species.

A third statement is drawn from the German Romantic tradition, and specifically from the Weimar circle associated generally with Goethe.[10] In this we see a close connection of the concept of nature with the form of dynamic pantheism that Johann Herder first developed by reworking some of Spinoza's monism and that was then expanded by Schelling in such works as *Von derWeltseele*.[11] In this form, Nature is conceived as a surrounding, vitalizing reality that animates and connects life and even serves to unite conscious experience and the material world. It is something known better through aesthetic awareness than by rational cognition. An essay on nature by Goethe's disciple Georg Christoph Tobler, dating from 1783, that was transcribed by Goethe at this time in his *Tiefurt Journal,* expresses this conception of nature well:

> Nature! We are surrounded and embraced by her—powerless to leave her and powerless to enter her more deeply. Unasked and without warning she sweeps us away in the round of her dance and dances on until we fall exhausted from her arms.

She brings forth ever new forms: what is there, never was, what was never will return. All is new, and yet forever old.

We live within her, and are strangers to her. She speaks perpetually with us, and does not betray her secret. We work on her constantly, and yet have no power over her. . . . She is the sole artist, creating extreme contrast out of the simplest material, the greatest perfection seemingly without effort, the most definite clarity always veiled with a touch of softness. Each of her works has its own being, each of her phenomena its separate idea, and yet all create a single whole. . . . There is everlasting life, growth, and movement in her and yet she does not stir from her place. She transforms herself constantly and there is never a moment's pause in her.

She thought and she thinks still, not as man, but as nature. She keeps to herself her own all-embracing thoughts which none may discover from her.

All men are in her and she in all.[12]

I have used these statements by Boyle, Buffon, and Tobler/Goethe as snapshots of three different conceptions of nature that were available to Darwin in his formative years. Boyle's distinctions, for example, constitute the primary content of articles in British reference works into the early nineteenth century.[13] Darwin worked through Buffon's *Natural History* in its English translation, containing the essays on nature, in the summer of 1840. The contact with strands of German philosophy of nature was made indirectly through the intermediary of Alexander von Humboldt.

Darwin first seems to have encountered Humboldt's writings during his years at Cambridge, probably through the influence of his mentor, John Stevens Henslow, who was teaching Humboldt's botanical geography in his descriptive and physiological botany course. Humboldt's impact on Darwin was particularly striking in the spring of 1831 during Darwin's last term in residence at Cambridge when he completed Henslow's course for the second time. At this time Darwin was also preparing for the expedition to the Canaries that was replaced by his appointment as naturalist to accompany the H. M. S. *Beagle.*[14] During Darwin's long intellectual isolation of nearly five years on the *Beagle,* he was confined to a library of approximately 275 books, with Humboldt's writings, particularly his *Personal Narrative,* serving as one of the few works of a reflective and philosophical character to which he returned again and again. Thus Humboldt's impact on Darwin in this period was much deeper than might have been the case if he had remained in England, immersed in the complex British discourse on issues in science, theology, and political economy that was taking place in the 1831–36 period. As Robert Richards has developed in chapter 8 of this volume and elsewhere, Humboldtianism was deeply imbedded in Darwin's thinking in these years.[15] Humboldt's conception of a holistic "physics of the earth," a science connecting diverse scientific inquiries together in a more comprehensive science of nature, seems to form the model for Darwin's own efforts to link geology and zoology, paleontology and stratigraphy, botany and zoology during the *Beagle* years.[16]

In Darwin's reflections during his formative *Beagle* period, "Nature," typically referred to in capitalized form, takes on many of the attributes of Humboldt's Goethean conception of nature as developed on these points in his early writings. Like Humboldt, Darwin replaces the theological language of "creation" or a creator-God with that of Nature. It is Nature that "chooses her vocalists out of other tribes" of birds in the forest.[17] The same Nature is the designer of landscape.[18] It is seen "in its grandest form in the regions of the Tropics."[19] Even in his remarkably rare uses of theological language in the writings of the *Beagle* years, Darwin speaks of the great primeval forests as "temples filled with the varied productions of the God of Nature."[20]

Philosophy of Nature and Darwinian Transformism

Darwin's transformist theory, conceptualized in its earliest formulation in the nine months following the return of the *Beagle* to England, elaborated in the B–E Notebooks that occupied the period from July 1837 to July 1839, and paralleled by the M–N Notebooks on mental powers running from July 1838 to the summer of 1839, was the result of a remarkable burst of creative work in which Darwin drew into a synthesis the empirical work he had carried out in functional biology, Lyellian geology, biogeography, and paleontology with some underlying metaphysical assumptions that were never made fully explicit. During this creative period, Darwin's natural philosophy was in no sense static. There was an intense interaction of his preexistent philosophical background, the *Beagle* empirical researches, and his remarkable travel experiences with the complex intellectual world of late 1830s Britain that he entered after the years of virtual isolation at sea. This was the world of post–Reform Bill Britain. The scientific landscape had altered remarkably, with new contacts with the Continent that had developed particularly with the foundation of the British Association for the Advancement of Science in the year of Darwin's departure. The foreign scientific authorities in the life sciences had shifted from the French authors to the new wave of Germans — Karl Ernst von Baer,

Christian Ehrenberg, Johannes Müller, Gotthelf Treviranus—and almost immediately after his return Darwin felt "he must learn German."[21]

In the broad and omnivorous readings of the early transformist period, which included extensive retrospective reading in numerous journals that had appeared in the 1831–36 period as well as a broad set of both technical and more general works, Darwin encountered an essay by Goethe's disciple Carl Gustav Carus, translated in Richard Taylor's *Scientific Memoirs* in 1837 under the title "The Kingdoms of Nature: Their Life and Affinity." This ambitious essay, drawing on the biological treatises of Goethe, Treviranus, Heinrich Steffens, and others, sought to display the inner harmony between the conscious and physical worlds, revealed by the "reciprocity of bodily action of various kinds with surrounding objects."[22] The overcoming of the duality of inner and outer worlds was achieved by an aesthetic awareness of "the *supreme* and *one,* which is alike the foundation of nature and mind."[23] This unifying reality, the source of life, consciousness, and the unity of all organic life, was both divine and natural:

> Every natural being must therefore appear, like nature in general, partly as a unity (in which light only it is an individual), and partly as a multiplicity, in which light it is infinitely divisible. . . . Universal nature is consequently to be considered as the highest, the most complete, the original organism; and *in* nature those individuals only are to be called organisms which, as unities under certain external conditions, that in their relation to other natural unities, continually develop themselves inwardly and outwardly into a real multiplicity. Among such organisms the most prominent are those bodies which, including our planet, constitute the system of the universe, and display themselves in continual motion and formation; those on our planet consist of plants and animals.[24]

Nature, for Carus, was the source of moral order, of the sense of beauty. It was the ground for the "indissoluble union as well as the beauty and regularity of the phaenomena surrounding man and existing within him." Grasp of this "harmony and purity" of universal nature "must necessarily stimulate us, not only to penetrate more deeply into the mysteries of science, but also to conform our own inward life to that harmony and purity which are presented by universal nature."[25]

Carus's essay supplied Darwin with a more explicit development of some of the themes of Humboldt's philosophy of nature that Darwin assimilated during the *Beagle* years. The resonance of Darwin's own thinking at this time with the themes in Carus is evident. Commenting on the Carus essay in an entry that immediately followed a series of comments in the C Notebook on the relation of "life" to the laws of chemical combination, Darwin wrote:

> After reading "Carus on the Kingdoms of Nature, their life & affinity" **in Scientific Memoirs** I can see that perfection may be talked of with respect to life generally.—where <">unity constantly develops multiplicity<"> [(his definition "constant manifestation of unity through multiplicity" this unity,—this distinctness of laws from rest of] universe<< which Carus considers big animal>> becomes more developed in higher animals than in vegetables. . . . It is very remarkable as shown by Carus how intermediate plants are between animal life & "*inorganic life*".—animals only live on matter already organized.—This paper might be worth consulting, if any Metaphysical speculations are entered in upon life. **Namely Carus.**[26]

In a later section of the C Notebook, still dated to around May–June 1838, he then entered the following comment:

> There is one living spirit, prevalent over this word [*sic* = world], (subject to certain contingencies of organic matter & chiefly heat), which assumes a multitude of forms <<each having acting principle>> according to subordinate laws.—There is one thinking <<<& Creat> sensible>> principle (intimately allied to one kind of organic matter.—brain. & which <prin> thinking principle. seems to be given or assumed according to a more extended relations of the individuals, whereby choice with memory. *or reason?* is necessary).—) which is modified into endless forms, bearing a close relation in degree & kind to the endless forms of the living beings.—We see thus Unity in thinking and acting principle in the various shades of <dif> separation between those individuals thus endowed, & the community of mind, even in the tendency to delicate emotions between races, & recurrent habits in animals.—[27]

This postulation of a general "living spirit" associated with organic matter that seems to unite forms of life, and the proposal of a distinct but related "thinking principle," allied to one form of organic matter—brain—that is the basis of consciousness, displays some of the distance between Darwin's thought in this period and the assumptions of inert materialism or Newtonian mechanism.

Vital and Physical Laws

The contact with Thomas Malthus's *Essay on Population* a few months following these entries in late September 1838 altered several aspects of Darwin's preexistent reflections on nature. It also supplied him with a new set of metaphors and a new framework for analysis of the natural world by which he could formulate the principle of natural selection. Entering his discourse from this date are a new set of images — metaphors of physical force, the pounding of hammers on wedges into an unyielding surface, the notion of a universalized struggle for existence. Malthus's emphasis on the relations of human population to food supply and living space was universalized by Darwin into a total struggle for existence in which all living beings, including even those things regarded by Malthus as food, were involved in an inexorable drive to increase to the maximum. This principle of population increase gave Darwinian theory an "inertial" principle, enabling him to shift attention away from a search for an explanatory theory of life that would explain causally the transformism of species, to an emphasis on the controls on this dynamic power of population increase. With Malthus, Darwin also encountered a new conception of nature closely associated with Malthus's analysis of the principle of population — nature as a selector, even a harsh schoolmaster, that has fashioned humankind toward hidden ends by means of a rigorous struggle for existence:

> Nature, in the attainment of her great purposes, seems always to seize upon the weakest part. If this part be made strong by human skill, she seizes upon the next weakest part, and so on in succession; not like a capricious deity, with an intention to sport with our suffering and constantly to defeat our labours; but like a kind though sometimes severe instructor, with the intention of teaching us to make all parts strong, and to chase vice and misery from the earth. . . ; [we] always find Nature faithful to her great object, at every false step we commit ready to admonish us of our errors, by the infliction of some physical or moral evil.[28]

The new images and language that enter with the Malthus encounter underpin much of the scholarly tradition that has localized Darwin's thought in the "common context" of British political economy, Newtonianism, natural theology, and empiricism.[29] This view requires revision to take full account of an unusual formative period that included his novel experiences at Edinburgh and Cambridge, followed by the limited reading carried out in the intense isolation of the *Beagle* years.[30]

We can follow the continuities and changes in Darwin's thinking in this complex period of his formation in the London years by looking closely at his appeals to "laws
of nature" or "laws of life" in his writings of the post-Malthus 1840s. The appeal
to the concept of natural laws is not a new phenomenon in Darwin's thought.
From the time of his encounter with John Herschel's *Introduction to the Study of
Natural Philosophy* at Cambridge in the spring of 1831, the language of "law" entered
his discourse. But his early discussions of "laws of life," such as we find opening the
first species transformation or B Notebook in July of 1837,[31] must be read against
the background of a distinction commonly being made in the biomedical literature of his day between two kinds of laws, those governing animate and inanimate nature.[32]

   That Darwin was recognizing such a distinction is evidenced by his most extended single discussion of the "laws of nature," which is found in a set of manu-
scripts later designated by Darwin "Old and Useless Notes." This is a two-column
back-and-forth entry that Darwin wrote around March 1840 in response to reading John Stuart Mill's discussion of Samuel Taylor Coleridge in the *Westminster Review* of March 1840[33] and in close conjunction to reading a summary of Jean Bap-
tiste Lamarck's discussion of biological laws in William Kirby's Bridgewater Treatise
of 1835.[34] Darwin's discussion begins with a distinction between two orders of
natural laws that are then drawn into association with one another:[35]

ʃHas any vegetable or animal *matter* been
formed by the union of *simple* non-organic
matter without action of vital laws.

Effects of Life in the abstract is matter united
by certain laws different from those, that govern in the organic World; life itself being the
capability of such matter obeying a certain &
peculiar system of movements different from
inorganic movements.—See Lamarck for the
definition given in full.—

Hence there are two great <worlds, inor.>
systems of laws <<in the world>> the
organic & inorganic—The inorganic are
probably one principle for connect[ion] of
electricity chemical attraction, heat & gravity
is probable.—And the Organic laws probably
have some unknown relation to them.—

According to the individual forms of living
beings, matter is united in different modification, peculiarities of external form impressed,
& different laws of movements.

   As this discussion proceeds, Darwin is concerned to clarify the relation of
the "vital" and "physical" laws. In very simple forms of life there is such a close
connection that he is even willing to identify the two.

This is true as long as movement of sensitive plant can be shewn to be direct physical effect of touch & not irritability, which at least shows a local will, though perhaps not conscious sensation.

In the simplest forms of living beings namely <<*one individual*>> vegetables, the vital laws are definitely (<like> <<as>> chemical laws) as long as certain contingencies are present, (contingencies as heat light &c).

During growth, <extres> tissue <c.o.> unites matter into certain form; invariable, as long as not modified by external accidents, & in such cases modifications bear fixed relation to such accidents.

But such tissue <must> bears relation to whole, that is enough must be present to be able to exist as individual.—

But the ensuing discussion reinforces, rather than diminishes, Darwin's earlier assumption that the principles governing the life world are in some important respects very different from those governing the physical world. Most notably, there are principles of consciousness and will that extend down into the most elementary forms of life:

I here omit the case (if such there are) of animals enjoying only movements such as sensitive plant. (But I include irritability for that require will in part ʃWhy more so than movement of sap. or sunflower to sun? ∵ I should think there was direct << physical >> effects of more or less turgid vessels; effect of heat, light or shade.)

_____

Joining two difficulties into one common one always satisfactory, though not adding to positive knowledge. lessening amount of ignorance

_____

ʃin Corallina are not two kinds of *life* vegetable and animal strictly united?

_____

In animals, growth of body precisely same as in plants, but as animals bear relation to less simple bodies, and to more extended space, such powers of relation required to be extended.

Hence a sensorium, which receives communication from without, & gives wondrous power of willing. These +willings <<+can the word *willing* be used without consciousness, for it is not evident, what animals have consciousness>> are common to every animal, instinctive and unavoidable.

These willings have relations to external contingencies, as much as growth of tissue and are subject to accident; the sexual willing comes on periods of year as much as inflorescence.—

The radicle of plants absorb by physical laws of endosmosis exosmic juices. arms of polypus, show either local or general will, & stomach likewise <<does>> .

The continuity of these discussions from the 1840s with the 1838 C Note-book reflections on the universal vital spirit that unites matter and consciousness indicates the persistence of important themes from the pre- to post-Malthusian period.[36] In the next section, my intent is to examine the evidence for the conti-nuity of these themes into the mature post-*Origin* period.

Darwin on Nature: Post-Origin Reflections

By the date of the publication of the *Origin,* several transformations of earlier for-mulations of issues in Darwin's writing had occurred that cannot be examined in detail here.[37] The published *Origin* abandoned the earlier reference to distinct "vital" laws that govern organic nature in contrast to inorganic physical laws. Only one category of law, without adjectival qualification, was employed. Close exami-nation of Darwin's usages of the concept of law in the *Origin* shows, however, that the majority of his references to "law" and "laws"—laws of "inheritance," "vari-ation," "reproduction," "transmission," "correlation of parts," "balance," and "em-bryonic development"—would earlier have been designated as "vital" laws. But the significance of this change in Darwin's language should not be overinterpreted. It does not mean a clear shift of Darwin's framework away from his earlier concep-tions to Newtonian or Comptean-positivist interpretations.[38] Even the well-known final paragraph of the *Origin* in all its editions maintains a subtle distinction between the fixed natural law of gravitation governing the solar system and the source of vi-tality that has brought living forms into existence, suggesting that it is a misread-ing to assume that Darwin claimed *inorganic* physical laws had created the various forms of life.[39]

As Robert Richards has developed in this volume, Darwin also employed highly intentional and teleological metaphors in his discussion of the actions of "natural" selection in the first edition of the *Origin.* As Darwin's theoretical statements moved from the private to the public sphere in 1859, however, critics and commentators quickly drew attention to the suspect intentional imagery in Darwin's strong par-allels between the selective action of human art and the more penetrating selection of a wiser nature, odd-sounding to the ears of 1860s scientists and philosophers in the era of Comte, Mill, Helmholtz, and Kelvin.[40] But Darwin promptly regretted these formulations and frequently claimed after 1860 that his point was being mis-understood and that the locution "natural preservation" was a better description of his main theoretical principle.[41] His clarifying passage inserted in the opening dis-cussions of chapter 4 in the third edition of the *Origin* of 1861 and maintained in all subsequent editions presumably laid the issue to rest:

Several writers have misapprehended or objected to the term Natural Selection. . . . It has been said that I speak of natural selection as an active power
or Deity; but who objects to an author speaking of the attraction of gravity
as ruling the movement of the planets? Every one knows what is meant and is
implied by such metaphorical expressions; and they are almost necessary for
brevity. So again it is difficult to avoid personifying the word Nature; but I
mean by Nature, only the aggregate action and product of many natural laws,
and by laws the sequence of events as ascertained by us. With a little familiarity such superficial objections will be forgotten.[42]

In this redefinition of nature as simply the aggregate of natural laws,[43] Darwin certainly removed the immediate warrant for assuming a strongly teleological reading of his principle of natural selection. The extensive revisions of the
fourth, fifth and sixth editions of the *Origin* and his later writings also display unequivocally his efforts to eliminate or qualify his usages of intentional metaphors
in relation to nature and natural selection. This process of revision can be seen in
table 11.1's comparison of word frequencies of the use of the term *nature* in the
first and sixth editions of the *Origin* with those in the *Various Contrivances of Orchids*
(1862), the *Descent of Man* (1871), and the *Expression of the Emotions* (1872).[44] The
culmination of this development was the insertion of the phrase "survival of the
fittest" as a synonym for "natural selection" in the fifth edition of the *Origin* of 1866.

TABLE 11.1. Word Frequencies Regarding Nature

	Nature Total Term Frequencies	*Natural* *Selection* Frequency	"Intentional" Metaphors Applied to Nature
Origin of Species, 1st ed. (1859)	253	271	21
Origin of Species 6th ed. (1872)	326	363	16
Various Contrivances of Orchids (1862)	40	11	5
Expression of the Emotions (1872)	47	4	0
Descent of Man (1871)	152	140	1

This dramatic abolition of "intentional" imagery from Darwin's discussions of nature after 1861, particularly in works actually drafted in this period, presents us with an interpretive problem. Removal of the references to an "intentional" selective nature clearly altered Darwin's meaning of "natural selection" to denote a *resultant* effect of a nonteleological process of sorting, rather than an intentional selection by "nature." But does this alteration also have implications for Darwin's preexistent philosophy of nature?

I have argued in this chapter and elsewhere that two primary layers in Darwin's conception of nature can be distinguished, the original "Humboldtian" holistic concept of the *Beagle* years and the subsequent "Malthusian" intentional and selective nature that entered after 1839.[45] The second layer has clearly been stripped away in the later works. Has the same happened to the earlier concepts? When Darwin later reflected back on his early views in the *Autobiography,* he spoke of how "grand scenes" of the natural world had "formerly excited in me" a sense of the sublime, "intimately connected with a belief in God."[46] But these are referred to in the *Autobiography* as rejected views, and I do not suggest that we can see Darwin simply returning to the nature philosophy of his youth.

Nevertheless, I suggest that a key to understanding Darwin's mature work is the recognition of the persistence of important dimensions of this earlier philosophy of nature even after the excision of the "intentional" layer that had been imposed upon it. I will develop my argument on this point by focusing upon some select aspects of Darwin's interpretation of the relation of the mental and the physical in his later works.

Darwin's discussions of these issues in the *Descent of Man,* for example, are based upon a complete parallelism of inner mental states with external anatomical continuities that extend throughout the animal kingdom and even into the plant world. Darwin's solution to the mind-body problem can best be characterized as a kind of vital monism. Darwin's "reductionism," if we can term it that, is always an "intermediate" reductionism, in which he explains higher mental properties by simpler states of life that exhibit the *same* properties in more elementary form. In the language of the comparative anatomy of his day, mental relations of humans to other animals, like those of external anatomy, can be considered relations of true "homology" and not merely relations of "analogy." Only degrees of inner complexity that parallel the complexity of physical structure provide a basis for the distinctions of "life," "instinct," and "consciousness" in his evolutionary schema.

Consequently the strongly anthropomorphic language used to describe animal behavior in the *Descent of Man,* extended even to the behavior of earthworms in his last published work, is not an inexact and merely metaphorical use of language.[47] It is a direct implication of the monism that resolves the body-mind relation for him. Ants "play" together, "chasing and pretending to bite each other, like so many puppies," in the same univocal sense of "play" that he applies to the behavior of his own children. In ants this behavior is just manifest at a simpler and more elementary level.[48]

Similarly, "life" for Darwin is not explained through the actions of a physical system. It remains a dynamic, self-creative power that needs no more elementary explanation. For this reason, Darwin cannot be seen as a reductionist in the German biophysics tradition of Hermann von Helmholtz and Emil DuBois-Reymond, or that of the more strident German materialism of Carl Vogt and Ludwig Büchner. This is displayed by his treatment of vitality in one of his last works, the *Power of Movement in Plants* (1880). In this work he returned to inquiries in plant physiology that had once occupied him as a student at Cambridge.[49] Darwin does not reduce plant movements to mechanical actions driven by physical causes in the sense developed by Jacques Loeb later in his theory of forced tropisms. The ability of plants to follow light is instead due to a kind of inner "sensitiveness," and "it is hardly an exaggeration to say that the tip of the radicle thus endowed, and having the power of directing the movements of the adjoining parts, acts like the brain of one of the lower animals."[50]

None of this explicitly develops a systematic philosophy of nature, and Darwin uses the term *nature* only infrequently in his later works. We must triangulate on his underlying assumptions more by seeing what he does than with what he explicitly says about these issues.

To illuminate this point I shall focus on the revealing correspondence with William Graham, the Irish mathematician and lecturer on mathematics at St. Bartholomew's Hospital in London, whose *Creed of Science* of 1881 Darwin read avidly in the last year of his life and recommended to others. It also formed the subject of a series of letters between Darwin and Graham,[51] generating a renewed discussion of issues of design and purpose in nature that Darwin had conducted much earlier with John Herschel and especially with Asa Gray. In Graham's view, Darwin had conveyed to him "the clearest statement I know of his attitude to the most central philosophical and religious questions."[52] In the longest of his two surviving responses to Graham's letters, Darwin commented:

> [Your chief claim] is that the existence of so-called natural laws implies purpose. I cannot see this. Not to mention that many expect that the several great

laws will some day be found to follow inevitably from some one single law, yet taking the laws as we now know them, and look at the moon, what the law of gravitation is and no doubt of the conservation of energy—of the atomic theory, &c. &c., hold good, and I cannot see that there is then necessarily any purpose. Would there be purpose if the lowest organisms alone, destitute of consciousness existed in the moon? But I have had no practice in abstract reasoning and I may be all astray. Nevertheless you have expressed my inward conviction, though far more vividly and clearly than I could have done, that the Universe is not the result of chance. But then with me the horrid doubt always arises whether the convictions of man's mind, which has been developed from the mind of the lower animals, are of any value or at all trustworthy. Would any one trust in the convictions of a monkey's mind, if there are any convictions in such a mind?[53]

We see in this quote Darwin's persistent belief in the rule of natural laws and his belief in the unity of science that he assumes will some day result in one great explanatory law. He also denies that in their specific actions natural laws display evident teleological purpose. At the same time he affirms that the universe is not the result of something he is calling "chance." Darwin's meaning of "chance" here is not immediately transparent. One reading can suggest that he is affirming that all natural events happen in accord with deterministic laws, a reading I accept as the most probable. Or it could suggest that he is endorsing a weak form of teleological purposiveness of nature as a whole, if not of its individual parts. This latter reading is supported by similar statements we can find in Darwin's writings as early as 1859, and it is the interpretation given to these discussions by Graham himself, who interprets Darwin to have opposed the Democritean claim that the world is "the upshot of chance."[54]

We also see Darwin raise in his letter to Graham the possibility of an ultimate epistemological skepticism that might follow from his theory, the consequence that the naturalistic development of mind by the processes of nature would in the final analysis undermine any claims to the validity of conclusions of reason. Darwin raises this option only to leave it suspended, and nothing he says in the letter adequately responds to this possibility. Yet in Graham's view, Darwin remained a "robust and cheerful optimist."[55] The potential for such radical skepticism is countervened, it seems, by an underlying belief in the continuity between a rational nature and consciousness, providing the ground for the assumption that science and genuine knowledge are possible even though human reflection is derived from the processes of evolution. This claim is neither explicated nor justified. It simply rests upon a deep belief in the underlying ground of a substantive "nature." If the claims of William Darwin that began this chapter can be accepted, "Nature" it seems, if no longer a pantheistic surround, or a wise and intentional selector, nonetheless is still able to dictate a moral code and to demand from Charles Darwin a respect harboring on reverence.

CONCLUSION

Unpacking some of the dimensions of Darwin's complex philosophy of nature should free us from a monochromatic reading of Darwin and his philosophical heritage that has been imposed by neoselectionist evolutionary theory. If the arguments of this chapter can be accepted, it can be seen that the question of teleology and the purposiveness of nature are complex issues in Darwin's thought. My central argument has been that we cannot plot Darwin's intellectual development along a line that moves from the external design-contrivance natural theology of Paley to a mature belief in a purposeless nature that undercuts all teleological principles. Rather, there is a persistence of a philosophy of nature, indebted to certain strands within German reflections on nature, that underpins the concept of natural laws and the relation of mind and matter and that even sustains some kind of ethical code. If Darwin's "nature" is not imbued with the explicit teleology of Aristotle, it has at the very least some interesting connections with nonmechanical views of nature that may assist us in overcoming some of the contemporary split between "is" and "ought."

NOTES

1. Darwin Papers, Cambridge University Library, MS DAR 112 (ser B), 3d verso–3e recto. Hereafter cited by DAR numbers. This and other direct quotations from the DAR archives with permission of the Syndics of Cambridge University Library.

2. David Depew and Bruce Weber, *Darwinism Evolving: Systems Dynamics and the Genealogy of Natural Selection* (Cambridge, MA: MIT Press, 1995), chap. 10.

3. Robert Richards, *The Romantic Conception of Life* (Chicago: University of Chicago Press, 2002), chap. 10. See also Robert Richards, "Darwin's Romantic Biology: The Foundation of His Evolutionary Ethics," in *Biology and the Foundation of Ethics,* ed. J. Maienschein and M. Ruse (Cambridge: Cambridge University Press, 1999), 113–53.

4. For useful discussions of this background I am indebted to J. E. McGuire, "Boyle's Conception of Nature," *Journal of the History of Ideas* 33 (1972): 523–42; Catherine Wilson, "De Ipsa Natura: Sources of Leibniz's Doctrines of Force, Activity and Natural Law," *Studia Leibnitiana* 19 (1987): 148–72; and Jean Ehrard's classic *L'idée de nature en France dans la première moitié du xviiie siècle* (1963; reprint, Paris: Michel, 1994).

5. Descartes, *Le monde* (1630), in *Oeuvres de Descartes,* ed. C. Adam and Paul Tannery (Paris: Cerf, 1909), 11:36–37.

6. Boyle's gives eight separate meanings to the concept of nature in his *Free Inquiry into the Vulgarly Received Notion of Nature,* ed. E. B. Davis and M. Hunter (Cambridge: Cambridge University Press, 1996). These eight definitions subsequently formed the substance of Ephraim Chambers's discussion of the concept of nature in his *Cyclopaedia: Or an Universal Dictionary of Arts and Sciences* (London: J. J. Knapton, 1728), 4:617–18. Via Chambers, Boyle's definitions formed the basis of the article "Nature," in Diderot's *Encyclopédie ou dictionnaire raisonnée* (Neuchâtel, 1765), 11:40–41.

7. On Boyle's opponents, see Edward Davis, introduction to Boyle, *Free Inquiry,* xxii. In his *True Intellectual System of the Universe* (1678), bk. 1, chap. 3, Ralph Cudworth had considered the domain of living beings in particular to require a conception of an inward principle of life and organization, separate from God's direct supervision, that could account not only for the regularity of things but also for monstrosity and abnormalities. I am using the selection from the *True Intellectual System* in *The Cambridge Platonists,* ed. C. A. Patrides (Cambridge: Cambridge University Press, 1980), 293.

8. Boyle, *Free Inquiry,* § 4, p. 36.

9. Buffon, "De la nature: Première vue," in *Histoire naturelle, générale et particulière,* vol. 12 (1764), reprinted in *Buffon: Oeuvres philosophiques,* ed. J. Piveteau (Paris: Presses Universitaires de France, 1954), 31. My translation.

10. Robert Richards's study *Romantic Conception,* esp. chap. 11, has clarified the relations between the *Naturphilosophie* of Schelling and that of the Romantic movement of the Weimar circle and some of the connections between Herder and the pantheism revival that predated Schelling. See also John Zammito, *The Genesis of Kant's Critique of Judgment* (Chicago: University of Chicago Press, 1991), chaps. 8–10. Because of the interactions that developed later between Goethe and Schelling, it is difficult to separate these two forms of the philosophy of nature after 1800. Richards's analysis of the rapid development in Schelling's project as it emerged from the influence of Fichte in the *Ideen zu einer Philosophie der Natur* (1797), through the *Von der Weltseele* (1798), the *Erster Entwurf eines Systems der Naturphilosophie* (1799), and the *Einleitung zu dem Entwurf eines Systems der Naturphilosophie* (1799), displays several strands within Schelling's own *Naturphilosophie,* with his latter expressions more important for the theme of this chapter. Consequently, my association of Darwin with German philosophy of nature via Humboldt, in general agreement with Richards (*Romantic Conception,* chap. 14), is not to claim that Darwin was developing the strand of systematic *Naturphilosophie* that can be followed from Schelling's earlier expressions into Lorenz

Oken, or as it was developed in British contexts by such biological theorists as Joseph Henry Green and Richard Owen. I have discussed some of these issues in greater detail in Phillip Sloan, "Whewell's Philosophy of Discovery and the Archetype of the Vertebrate Skeleton: the Role of German Philosophy of Science in Richard Owen's Biology," *Annals of Science* 60 (2003), 39–61.

11. I have discussed aspects of this monistic concept of nature in Phillip Sloan, "The Sense of Sublimity: Darwin on Nature and Divinity," *Osiris* 16 (2001): 251–59.

12. George Christoph Tobler, "Nature," as transcribed by Goethe in his journal, in *Goethe: Scientific Studies,* ed. and trans. D. Miller (New York: Suhrkamp, 1988), 3. In his later commentary on this essay in 1828, following his more systematic study of Kant and Schelling, Goethe acknowledged that it accurately captured his conception of nature in his early years but that from his perspective of 1828, it was incomplete in its failure to deal sufficiently with his notions of polarity and intensification (Goethe, *Scientific Studies,* 6–7).

13. See the entries on "Nature" in the *Encyclopedia Britannica* from the first edition of 1771 through the fourth edition of 1810. In altered form, Boyle's formulations still survive in the entry of the seventh edition of 1842. Boyle's definitions also continue to form the basis of the articles in William Nicholson's *The British Encyclopedia or Dictionary of the Arts and Sciences* (London: Longmans, Hurst, Rees, Orme, 1809) and Abraham Rees's *Cyclopedia, or Universal Dictionary of Arts, Sciences, and Literature* (London: Longman, Hurst, Rees, Orme and Brown, 1819).

14. For more detailed development of this point, see Sloan, "Sense of Sublimity," esp. 252–61, and "The Formation of a Philosophical Naturalist," in *The Cambridge Companion to Darwin,* ed. J. Hodge and G. Radick (Cambridge: Cambridge University Press, 2003), 17–39.

15. See Richards, *Romantic Conception,* chap. 10. The style of Darwin's letters home had become so "Humboldtian" in character that his family remarked negatively on his adoption of Humboldt's style. See letter of Caroline Darwin to Charles Darwin, October 28, 1833, in *The Correspondence of Charles Darwin,* ed. F. Burkhardt and S. Smith (Cambridge: Cambridge University Press, 1985–), 1:345. I thank Robert Richards for this reference.

16. This point is developed in more detail in Sloan, "Formation." On Humboldt's conception of science, see especially M. Dettelbach, "Global Physics and Aesthetic Empire: Humboldt's Physical Portrait of the Tropics," in *Visions of Empire:Voyages, Botany, and Representations of Nature,* ed. D. P. Miller and P. H. Reill (Cambridge: Cambridge University Press, 1996), 258–92.

17. Richard Keynes, ed., *Charles Darwin's Zoology Notes and Specimen Lists from H.M.S. Beagle* (Cambridge: Cambridge University Press, 2000), entry for June 1832, 52.

18. See Darwin, *The Diary of the Voyage of the H. M. S. Beagle,* ed. N. Barlow (1933), reprinted in *The Works of Charles Darwin,* ed. Paul H. Barrett and R. B. Freeman (London: Pickering, 1986–89), 1:334.

19. Ibid., 46.

20. Ibid., 388.

21. Letter of Elizabeth Wedgwood to Hensleigh Wedgwood, November 16, 1836, in Darwin, *Correspondence,* 1:520.

22. C. G. Carus, "The Kingdoms of Nature, Their Life and Affinity," in *Scientific Memoirs,* 4 vols., ed. R. Taylor (1837; reprint, New York: Johnson Reprint, 1966), 1:223.

23. Ibid., 223. Emphasis in original.

24. Ibid., 225–26.

25. Ibid., 254.

26. Darwin, C Notebook, fols. 103–4, as transcribed in Charles Darwin, *Charles Darwin's Notebooks, 1836–1844,* ed. P. H. Barrett et al. (Cambridge: Cambridge University Press, 1987), pp. 269–70. I will use transcription conventions as employed in the Barrett edition: < > indicates crossouts; << >> indicates interlineations; boldface indicates marginal additions. The bracketed comment [his definition . . . rest of] is Darwin's own. Ellipses indicate my deletion.

27. Ibid., fols. 210e–211, p. 305. Compare to Carus, "Kingdoms of Nature," 1:227. This passage also seems to relate to his reading of William Kirby's *On the History, Habits and Instincts of Animals,* 2 vols., Bridgewater Treatise 7 (London, 1835), 2:247. See "Old and Useless Notes" (DAR 91, fol. 36) in Darwin, *Charles Darwin's Notebooks,* 613. Although Kirby refers to a universal instinct "modified according to species," Darwin alters this instinct into a "thinking" principle inherent in matter.

28. T. Malthus, *An Essay on the Principle of Population,* 1803 edition, 2 vols., ed. P. James (Cambridge: Cambridge University Press, 1989), 2:117. Darwin read Malthus in the sixth edition of 1826. This passage is unaltered in the 1826 edition.

29. See, e.g., essays in R. M. Young, *Darwin's Metaphor: Nature's Place in Victorian Culture* (Cambridge: Cambridge University Press, 1985).

30. See Sloan, "Formation," 26–37.

31. Darwin, B Notebook, fol. 229, *Charles Darwin's Notebooks,* 228.

32. For example, Darwin's early teacher in physiology at the University of Edinburgh during his medical student days (1825–27), W. P. Alison, expounded his physiology in terms of the distinction between "vital" laws governing life and normal physical laws governing inorganic nature. See W. P. Alison, *Outlines of Physiology and Pathology* (Edinburgh: Blackwood, 1836), "Preface" esp. vii–ix, and "Supplement" on vital action. For additional discussion of the distinction of vital and inorganic laws of nature in the period surrounding Darwin's writings, see W. D. Clark, "Report on Animal Physiology," *Report of the Fourth Meeting of the British Association for the Advancement of Science* (London: Taylor, 1835), 95–142. See also W. B. Carpenter, "On the Differences of the Laws Regarding Vital and Physical Phenomena," *Edinburgh New Philosophical Journal* 24 (1838): 327–53. For Carpenter, special properties of life are transmitted in a lawlike way by generation. Word frequencies for uses of *law*" in Darwin's combined Notebooks are "Law" (89) and "Laws" (88). Of these 72 (38 + 34) are in some way references to "vital" or specifically biological laws. There are

seven occurrences of "law(s)" of nature. Word frequencies have been determined from D. Weinshank et al., *A Concordance to Charles Darwin's Notebooks, 1836–1844* (Ithaca, N.Y.: Cornell University Press, 1990).

33. John Stuart Mill, "Review of the Works of Samuel Taylor Coleridge," *Westminster Review* 33 (1840): 257–302.

34. Kirby, *On the History.* Kirby is summarizing Lamarck's discussion in the "Discours préliminaire" of *Histoire naturelle des animaux sans vertèbres,* vol. 1 (Paris, 1815).

35. Darwin, DAR 91, fols. 34–35. A complete transcription of this document is given in *Charles Darwin's Notebooks,* 610–11. I have made some slight modifications from this published transcription on the basis of my own reading of the original manuscript.

36. This continuity is also in evidence in a concluding note to his first comprehensive drafting of his theory of natural selection, the thirty-five-page pencil sketch of 1842: "The supposed creative spirit does not create either number or kind, wh. frm. analogy adapted to site (viz. New Zealand) it does not keep them <<all>> permanently adapted <<to>>any country—it works on spots or areas of creation—it is not persistent for geol periods—it creates form of same group in same regions, with no physical similarity—its power seems influenced or related to the range of other species wholly distinct of the same genus.—"Darwin, "Draft of 1842," DAR 6, fol. 35. My transcription of the MS differs slightly from the printed version in Francis Darwin, ed., *Foundations of the Origin of Species* (Cambridge: Cambridge University Press, 1909), 52 n.

37. See Sloan, "Sense of Sublimity," for details on the intermediary period.

38. For a contrasting view, see Depew and Weber, *Darwinism Evolving,* chaps. 4–5, in which they argue that Darwin transferred the Newtonianism of British political economy to the living world. See also S. Schweber, "The Origin of the *Origin* Revisited," *Journal of the History of Biology* 10 (1977): 229–316.

39. The insertion of the phrase "by the Creator" into the final paragraph in the second (1860) and all later editions does not significantly alter the implicit distinction of vital laws as the secondary means.

40. Lyell seems to have been the first to raise this objection. See Darwin, *Correspondence,* 8:250. More extended objections on this point were raised by William Henry Harvey in a letter of August 24, 1860 (Darwin, *Correspondence* 8:329).

41. Ibid., 8:371, 389, 397, 403, 416.

42. This passage is unchanged through all subsequent editions. The quote is from Charles Darwin, *Origin of Species,* 6th ed. (New York: Modern Library, 1934), 64. The substance of this passage was later incorporated into the opening discussions of Darwin's *Variation of Animals and Plants under Domestication,* 2nd ed. (New York: Appleton, 1874), 1:6–7.

43. His meaning here is most clearly illuminated by his one explicit definition of the concept of nature prior to the publication of the *Origin* that appears in the "Natural Selection" manuscript of 1856–58 in the critical edition of R. Stauffer, ed., *Charles Darwin's Natural Selection* (Cambridge: Cambridge University Press, 1975). In an insertion made

following the phrase "See how differently Nature acts!" Darwin has added: "By nature, I mean the laws ordained by God to govern the Universe" (ibid., 224). In terms of the previous meanings of the concept of nature outlined in this chapter, there are the strongest similarities between Darwin's 1856–58 definition and that given by Buffon in 1764. Darwin worked through the English edition of Buffon's *Natural History* containing his essays on nature in the summer of 1840.

44. The word frequencies have been derived from the concordance to the first edition of the *Origin* (P. H. Barrett, D. J. Weinshank, and T. T. Gottleber, *A Concordance to Darwin's "Origin of Species," First Edition* (Ithaca, N.Y.: Cornell University Press, 1981) and from the similar concordances to the first edition of the *Descent of Man*—Paul H. Barrett et al., *A Concordance to Darwin's "The Descent of Man, and Selection in Relation to Sex"* (Ithaca, N.Y.: Cornell University Press, 1987)—and the *Expression of the Emotions*—P. H. Barrett et al., *A Concordance to Darwin's "The Expression of the Emotions in Man and Animals"* (Ithaca, N.Y.: Cornell University Press, 1986). The frequencies for the sixth edition of the *Origin* and the *Various Contrivances* have been determined by word search on the electronic CD-ROM version, supervised by M. Ghiselin (California Academy of Sciences/ Lightbinders, 1994).

45. Sloan, "Sense of Sublimity."

46. Francis Darwin, ed., *The Autobiography of Charles Darwin and Selected Letters* (1892; reprint, New York: Dover, 1958), 65.

47. See Charles Darwin, *The Formation of Vegetable Mould through the Action of Worms, with Observations on Their Habits* (1881), in *Works*, vol. 28, esp. pp. 11, 15, 29.

48. Charles Darwin, *The Descent of Man and Selection in Relation to Sex* (facsimile reprint of 1871 London first edition; Princeton, N.J.: Princeton University Press, 1981), 39.

49. His mentor, John Stevens Henslow, had treated these issues at some length in his physiological botany lectures of the 1830s that Darwin attended at least twice. Henslow's lectures from the mid-1830s are available as his *Descriptive and Physiological Botany* (London: Longman, Rees, Orme, Brown, Green & Longman, 1836). I have discussed Henslow's views on vital powers in Phillip Sloan, "Darwin, Vital Matter, and the Transformism of Species," *Journal of the History of Biology* 19 (1986): 369–445, esp. 373–96.

50. Charles Darwin, *The Power of Movement in Plants* (New York: Appleton, 1888), 573.

51. Graham speaks of four letters between them in his letter to Francis Darwin (Graham to Francis Darwin, DAR 144). Only a press copy of Graham's letter to Francis and a press copy of Darwin's letters of July 3 and August 5 of 1881 to Graham in copies in Graham's hand survive in the DAR materials.

52. Graham to Francis Darwin, n.d., DAR 144, Graham letters, item 1, fol. 1.

53. Charles Darwin to William Graham, July 3, 1881, DAR 144, item 2, fols. 1–2. This is a press-copy in another hand. I have made slight corrections on the basis of the manuscript from the version published in *The Life and Letters of Charles Darwin, Including an Autobiographical Chapter* (1887; reprint, New York: Appleton, 1888), 1:284–85.

54. Graham to Francis Darwin, n.d., DAR 144, Graham letters, fol. 2. The letter accompanies his transmission to Francis of copies of his correspondence with the elder Darwin after Darwin's death. This interpretation closely connects to similar statements in Darwin's correspondence of earlier years. See, e.g., the letter of Darwin to Asa Gray, November 26, 1860, in Darwin, *Correspondence,* 8:496.

55. Letter of Graham to Francis Darwin, n.d., DAR 144, fol. 2. The full statement reads: "As regards Evolution and the progress of our species the words of your father in the letters referred to are optimistic as they are in his books, and it is a good thing to find a robust and cheerful optimist in these days like your father and Herbert Spencer."

CHAPTER 8

Darwin's Metaphysics of Mind

Robert J. Richards

My theory would give zest to recent & Fossil Comparative Anatomy, & it would lead to study of instincts, hereditary & mind hereditary, whole metaphysics.

Darwin, Notebook B, 1837

Our image of Darwin is hardly that of a German metaphysician. By reason of his intellectual tradition—that of British empiricism—and psychological disposition, he was a man of apparently more stolid character, one who could be excited by beetles and earthworms but not, we assume, by abstruse philosophy. Yet Darwin constructed a theory of evolution whose conceptual grammar expresses and depends on a certain kind of metaphysics. During his youthful period as a romantic adventurer, he sailed to exotic lands and returned to construct a theory that attacked the citadels of orthodoxy. In the long process of theory construction, he explored difficult philosophical questions—for instance, the nature of reason and the mind-body problem. Moreover, he founded that theory on something like a concept of absolute mind, echoing from afar ideas propounded by such German Romantic scientist-philosophers as Friedrich Schelling and, more proximately, Alexander von Humboldt.

In this essay, I will explore the metaphysical grammar that underlay Darwin's theory. This grammar structured the way he joined the various parts of his conception and reveals itself most perspicuously in the metaphors that he constantly deployed to articulate his ideas. He used these tropes, certainly, to make his ideas come alive for his readers. But as he constructed his theory, he also employed the more significant of them to explain to himself the nature of his slowly developing notions. In particular, he came to understand human mind, productive nature, and his special explanatory device of natural selection all with the indispensable aid of particular metaphors and similes. In what follows, I will first consider specifically his developing ideas about rational mind, its animal origins and human embodiment, and then turn to what might be called the concept of absolute mind, a concept that structured his general theory of evolution by natural selection.

HUMAN MIND

Darwin recognized from the beginning of his theorizing about the transmutation of species that he would have to include human beings within the ambit of his considerations. Should he allow man to escape the net of his hypothesis, our species would drag the Creator back into the picture—something that Darwin had no intention of permitting. This is not to say that he initially denied the work of a Creator God. He recognized the need for a divine shove to get the world spinning; but during his travels in South America he was conscious of the absence of a benevolent personal power, when, for instance, he watched young Indian women being slaughtered by ignorant Spanish soldiers because "they breed so." This kind of experience caused him to ask himself: "Who would believe in this age in a Christian civilized country that such atrocities were committed?"[1] Not that Darwin had a particularly high estimate of the intellectual or moral worth of the natives. As he viewed the naked, paint-smeared Fuegians, whose language seemed little better than the guttural cries of a beast, he reflected that "one can hardly make oneself believe that they are fellow creatures placed in the same world."[2] From his experience of the extremes of human behavior during his *Beagle* voyage (1831–36), Darwin came to perceive a gradation among human beings, from the Fuegians who seemed so like the animals, through the hardly more civilized gauchos, to his own messmates aboard the *Beagle*. Yet while on the voyage, he had not yet come to the view that there might be a transition from animals to man.

During the few months after he returned from his journey, Darwin did become convinced that species were not stable; and with this conviction, he quickly began to explore questions of instinct, mind, and, as he termed it, the "whole metaphysics."[3] He moved fairly rapidly from considerations of anatomical similarities among animals to cognitive comparisons between animals and man; and so he came to rest the divinity of man on the shoulders of a monkey: "He is Mammalian—his . . . origin has not been indefinite—he is not a deity, his end <<under present form>>will come, (or how dreadfully we are deceived) then he is no exception.—he possesses some of the same general instincts, <as> & moral feelings as animals.—they on other hand can reason—but Man has reasoning powers in excess. Instead of definite instincts—this is a replacement in mental machinery—so analogous to what we see in bodily, that . . . it does not stagger me."[4]

In coupling human mind in train with animal mind, Darwin followed out one deeply embedded track of the philosophical legacy of British empiricism, namely the contention that rational activity consisted in the manipulation and association of faint sensory images, something of which even animals were capable. His grandfather, in *Zoonomia,* endorsed this sensationalist view of reason, and Darwin followed in his progenitor's footsteps. Also like Erasmus Darwin, Charles became a reader of David Hume.[5]

Darwin picked up Hume's *Inquiry concerning Human Understanding* in August of 1838, just after he had read Malthus's *Essay on Population.* In accord with Hume and his grandfather, Darwin considered, as he jotted in his Notebook N, that "[r]eason in simplest form probably is simple comparison by senses of any two objects— they by VIVID power of conception between one or two absent things—reason probably mere consequence of vividness & multiplicity of things remembered & the associated pleasure as accompanying such memory."[6] If reason were only the comparison of recalled sensory images, then animals would be as capable of reason— at least in a rudimentary way—as any Fuegian. And looking in the other direction, reason in its elemental form seemed little different than animal instinct. As a mental disposition, instinct was a set of cognitive impulses that led to stereotyped behavior in a species—for example, weaverbirds that innately knew how to tie the peculiar knot that held their nests together. Instinct might then be considered a rather rigid pattern of mental structures and reason a more flexible pattern that allowed an animal to accommodate to environmental contingencies— or at least this is the way Darwin thought about these mental powers.[7]

If animals possessed mental faculties—both instincts and modest rational abilities—the question became significant: What was the connection between mind and brain? After all, a Humean analysis of reason takes place on a phenomenal plane, but Darwin was also deeply interested in reason's assumed biological substructure. Instinct seemed clearly a biological trait, since it could be inherited and modified through breeding. Thus when two varieties of dog, each with different hunting instincts, were mated, their offspring would show not only anatomical features common to both parents but also a mixed repertoire of instincts.[8] Such evidence demonstrated that mental structures were heritable in a way no different than anatomical structures. But if heritable, they had in the first instance to be located in some physical traits. The brain, Darwin assumed, was their locus.

In this early period of theorizing (1838–42), Darwin attempted to clarify exactly the relation between mind and its neural substrate. He reflected, for instance, on the way in which drunks suffering from delirium tremens could still perform wobbly mental acts without much consciousness—an example that might be familiar to one who had voyaged across oceans and messed with rough-hewn sailors. But even considering what he himself might do out of well-worn habit suggested to Darwin that behavior could be mentally controlled without conscious attendance. Evidence of this kind indicated that instinct—which was, after all, only inherited habit—resulted from certain brain patterns. This meant that the extraordinary instincts of animals could be wrested from the Empyrean, where many natural theologians found their origin, and brought back to earth. Instincts were due to the modifications of a quite mundane brain. "A train of thought, action &c will arise from physical action on the brain," and so, as he concluded, this "renders much less wonderful the instincts of animals."[9]

Despite the close connection of instinct and thought with the brain, Darwin yet regarded their mental existence as distinct from their source in brain. But how to conceive it? He tried several metaphors to help conjure with the connection between mind and body. The brain, for example, might secrete thoughts as the liver secreted bile.[10] Or better, thought might be a *force* of the brain comparable to the gravitational attraction that bodies exerted on one another.[11] Darwin rather liked the proportional comparison of thought to brain with gravitational force to matter. He performed several thought experiments—so to put it—to stabilize the metaphor further. For instance, he let free-floating ideas drift through his mind; he then would try to hold one simple idea in consciousness (say, the idea of the color red); and finally he would engage in abstract thought to solve a problem. In this progression, he noted that these tasks required increasing effort.[12] Since he felt a sliding degree of effort while moving from one kind of cognitive performance to another, this suggested that mentality was indeed like a force of the brain. And so in this respect, ideas were quite analogous to other forces in the physical world.[13]

Such thought experiments led Darwin along a path toward a traditional conundrum: How can we distinguish the mental state of dreaming from that of being consciously awake? His metaphor of force gave him a tentative solution. He decided that dreaming required virtually no effort but that consciousness demanded a continuing effort to compare present with past ideas. The active effort of comparing indicated that real work was being done. Hence degree of effort might well be a sign that Descartes's daemon was not beguiling us.[14]

Darwin's thought experiments also led him to ponder the nature of the imagination, which almost seemed like dreaming. Yet, on reflection, he recognized that his own thinking about various aspects of his theory abounded in imaginative constructions; they were actually doing work. He came to regard such "castles in the air" as a propaedeutic to real scientific discovery. This, I believe, is an important aspect of Darwin's own science and worth lingering over. As I will show in a moment, imaginative constructions lie at the root of his thought about evolution and decisively control aspects of his theory. In his Notebook M, he reflected, with a modicum of humor, on the nature of these castles in the air and their relation to the development of scientific hypotheses: "Now that I have a test of hardness of thought, from weakness of my stomach, I observe a long castle in the air is as hard work . . . as the closest train of geological thought—the capability of such trains of thought makes a discoverer, & therefore . . . such castles in the air are highly advantageous, before real train of inventive thoughts are brought into play."[15]

Darwin may have derived his attitude about the importance of imaginative constructions from reading both Charles Lyell's *Principles of Geology* and Humboldt's *Personal Narrative of Travels to the Equinoctial Regions of the New Continent*. Both treatises employ imaginative scenarios to provide reasonable suppositions concerning events of the past. And through the dexterity of their telling, these imaginative stories gradually become insinuated in the reader's mind as sound evidence. For instance, in the space of a few pages of the second volume of the *Principles,* Lyell begins virtually every paragraph with such locutions as "Let us next imagine a few cases of the elevation of land . . . ," "Let us next suppose . . . ," "We will imagine the summits . . . ," and so on throughout his volumes.[16] By use of these inviting castles in the air, Lyell argued ingeniously for his theory of geological uniformitarianism, a theory that Darwin adopted to secure his own proposals. Lyell's powerful example could not but convince the young naturalist of the importance of imagination for guiding reason. His own practice in the *Origin* would demonstrate how such imagination not only stabilized reason but also gave it a distinct trajectory.

Darwin's construction of imaginative thought as a quasi-physical force brought him only so far in his effort to understand the mind-brain relationship. He, like many philosophers before him, arrived at an impasse. He could not go any further in parsing the relationship—though for his purposes, he need not have gone any further. He concluded that all we could really say about the mind-body connection was that "thought & organization run in a parallel series."[17]

DARWIN'S "MATERIALISM"

During the late 1830s, while constructing the fundamental features of his theory, Darwin was quite aware of the philosophical and theological implications of his conception of mind. Since human mental traits were comparable to those of animals, differing only in degree, he felt assured that his theory of transmutation could indeed bring humans within its purview.[18] By employing the resources of a moderate Humean perspective, which regarded both instinctive cognition and reason as constituted by sensory images in more or less fixed patterns, he could smooth the way for charting the development of the human species out of animal species. But this meant that certain kinds of traditional language used to describe the human mind would no longer be applicable, particularly language that carried a theological burden—the term *soul,* for instance. Since such locutions were not used for animals, there was no longer justification, Darwin concluded, for using them to refer to human beings.[19]

Darwin recognized that his conception of human mind would be labeled by some as materialistic, a designation that was often used synonymously with *atheistic.* In this early period, Darwin was certainly no atheist, nor even yet an agnostic; and he did not want to leave himself liable to the charge of irreligion. In his notebooks, he cautioned himself: "To avoid stating how far I believe in Materialism, say only that emotions, instincts, degrees of talent, which are hereditary are so because brain of child resembles parent stock."[20] Not only was this modest conclusion safe, but it recognized his own unsettled beliefs about mind. At the phenomenal level, he maintained that little difference existed between animal instinct and animal reason and that animal reason differed only in degree of complication from human reason. He did affirm that ideas were related to brain in a deterministic fashion—maybe the relationship was like force to matter; yet the exact nature of the connection eluded him, as it still does us.

From this early period through the first few years after the publication of the *Origin of Species* in 1859, Darwin remained a theist, if a rather weak-kneed one.

He did not think his sort of materialism of mind necessarily led to atheism or even precluded an afterlife—though he did not explore these issues with any logical rigor. Rather, his whole theory, as he initially constructed it, had a teleological orientation, an orientation that gave succor to the more orthodox conception of a God-driven universe, in which he still believed. As he put it around 1839 in a rather inchoate passage: "This Materialism does not tend to Atheism . . . we are steps towards some final end—production of higher animals—perhaps, say attribute of such higher animals may be looking back. ∴ Therefore, consciousness, therefore reward in good life."[21] Darwin thus held that the growth of consciousness—that is, the production of the higher animals—was the final cause of the transmutational process. A higher consciousness, according to the above quoted passage, may look back to its roots in animal life; but it may also look forward to a divine reward for a good life. The notion that we would be rewarded in an afterlife slowly faded from Darwin's system of belief. What did not fade, however, was his assumption of a teleological orientation for the evolutionary process. The last few lines of the *Origin of Species* reiterate the idea that the purpose of the "war of nature" has been "the production of the higher animals."[22] For Darwin, it was what was left of divine purpose after God took leave. Nature came to inherit the mantle of the recently departed deity, displaying almost all of the divine powers—omniscience, benevolence, creativity, and wisdom. This is but to say that Darwin came to conceive of nature as possessed of something like absolute mind.

ABSOLUTE MIND AND THE EVOLUTIONARY PROCESS

From the beginning of his theorizing, Darwin employed mind as a model for understanding the evolutionary process. In the initial pages of his first transmutation notebook, begun in 1837, he queried himself: "Each species changes. Does it progress. Man gains ideas. The simplest cannot help—become more complicated; & if we look to first origin there must be progress."[23] Here the progressive character of mind became a way of understanding the progressive character of species development.[24]

Also in this early notebook, Darwin employed a cognitive device to explain species adaptation. He had, in his pre-Malthusian work, proposed that species alteration would occur as the result of the direct effects of the environment—hence animals in colder climates would produce heavier coats. He quickly realized, however, that direct environmental impact could hardly shape organisms to their surroundings in an intricate manner. In mid-1838, he suggested, alternatively, that animals might develop habits that, if practiced over many generations, would become instinctive — that is, manifested without learning. These instincts, he believed, would then come slowly to change anatomy. In this way, mind — expressed in acquired habit and instinct — could produce adaptations that more finely fitted an animal into its environment. So, for example, Darwin considered how habit might modify the foot of the jaguar: "Fish being excessively abundant & tempting the Jaguar to use its feet much in swimming, & every development giving great vigour to the parent tending to produce effect on offspring — but whole race must take to that particular habit.— All structures either direct effect of habit, or hereditary & combined effect of habit."[25] Animal mind, then, could adapt individuals of a species to their particular surroundings, and these adaptations would be delivered to descendants through inheritance. Even after having arrived at his principal device of evolutionary change — natural selection — Darwin never abandoned inherited habit as an auxiliary mode of alteration, and he also advanced it as a source of variation on which natural selection might operate.

It is often assumed that when Darwin read Malthus in late September 1838, the idea of natural selection dropped fully formed from his brain. Certainly the idea of natural selection had enough analytic simplicity to make it seem an all-or-nothing affair. Thomas Henry Huxley, on reading the *Origin of Species,* exclaimed, "How extremely stupid not to have thought of that!" The apparent simplicity of the idea has often led critics to identify Darwin's natural selection with later formulations by neo-Darwinians. But the idea came slowly to Darwin, and then in the guise of a kind of preternatural intelligence, a kind of absolute mind.

The best way to show this and to indicate how natural selection was modeled, not on some machinelike operation, but on mind, is to begin with its formulation in the *Origin* and then excavate its deeper strata. In what follows, I will move back to the manuscript Darwin was working on when he received that fateful letter from Alfred Russel Wallace in 1858 and then further back to the essays of 1842 and 1844, when he first sketched out his ideas at length.

In the fourth chapter of the *Origin,* Darwin describes natural selection in contrast to man's selection:

Man can act only on external and visible characters: nature cares nothing for appearances, except in so far as they may be useful to any being. She can act on every internal organ, on every shade of constitutional difference, on the whole machinery of life. Man selects only for his own good; Nature only for that of the being which she tends. . . . If may be said that natural selection is daily and hourly scrutinizing, throughout the world, every variation, even the slightest;

rejecting that which is bad, preserving and adding up all that is good; silently and insensibly working, whenever and wherever opportunity offers, at the improvement of each organic being in relation to its organic and inorganic conditions of life.[26]

Several aspects of this description of natural selection reveal a deeper conception, namely something like absolute mind as operative in nature. First, note that when Darwin says nature seeks "the good of the being which she tends" and aims at the "improvement of each organic being," we want to know what or who is the *being* to which he refers. It cannot be the individual organism, which, strictly speaking, is not improved by nature, though its offspring might be. Moreover, does nature seek the improvement of *each* organic being, as Darwin says? Hardly, since most organic beings will be destroyed by natural selection. Nature, at least as the language of this passage suggests, must be aiming at an ideal end that transcends the individual—something only a mind might do. Throughout the *Origin,* but particularly in the last paragraph of the book, Darwin indicates this ideal end, this final cause of the whole evolutionary process, to be the production of the higher animals. This means that death and destruction are the agents, the necessary agents, for realizing the greater good that nature *intends.*

Natural selection, in the passage just quoted, has a vision that can penetrate into the very fabric of life, detecting the slightest variation and then selecting that variation for the good of the creature—or at least its descendants. This is, as Darwin portrays it, an altruistic act, unlike man's selfish choices. The actions of natural selection are thus hardly that of a machine, even a very powerful Manchester spinning loom—or of mute, blind causal forces.

The passage in the *Origin* has its progenitor in the manuscript that Darwin put aside in 1858 when he got Wallace's letter describing a similar theory of species transmutation. In that manuscript, which was to be called "Natural Selection," his formulation of the comparable passage ran:

[Man] selects any peculiarity or quality which pleases or is useful to him, regardless whether it profits the being. . . . See how differently Nature acts! She cares not for mere external appearance; she may be said to scrutinize with a severe eye, every nerve, vessel & muscle. . . . Can we wonder then, that nature's productions bear the stamp of a far higher perfection than man's product by artificial selection. With nature the most gradual, steady, unerring, deep-sighted selection,—perfect adaptation to the conditions of existence.[27]

As we move down through the compositional strata, we see ever more clearly that natural selection takes on intellectual qualities. Here selection is "deep-sighted" and able to produce perfect adaptations—again, certainly out of the range of possibility for any machine.

The *Origin*'s description of natural selection has yet further depths, which are revealed in the essays of 1842 and 1844, the first extensive renditions of his theory. Darwin had a fair copy made of his 1844 essay, so as to bequeath it to posterity should he die before having a chance to write a proper book. At this level, the metaphorical formulation stands clear—but a formulation whose grammar controlled the structure of the mature theory. Darwin wrote:

> Let us now suppose a Being with penetration sufficient to perceive differences in the outer and innermost organization quite imperceptible to man, and with forethought extending over future centuries to watch with unerring care and select for any object [i.e., for any purpose] the offspring of an organism produced under the foregoing circumstances; I can see no conceivable reason why he should not form a new race (or several were he to separate the stock of the original organism and work on several islands) adapted to new ends. As we assume his discrimination, and his forethought, and his steadiness of object, to be incomparably greater than those qualities in man, so we may suppose the beauty and complications of the adaptations of the new races and their differences from the original stock to be greater than in the domestic races produced by man's agency.[28]

A comparable image can be found in Darwin's first essay (1842) elaborating his theory.[29] He initially used these tropes of a powerful intelligence to explain natural selection to himself, to work out his understanding of this burgeoning idea. And it is patent from their residual expression in the manuscript "Natural Selection" and in the *Origin of Species* that the structure of the metaphor still controlled his understanding and the development of his general theory. Indeed, so embedded in the theory were the implicit features of the metaphorical image that when Wallace suggested to him that he replace the term *natural selection* with Spencer's formulation "survival of the fittest," he declined.[30]

One could argue, of course, that Darwin's metaphor of a powerful mind doing the selecting in nature was only a rhetorical device meant to make the idea easier for his readers to digest. One might, therefore, wish to distinguish Darwin's expression in the *Origin* from what the theory *really* entails. This would be to suggest, however, that the real theory was some timeless Platonic entity rather than a creature of history. I believe that Darwin's theory is enmeshed in his expression of it, not floating in some third world that only the likes of Karl Popper or Imre Lakatos — or Michael Ruse — might communicate with.[31] And, of course, it is the idea of natural selection grounding its actual expression that has controlled the further articulation of the general theory.

If Darwin's metaphorical construction of natural selection is more than a *façon de parler,* then it should make a difference in his theory, distinguishing his actual theory from what it might have been if natural selection were to be understood in the way a neo-Darwinian might construe it. There are, I believe, four general features of the theory that might have looked rather different had Darwin rendered natural selection in the denuded, anemic terms of the modern scientist.

The first difference concerns Darwin's notion of creation through law. From the early notebooks through the essays and the *Origin,* Darwin held that the rise and development of creatures occurred through natural laws. Yet in the essay of 1842, he emphasized the difficulty of conceiving of natural law as having the requisite power to fashion the most intricate contrivances. Law could have this power, however, if it were the legislation of a superior mind. As he put it in the essay: "Doubtless it at first transcends our humble powers to conceive laws capable of creating individual organisms, each characterized by the most exquisite workmanship and widely extended adaptations. It accords better with [our modesty] the lowness of our faculties to suppose each must require the fiat of a creator, but in the same proportion the existence of such laws should exalt our notion of the power of the omniscient Creator."[32] The point here is that the only kinds of laws capable of producing the infinitely fine adaptations exhibited by creatures — insofar as we could comprehend such laws — were those established by mind.

Phillip Sloan (in chapter 7 of this volume) has observed the distinction Darwin made in his early theorizing between the laws governing the inanimate universe and those operative in the organic world. The basis of the distinction was ultimately located for Darwin in the intentions of the divine mind. While Darwin muted the distinction in the *Origin,* Sloan clearly shows that it still continued to operate in the actual articulation of the theory. Thus the entire evolutionary system, as Darwin proposed it in the first edition of the *Origin,* was predicated on mind, on intelligence.

The second feature of Darwin's general theory that the metaphorical rendering of natural selection has generated is the conception of selection as operating gradually and by minute increments. If natural selection performed like a Manchester spinning loom, the product would not have been fine damask—or the exquisite eye of the vertebrate. But if natural selection had preternatural intelligence and could see into the very depths of a creature, was ever watchful, and always selected the best variations, no matter how small, then something like the vertebrate eye might gradually result. Thomas Henry Huxley, Darwin's great friend and champion, insisted on the machinelike character of selection—and he also maintained, consequently, that the evolutionary process occurred hesitatingly and saltationally. Darwin, by contrast, assumed the process to be slow, gradual, and fine.[33] The machine analogy simply did not form part of Darwin's initial conception of the evolutionary process—indeed, the very word *machine* in any of its forms appears only once in the *Origin*, hardly what you would expect if mechanism were a fundamental assumption for understanding the operations of living nature.

A third contribution of Darwin's metaphor to his theory has to do with the contrast with artificial selection. The absolute intelligence implicated in the metaphor helps explain a very curious claim Darwin made about speciation in the *Origin*, one that no modern evolutionist would accept, namely that large numbers of a given species in one location promote, per se, faster evolution. This claim is based on the successful practice of breeders. Darwin observed in the *Origin* that artificial selection would work more swiftly if breeding stocks were large, since "variations manifestly useful or pleasing to man appear only occasionally, [so] the chances of their appearance will be much increased by a larger number of individuals being kept."[34] Some pages later, he reintroduced this condition as one necessary for the success of natural selection: "A large number of individuals, by giving a better chance for the appearance within any given period of profitable variations, will compensate for a lesser amount of variability in each individual, and is, I believe, an extremely important element of success."[35] Of course, Darwin is right. With large flocks, the absolute number of favorable variations ought to increase. But the proportion of favorable to unfavorable (or neutral) variations should remain constant; and, indeed, large flocks will be even more subject to the phenomenon of swamping out (when favorable varieties breed with unfavorable). Only if natural selection acts intelligently and with foresight will large numbers avail. And this is what Darwin assumes.

Other aspects of Darwin's theory would likely be different were natural selection really the result of blind mechanism. Let me conclude with one final feature of the theory, already adumbrated, that clearly demonstrates the intentional, mental character that Darwin ascribed to natural selection. If natural selection were endowed with supreme intelligence, we would expect it to act for ends, for goals.

And this was precisely the way Darwin understood it to operate. In the preceding section, I cited his remarks, in 1839, to the effect that the evolutionary process was "stepping towards some final end—production of higher animals." He retained this conception during the two decades he worked on his theory. In the concluding paragraph of the *Origin,* he exclaimed that all the death and destruction imposed by natural selection aimed at a final cause, which was "the most exalted object which we are capable of conceiving, namely the production of the higher animals."[36] Evolution for Darwin was progressive and goal directed, which it certainly would not be if it were merely the result of blind causes clashing by night.

In later years, Darwin's metaphysical assumptions—or at least their overt expression—faded as critics teased them out and held them up for inspection (see chapter 7 of this book). But during the formative period of his theory construction, when the fundamental features of that theory were established, those assumptions formed the deep grammar of his conception, controlling what the theory was capable of asserting. Today, we understand the evolutionary process differently. Darwin's formulation of the operations of natural selection is not ours. We are neo-Darwinians, which, needless to say, Darwin was not.

NOTES

1. Charles Darwin, *Charles Darwin's Beagle Diary,* ed. R. D. Keynes (Cambridge: Cambridge University Press, 1988), 180.

2. Ibid., 222.

3. Charles Darwin, Notebook B (MS p. 228), in *Charles Darwin's Notebooks, 1836–1844* (hereafter *Notebooks*), ed. Paul Barrett et al. (Ithaca, N.Y.: Cornell University Press, 1987), 227.

4. Darwin, Notebook C (MS pp. 77–78), in *Notebooks,* 263. Single-wedge quotes indicate Darwin's deletions, double-wedge quotes his insertions.

5. I have discussed the sensationalist epistemology of Erasmus Darwin and Charles's debt to this grandfather in greater detail in Robert J. Richards, *Darwin and the Emergence of Evolutionary Theories of Mind and Behavior* (Chicago: University of Chicago Press, 1987), 31–40, 105–6.

6. Darwin, Notebook N (MS p. 21e), in *Notebooks,* 569.

7. Darwin, Notebook D (MS p. 118), in *Notebooks,* 371: "It is less wonderful that childs nervous system should build up its body, like its parent, than that it should be provided with many contingencies how to act—so with the mind. The simplest transmission is direct instinct & afterwards enlarged powers to meet with contingency."

8. Darwin, Notebook N (MS pp. 43e–44e), in *Notebooks,* 574–75.

9. Darwin, Notebook M (MS pp. 81), in *Notebooks*, 538.

10. Darwin, "Old and Useless Notes" (MS pp. 37), in *Notebooks*, 614.

11. Ibid. (MS pp. 37, 39), in *Notebooks*, 614, 616.

12. Darwin, Notebook M (MS pp. 89–92), in *Notebooks*, 540.

13. Darwin, "Old and Useless Notes" (MS p. 41v), in *Notebooks*, 618

14. Darwin, Notebook M (MS p. 103), in *Notebooks*, 544.

15. Ibid. (MS pp. 34–35), in *Notebooks*, 527.

16. Charles Lyell, *Principles of Geology*, 3 vols. (London: Murray, 1830–33). For the sentences cited, see 2:163–67.

17. Darwin, "Old and Useless Notes" (MS p. 41v), in *Notebooks*, 618.

18. He had a more difficult time with man's moral character but finally came to several ingenious resolutions to the problems of moral evolution. See Richards, *Darwin*, chs. 2 and 5.

19. Darwin, "Old and Useless Notes" (MS p. 36), in *Notebooks*, 613–14.

20. Darwin, Notebook M (MS p. 57), in *Notebooks*, 532–33.

21. Darwin, "Old and Useless Notes" (MS p. 37), in *Notebooks*, 614.

22. Charles Darwin, *On the Origin of Species* (London: Murray, 1859), 490.

23. Darwin, Notebook B (MS p. 18), in *Notebooks*, 175.

24. That Darwin's mature theory was progressive I have argued at some length in Robert J. Richards, *The Meaning of Evolution: The Morphological Construction and Ideological Reconstruction of Darwin's Theory* (Chicago: University of Chicago Press, 1992), ch. 5.

25. Darwin, Notebook C (MS p. 63), in *Notebooks*, 259.

26. Darwin, *Origin of Species*, 83–84.

27. Charles Darwin, *Charles Darwin's Natural Selection: Being the Second Part of His Big Species Book Written from 1856 to 1858*, ed. R. C. Staufer (Cambridge: Cambridge University Press, 1975), 224–25.

28. Charles Darwin, *The Foundations of the Origin of Species: Two Essays Written in 1842 and 1844 by Charles Darwin*, ed. Francis Darwin (Cambridge: Cambridge University Press, 1909), 85.

29. See ibid., 6.

30. Charles Darwin to Alfred Russel Wallace (July 5, 1866), in *Life and Letters of Charles Darwin*, ed. Francis Darwin, 2 vols. (New York: D. Appleton, 1891), 2:229–30. Darwin did introduce the designation "survival of the fittest" in the fifth edition of the *Origin*, though without abandoning "natural selection." See *The Origin of Species by Charles Darwin: A Variorum Text*, ed. Morse Peckham (Philadelphia: University of Pennsylvania Press, 1959), 163–65.

31. I have discussed the liability of the Platonic interpretation of theories in Robert J. Richards, "The Epistemology of Historical Interpretation: Progressivity and Recapitulation in Darwin's Theory," in *Biology and Epistemology*, ed. Richard Creath and Jane Maienschein (Cambridge: Cambridge University Press, 2000), 64–88.

32. Darwin, *Foundations*, 52.

33. Thomas Henry Huxley, "The Origin of Species" (1860), in his *Darwiniana*, vol. 2 of *Collected Essays* (London: Macmillan, 1893), 77: "Mr. Darwin's position might, we think, have been even stronger than it is if he had not embarrassed himself with the aphorism, '*Natura non facit saltum*,' which turns up so often in his pages. We believe . . . that Nature does make jumps now and then, and a recognition of the fact is of no small importance in disposing of many minor objections to the doctrine of transmutation." Huxley remarked on his differences with Darwin in a letter to William Bateson: "I see you are inclined to advocate the possibility of considerable 'saltus' on the part of Dame Nature in her variations. I always took the same view, much to Mr Darwin's disgust, and we used often to debate it." See Thomas Henry Huxley to William Bateson (February 20, 1894) in *Life and Letters of Thomas H. Huxley*, 2 vols., ed. Leonard Huxley (New York: D. Appleton, 1900), 2:394.

34. Darwin, *Origin of Species*, 41.

35. Ibid., 102. See similar claims made on 105, 107, and 125.

36. Ibid., 490.

CHAPTER 9

Are There Metaphysical Implications of Darwinian Evolutionary Biology?

Jean Gayon

GENERAL FRAMEWORK: PHILOSOPHICAL VERSUS METAPHYSICAL IMPLICATIONS OF DARWINISM

In her *Philosophical Testament* (1995), Marjorie Grene proposes to evaluate the philosophical implications of the fact of evolution through Kant's famous phrases about the four questions asked by philosophy: "What can I know?" "What ought I to do?" "What may I hope?" and, summarizing the others, as Kant says in his *Lectures on Logic,* "What is Man?" (Grene 1999, 107). The first three questions echo the modern debates over evolutionary epistemology, evolutionary ethics, and the relationship between religion and evolution. Kant's fourth question, which reduces philosophy to anthropological issues, makes even clearer why evolution should be important for philosophers: evolutionary biology has much to tell about humans (Grene 1995, 89–112).

In my essay I will also rely upon Kant, but in a different way. My intention is to take seriously the title of this conference, "Darwinism and Metaphysics." *Darwinism* is more specific than *evolution.* *Metaphysics* is also more specific than just *philosophy.*

If we want to examine the metaphysical implications of Darwinism, we can hardly avoid a minimal characterization of what we mean by *metaphysics.* *Metaphysics* is a rather frightening word for a philosopher today. Historians of philosophy may be at ease with it because they are knowledgeable about the exquisite variety of senses that it has had in the past. But for most philosophers engaged in the discussion of today's philosophical problems, it is an obsolete and confusing word. In fact, I do not think it is possible to unify the various major historical meanings of the term. However, if we want to answer the question raised by this conference, we need at least some notion of metaphysics.

I have chosen to rely on two different perspectives on metaphysics. I borrow them from two philosophers who proposed them at a time when the employment of this word was still natural for philosophers, although it was also beginning to become problematic for a number of them. These two philosophers are Auguste Comte and Immanuel Kant. Each of them can give us a notion of "metaphysics" that may help us assess the relationship between Darwinism and metaphysics. To put it simply, Auguste Comte gives us a negative view of metaphysics, and Kant gives us a positive view. Each of these perspectives will be useful to us, but for different reasons.

Comte's negative view of metaphysics is related to his "law of the three states," which stipulates that any area of knowledge will typically pass through three successive stages: the "fictitious" (or "theological") state, the "abstract" (or "metaphysical") state, and the "positive" (or "scientific") state. The metaphysical regime of discourse consists in explaining the phenomena with the help of abstract entities, supposedly inherent in the natural beings. By this usage, *metaphysics* is the name for a regime of knowledge founded upon verbal abstractions. This pejorative sense of *metaphysics* makes sense only in comparison with the alleged better state of knowledge that must follow: the state of positive science, which consists in the discovery of the laws on the basis of both observation and reasoning (Comte 1975, 21–22). The pejorative sense of metaphysics as verbal knowledge is useful to us. It can illuminate the special relationship between Darwinism and metaphysics that can emerge spontaneously within the most scientific aspects of evolutionary biology: like any scientific inquiry, evolutionary biology involves conceptual difficulties resulting from implicit preconceptions, verbal obscurities, and so on.

The positive view of metaphysics that I will use is more complex. In its most traditional sense, *metaphysics* is synonymous with Aristotle's "first philosophy," which includes the knowledge of divine things and the knowledge of the ultimate principles of action and of the sciences. More widely, on a large historical scale, metaphysics embodies the notion of a kind of knowledge more essential and fundamental than any other, both in terms of its domain of objects and in terms of the mode of thinking that it involves. I said earlier that I would not try to propose an all-encompassing definition of metaphysics. I will thus move directly to Kant. Kant's reflection upon metaphysics has two advantages for us. First, it is one of the most systematic categorizations of the domains of metaphysical investigation ever proposed. Second, this categorization will prove useful for classifying the various kinds of interactions between Darwinism and philosophy.

For Kant, *metaphysics* is a synonym for "*pure philosophical cognition*" (*Prolegomena,* § 1) or "the whole of pure philosophy" (*Critique of Pure Reason,* "Architectonics of Pure Reason" § 1). It includes all a priori knowledge drawn from pure understanding or reason alone, with no help of experience. It includes both critical knowledge—that is, the definition of the powers and limits of reason—and the systematic and a priori knowledge that we can have in the domains of both nature and ethics (*Critique of Pure Reason,* "Architectonics of Pure Reason"). This definition means that the word *metaphysics* can be applied to four different areas of philosophical discussion, which are explicitly enumerated in the "Architectonics of Pure Reason":

1. "Transcendental philosophy of pure reason," or the examination of the a priori conditions of any knowledge. It is concerned with the formal conditions of all knowledge, not its "matter" or content. This sense of the word *metaphysics* corresponds to what was later called epistemology (or theory of knowledge).

2. The "hyperphysical study of nature," or dogmatic metaphysics. It consists of a purely rational speculation about things that are absolutely beyond experience: the Soul, the World, and God. Kant believed that this kind of speculation was condemned to illusion and thus was not a genuine knowledge.

3. The "metaphysics of nature," especially "metaphysics of corporeal nature." Kant says that the metaphysics of corporeal of nature is commonly called "physics." It presupposes the empirical concept of matter, but insofar as it is "metaphysical," it consists of all that can be known about matter without any other empirical principle. Kant's *Metaphysical Foundations of Natural Science* is a good example: this book is entirely devoted to the concept of matter. The other component of the metaphysics of nature is the "metaphysics of the thinking nature"; however, Kant believed that it was a hopeless philosophical project (*Metaphysical Foundations of Natural Science,* Preface).

4. "Metaphysics of morals," or the science of the a priori principles of the practical use of pure reason.

The first two senses are often characterized by Kant as belonging to logic because they deal with formal issues and the last two as specifically metaphysical because they bear on material issues. But Kant oscillated between two uses of the word *metaphysics*, a broad use and a restricted use. Since my purpose is not to comment on Kant, I will not discuss this problem. Kant's broader notion of metaphysics will be taken as a convenient classification of the possible senses of the word. As I will show shortly, each of these senses provides a dimension for the description and assessment of the relationship between Darwinism and metaphysics. If we now add Comte's negative conception of metaphysics, we are left with five different types of questions that, I believe, constitute the domain of the Darwinism-metaphysics debate.

Question 1 is suggested by Comte's pejorative notion of metaphysics. It concerns the methods and doctrines of evolutionary biology as a positive science. Like any positive science, evolutionary biology encounters a number of conceptual puzzles. Philosophers help to solve them by clarifying, and possibly eradicating, the naive metaphysical presuppositions embodied in current theories and methods. This first interaction between Darwinism and metaphysics corresponds more or less to the industry of the modern "philosophy of biology."

The four other questions correspond to Kant's four areas of metaphysical investigation. I take them in order.

Question 2 has to do with the philosophical foundations of knowledge in general, or what Kant called transcendental philosophy. Most of modern epistemology since the end of nineteenth century has consisted in developing this foundational enterprise in various ways. The issue is whether a Darwinian anthropology challenges this enterprise or can be helpful to it.

Question 3 concerns the kind of metaphysics that Kant believed to be discredited: rational psychology, rational cosmology, and rational theology. I will argue that the Darwinian conception of evolution has not generated anything new in this philosophical debate. At best, Darwinism has indirectly fortified Kant's criticism of speculative metaphysics.

Question 4 is: Has Darwinism had any repercussion on the metaphysics of nature? The answers to this question have differed sharply. Major philosophers of the twentieth century answered negatively; others have argued that evolution is crucial for the metaphysics of nature, but most of them have been non-Darwinian.

Question 5 is about ethics. Evolutionary ethics, and especially Darwinian ethics, has been popular recently among philosophers. The question is whether evolutionary ethics is compatible with a foundational approach to ethics. Here again, the answers differ sharply: some philosophers think that a Darwinian ethics implies a total renunciation of any metaphysics of morals, others not.

Such are the five dimensions along which Darwinism can be evaluated if metaphysics is at stake. Of course, I will not provide a detailed treatment of each of these problems. My objective is to propose a categorization of the problems. The rest of this essay will make a little more explicit the five questions I have raised.

QUESTION 1: PHILOSOPHY OF BIOLOGY

The first dimension of the relationship between Darwinism and metaphysics in the philosophical literature is treated in the current philosophy of biology. Since the 1970s, the philosophy of biology has developed as a special branch of philosophy of science. In its dominant anglophone version, it has been concerned largely, though not exclusively, with conceptual and methodological difficulties found in the common practice of evolutionary biology (e.g., the meaning of species and other taxonomic categories, phylogenetic inference, and ontological or methodological problems related to the uses of concepts such as natural selection, fitness, adaptation, environment, and chance). Most of the authors involved in this special branch of philosophy of science do not make current use of the word *metaphysics*. Their objective is to clarify some aspects of evolutionary biology (or other aspects of biological science) that can manifestly benefit from the help of the specialists of metadiscourse, namely philosophers.

A particular relationship between metaphysics and Darwinism can be identified at this level. Insofar as modern philosophers of biology discuss the kind of problems that I have just evoked, they do not claim to propose a special metaphysical doctrine or to discuss any classical metaphysical problem. They do not propose a general theory of the foundations of knowledge; they do not take a stance on the simplicity or immortality of the soul; they do not build an argument about the ultimate signification of evolution for the metaphysics of nature; they do not discuss the ultimate foundations of morality or law. In brief, they do not engage themselves in the domain of metaphysics in any of the senses that were pointed out by Kant. But a certain kind of metaphysical preoccupation does appear in their theoretical activity. They are concerned with a number of linguistic difficulties generated by the particular language of biology. They are concerned with verbal abstractions that sometimes obscure the signification of a particular concept of evolutionary theory: for example, tautologies associated with the use of concepts such as "fitness" and "adaptation" and preanalytic representations of notions such as "type," "logical class," "history," "origins," "individual," and "contingency." Such notions, and others, play a significant role in the construction of evolutionary theories, in testing procedures, in the choice of hypotheses, and in the definition of acceptable research programs. At that point, philosophers' familiarity with the history of metaphysics can be useful. They can make the evolutionary biologist aware of a number of naive metaphysical presuppositions involved in his or her scientific practice and discourse. Sometimes they even propose solutions. Metaphysics, then, functions as a sort of conceptual toolbox that helps positive science to clarify of its own theories.

Some authors do assume the use of the word *metaphysics* in the context of this particular enterprise of clarification of scientific concepts. This is obvious, for example, in the case of Michael Ghiselin's book *Metaphysics and the Origin of Species* (1997). At the beginning of his book Ghiselin says, "Metaphysics is usually considered a branch of philosophy, equal in autonomy—perhaps in dignity—to ethics, aesthetics, epistemology and logic. Here, however, it will be treated as one of the natural sciences" (12). But why is metaphysics fundamental for the natural sciences? Ghiselin explains that "metaphysical preconceptions profoundly influence the course of scientific investigation" (14). However, Ghiselin does not appear to be proposing a general metaphysics of nature, even one that would be deeply motivated by the Darwinian view of evolution. Rather, he is using a variety of traditional philosophical notions to clarify biological concepts or methods such as species and classification. Therefore, Ghiselin does not accomplish something genuinely different from the current activity of philosophy of biology. He uses the word *metaphysics* where others tend to avoid it. But this is just a matter of subjective preference. I do not see a crucial difference between his attitude and that of other philosophers of biology.

Auguste Comte's positivist interpretation of metaphysics is relevant for the philosophical qualification of this kind of activity: positive science is not metaphysics, but it always embodies tacit metaphysical preconceptions embodied in its own discourse. Here, of course, it is metaphysics in a weak sense of the word that comes to interact with Darwinism.

I now turn to stronger interactions, which involve one or another of the strong senses of metaphysics that I described earlier with the help of Kant.

QUESTION 2: TRANSCENDENTAL METAPHYSICS
(OR "EPISTEMOLOGY")

Kant considered transcendental philosophy to be a "propaedeutics" for any future metaphysics. The role of transcendental philosophy was to define the foundations and the limits of all knowledge, and more especially of the kind of knowledge that relies on the powers of reason alone, prior to or independently of any experience. Because the transcendental part of philosophy was a propaedeutics, Kant oscillated between two ways of presenting its relationship to metaphysics. Sometimes he stated that transcendental philosophy was concerned with formal questions, whereas metaphysics dealt only with particular contents. This entailed that transcendental philosophy would come before metaphysics without being itself a part of metaphysics. In other circumstances, Kant asserted that the elucidation of the foundations and limits of knowledge was in fact the most important part of metaphysics (on this ambiguity, see *Critique of Pure Reason,* "Architectonics of Pure Reason," and *Prolegomena,* § 5). Thus *epistemology* can be considered a modern term for what Kant considered to be the most important part of metaphysics (I take *epistemology* as the common English term for what continental philosophers usually call theory of knowledge).

The relationship between epistemology and Darwinian evolution has been the subject of a huge literature. I will merely summarize a few aspects of this story. Anglophone students can familiarize themselves with epistemology through a book entitled *Companion to Epistemology* (1992), edited by Jonathan Dancy and Ernest Souza. An odd characteristic of this book is that it contains no article on "epistemology." Furthermore, the introduction provides no definition of epistemology. But the index mentions one article that offers a straightforward definition of epistemology. That article, ironically, is entitled "Death of Epistemology." Here are the key sentences: "The task of epistemology . . . is to determine the nature, scope and limits, indeed the very possibility of human knowledge. Since epistemology determines the extent to which knowledge is possible, it cannot itself take for granted the results of any particular forms of empirical enquiry. Thus epistemology purports to be a non-empirical discipline, the function of which is to sit in judgment on all particular discursive practices with a view to determining their cognitive status" (Williams 1992, 89).

This very much resembles Kant's "transcendental philosophy." In particular, Kant would have fully endorsed the claim that this kind of investigation should not rely upon any form of empirical inquiry. Now, if we look at the history of the twentieth century on a large scale, we can observe that two major philosophical traditions have also endorsed this position. The first is Husserl's phenomenology, which is explicitly "transcendental." It is defined as the science that elucidates the intentional structures, the originating acts of consciousness, involved in any kind of knowledge. As such, phenomenology provides the ultimate foundation of science in general. Scientific theories, for Husserl, are not direct descriptions of nature; they are interpretations of nature by conscious subjects. Consequently, no scientific theory, physical, biological, or psychological, can be relevant for transcendental philosophy. Evolutionary theory has no privilege in that respect. Husserl also used a more special argument against the relevance of evolution for philosophy: any scientific investigation presupposes logic, and evolutionary biology, like any other scientific theory, presupposes logic. If the truths of logic were dependent upon a theory of the origins of mental powers in the human species, then all kinds of science would be impossible, including the science of evolution. (For a detailed analysis of the relationship between phenomenology and Darwinism, see Cunningham 1996.)

Similar arguments against the relevance of biological evolution for epistemology can be found in the philosophy of Bertrand Russell. Russell, of course, did not share Husserl's antiobjectivist view of knowledge. But he also could not admit the dependence of logic on a particular empirical theory. Logic was not the science of the laws of effective human thought but a kind of a priori knowledge that was necessary for any empirical science. Consequently, if logic was the essence of philosophy, philosophy itself was independent from evolution. (Cunningham 1996 also provides an excellent analysis of this historical episode.)

We see, therefore, that the two major epistemological traditions of the twentieth century, namely phenomenology and analytic philosophy, were prompt to exclude evolution. It took a while before this exclusion was seriously challenged. I hardly need to recall here the role of Quine's (1969) arguments about the naturalization of epistemology. For Quine, the question of the conditions for the possibility of knowledge can be construed as an empirical question. Some particular sciences, such as psychology and evolutionary biology, can help us understand why humans may reasonably think that they have accurate beliefs about the world. However, a more ancient source of evolutionary epistemology must be mentioned. In 1941, Konrad Lorenz proposed a Darwinian interpretation of Kant's notion of a priori. Lorenz proposed to explain the existence and diversity of the cognitive systems in animals on the basis of their survival value in particular ecological niches. According to Lorenz and his successors (e.g., Vollmer 1987), such a research program should be able to inform us about the origins, the reliability, and the limits of our cognitive system.

Evolutionary epistemology has generated a lot of discussion among philosophers. It is unclear whether it has seriously discredited the classical foundationalist project. Marjorie Grene (1995), whom I invoked at the beginning of this essay, makes an interesting observation on this question. She agrees that it is important to be aware of the phylogenetic background of our cognitive powers and of their inevitable limitations: "[W]e are real animals trying to find their way in a real world," not idle and disinterested spectators. But this kind of information is not enough for a philosophical account of human knowledge. We do not want to know only about the necessary conditions that limit our effective cognitive powers. We also want to decide what are the sufficient conditions for the satisfaction of our cognitive needs (Grene 1995, 109). This is a normative issue that involves conflicting views among historical actors. In that sense, it might well be hasty to conclude that Darwinism has discredited the sort of metaphysics that Kant characterized as transcendental philosophy.

QUESTION 3: DARWINISM AND "SPECIAL METAPHYSICS" (SOUL, WORLD, GOD)

I come to the commonest sense of *metaphysics:* a purely rational kind of theoretical knowledge bearing on objects that are absolutely beyond any possible experience—Soul, World, and God. What are the implications of Darwinism for this kind of metaphysics? I will be brief on this subject. It seems to me that the interaction has been and could only be weak. Kant had an extremely restrictive conception of the theoretical interest of the ideas of reason for empirical science. He thought that these ideas had absolutely no usefulness for empirical science (*Prolegomena,* § 44). For instance, if we want to explain mental phenomena, it is irrelevant whether the soul is a simple substance and whether it is immortal. Similarly, speculations about the beginning or eternity of the world are of no interest for the explanation of any event in the history of the world. Finally, concerning rational theology, Kant wittily remarked that "we must avoid any explanation of the order of nature drawn from the will of a supreme being, because this is no longer philosophy of nature, but rather the end of it" (*Prolegomena,* § 44). Conversely, Kant thought that natural science offered no help for the debate over the big issues of special metaphysics, such as the mortality or immortality of the soul, the spatial and temporal finiteness or infiniteness of the world, or the existence of God.

I do not think that Darwinian evolution gives us any serious reason to modify this verdict. I can hardly see how the standard theory of biological evolution could affect our speculations over the immortality of the soul, the finiteness of universe, or the existence of God. Take the existence of God. At first sight, this seems to have been a major occasion of an interaction between metaphysics and Darwinism. But this is questionable. Long before Darwin, theologians and philosophers had debated the value of the physical-theological proof of the existence of God. This proof was traditionally seen as the weakest, although it was the most popular. As Kant recalled in the *Critique of Pure Reason,* such a proof could perhaps demonstrate the existence of an architect of the world but not the existence of a creator of the world.

In reality, the interaction between Darwinism and religion has been principally a cultural issue, not a serious problem for nineteenth- and twentieth-century philosophers. Furthermore, even from the point of view of the cultural history of science, Darwinism has been in practice compatible with a very large array of religious preferences. To sum up this section, the interaction between Darwinism and what was formerly called "special metaphysics" has been rather poor, although highly significant from the larger viewpoint of cultural history.

QUESTION 4: DARWINISM AND THE METAPHYSICS OF NATURE

The impact of Darwinism on the metaphysics of nature is a more interesting issue. Kant thought that it was impossible to elaborate a purely rational cosmology, but he believed in the possibility of developing a metaphysical theory of nature in the special case of the corporeal nature. The book that he devoted to this question *(Metaphysical Foundations of Natural Science)* is entirely devoted to an exposition of the a priori determinations of the concept of matter. This concept includes an empirical element, but Kant argued that an a priori knowledge could be developed on this basis. I will not discuss further Kant's characterization of this science. I just take it as an illustration of the project of a metaphysical theory of nature. For Kant, such a theory could be developed only in the case of the concept of matter. But we can broaden the notion and consider the possibility of a kind of metaphysics of nature—an exploration of general features of the natural world—that would have other bases. German *Naturphilosophie* was a major historical example of such an enterprise: it consisted in thinking about nature on the basis of the concept of the organism. The question is whether something similar might be done on the basis of the concept of evolution.

Russell, in his 1926 book *Our Knowledge of the External World,* developed an impressive argument against any project of this sort. His argument was that the evolution of life was restricted to so small a portion of the spatiotemporal world that it could not have a philosophical interest. The argument is summarized in his *Autobiography*: "What biology has rendered probable is that the diverse species arose by adaptation from a less differentiated ancestry. This fact is in itself extremely interesting, but it is not the kind of fact from which philosophical consequences follow. Philosophy is general, and takes an impartial interest in all that exists. The changes suffered by minute portions of matter on the earth's surface are very important to us active sentient beings; but to us as philosophers they have no greater interest than other changes in portions of matter elsewhere" (Russell 1978).

Biological evolution, then, had no philosophical interest for Russell because it belonged to a science that did not apply to all that existed in nature. Suzanne Cunningham, who quotes this text in her book *Philosophy and the Darwinian Legacy* (1996), underlines another reason why Russell could not consider evolution as a scientific model for the philosopher's reflection on nature. First, Russell thought that philosophical doctrines had to be not only general but also true in all possible worlds. In that respect, it is hard to imagine a model worse than biological evolution, especially Darwinian evolution: historical contingency is exactly the opposite of properties that could be verified in all possible worlds.

The example of Russell is extreme. Its main interest is to show why biological evolution can be problematic for the project of a metaphysics of nature. Of course, some philosophers have adopted a very different attitude. Spencer, Bergson, Peirce, and Teilhard de Chardin are good examples of philosophers who tried to construct a metaphysical theory of nature founded upon the model of biological evolution. All of them were fascinated by a cosmic conception of evolution; all of them were also fascinated by the emergence of novelty. However most of them were also reluctant to accept the Darwinian view of evolution. With the exception of Peirce, who emphasized the role of chance, philosophical evolutionism involved a preference for a progressionist view of the history of the universe.

On the whole, then, modern philosophy has not particularly favored the construction of a metaphysical science of nature founded on the Darwinian view of evolution.

QUESTION 5: METAPHYSICS OF MORALS

The last sense of *metaphysics* that deserves being confronted with Darwinism is metaphysics of morals, which Kant considered to be an investigation of the supreme principle of morality. This principle had to be defined a priori and based upon the representation of man as a rational being. Kant refused any attempt to derive this principle from empirical premises: "[T]he ground of obligation here must not be sought in the nature of man or in the circumstances in which he is placed, but sought a priori solely in the concepts of pure reason" (*Foundations of the Metaphysics of Morals,* Preface). His well-known answer to the problem of the foundation of morality was that a moral action was such that its maxim could be given the value of a universal law.

How can Darwinian evolution affect the philosophical project of a metaphysical theory of morals? At first sight, it seems that there can be no serious interference between the two. Darwin did indeed develop a theory of the origins of moral behavior, but there is a gulf between a natural history of morals and a philosophical theory of the possible foundations of morals. It is one thing to explain how humans came to develop a moral behavior and another thing to provide a justification for any possible moral action.

Evolutionary ethics, under the form of Darwinian ethics, has generated an interesting philosophical debate. Since the end of the nineteenth century, there has been intense debate over the possibility, or impossibility, of constructing a normative theory of ethics on the basis of Darwin's explanation of the origins of moral behavior in the human species. I will not describe the history of this debate, with its unending discussions over the "naturalistic fallacy." I will merely evoke two relatively recent positions that illustrate in modern terms the impact of Darwinism over the old problem of the metaphysics of morals. Each of the two authors has written an article with exactly the same title, "A Defense of Evolutionary Ethics." Robert Richards's article appeared in one of the first issues of *Biology and Philosophy* (1986); Michael Ruse published his own article in French in 1993 ("Une défense de l'éthique évolutionniste"). But under the same title the two authors propose two radically opposed defenses of evolutionary ethics. As we shall see, this disagreement has a direct bearing upon the issue of the metaphysics of morals, or, to put it in more modern terms, foundational ethics.

Ruse (1993) argues that the adoption of a Darwinian explanation of morals forces the philosopher to completely reject any kind of foundational ethics. Morals are an adaptation, just like our eyes, teeth, and hands. It is an illusion to think that an adaptation could be given a foundation. In particular, the most naive illusion consists in believing that the natural evolution of morals has been directed in some kind of progress, a progress that miraculously fits with the current justification of morals among philosophers. According to Ruse, "There is absolutely no reason for thinking that evolution involves progress. . . . Therefore, traditional evolutionary ethics is based on an erroneous view of evolutionary progress" (60). For Ruse, then, it is hopeless to try to reconcile Darwinism with a foundationalist theory of morals.

Richards's (1986, 272) argument in his own defense of evolutionary ethics goes in a different direction. He supports what he calls a revised version of Darwin's original conception of morals. On the one hand, he admits that the moral sense is the name for "a set of innate dispositions that, in appropriate circumstances, move the individual to act in specific ways for the good of the community." He insists that for Darwin the end of moral acts is not individual or general happiness but the "general good," understood as the vigor and health of the group. Morality has in fact evolved as the result of "community selection." On the other hand, Richards contends that this empirical explanation of the origins of moral behavior can be used to justify moral judgments. Richards insists that he is not committing the naturalistic fallacy. All moral theories involve empirical presuppositions, even if they do not explicitly recognize this point. In the case of Darwinian ethics, the argument consists in saying that *because* we have evolved as altruistic creatures, we are justified to construct a moral theory on this basis and say that we *ought to* heed the general good. The precise content of this idea of the general good is an open question that depends on the material and cultural conditions in which groups are placed and on the rational capacities of humans — another result of evolution through natural selection.

I am not sure whether I am convinced by this argument. The problem is not so much the naturalistic fallacy as the criterion of morality proposed by Richards. It is not at all obvious that morality should be equated with altruistic behavior. Many if not all human cultures are aware that certain altruistic behaviors are utterly immoral, especially when they involve a nonrecognition of the dignity of other groups. Group selection of altruistic behavior may have been important for the genesis of the human moral sense, but I can hardly be satisfied with the idea that morality is just altruism. Some cultural anthropologists have pointed out that most human cultures have a word expressing the idea of good, of something that is purely and simply good in the sense that it is not restricted to a particular group. My conviction is that biological evolution alone is insufficient for both explaining and justifying moral behavior as a kind of behavior motivated by such a value. But this is just a subjective reaction.

The obvious conclusion of this section is that Darwinian ethics is hardly compatible with a metaphysical or foundational theory of morals. Both Ruse and

Richards illustrate this point, though with different arguments. Personally, however, I do not think that Darwin's natural history of the moral sense can secure a sufficient basis for the elaboration of a moral theory. To quote Marjorie Grene (1995, 111) again, just as in the case of epistemology, Darwinism sets conditions for our ethical speculations, but it does not answer the ethical question.

CONCLUSION

If we really want to use the word *metaphysics* rather than just *philosophy,* there have been many kinds of interactions between Darwinism and metaphysics. I have proposed a classification of these interactions. For the majority of them, however, I have used the word *metaphysics* only for the purposes of a retrospective interpretation. In fact, it is rather hard to imagine what a Darwinian metaphysics might be. Intuitively, the two words seem to conflict. As I see it, there are three major reasons for this situation.

The first reason is that the major philosophical traditions of the twentieth century emerged on the basis of a common rejection of traditional metaphysics and naturalism. Analytic philosophy, phenomenology, and existentialism are obvious examples. In such a context, Darwinism is irrelevant for philosophy, which claims to have freed itself from metaphysics. In such philosophies, the question of a relationship between metaphysics and Darwinism does not exist.

The second reason is related to those few openly metaphysical systems that have been constructed with explicit reference to evolution. Since the end of the nineteenth century, such philosophical systems have received the name *evolutionism.* Spencer, Bergson, and Teilhard de Chardin are good examples. Each of the authors developed a metaphysical theory of nature, and each of them did it with a very different orientation in terms of traditional philosophy (materialism, spiritualism, and theism respectively). But all of them were skeptical with regard to Darwin's theory of evolution. Thus, when metaphysics and evolution met, Darwin was not invited.

The third reason for the difficulty of a marriage between Darwinism and metaphysics results from the kind of challenge that Darwinian science addresses to philosophy. Darwinism challenges philosophy when philosophy focuses on the human condition. Darwinism has affected almost all areas of anthropology. If the central question of philosophy is "What is man?" then it is hard for philosophy to ignore Darwinism. That is the main reason why there are so many obvious philosophical implications of Darwinism and so few obvious metaphysical implications.

REFERENCES

Comte, Auguste. 1975. *Cours de philosophie positive.* Paris, Hermann.

Cunningham, Suzanne. 1996. *Philosophy and the Darwinian legacy.* Rochester, N.Y.: University of Rochester Press.

Dancy, Jonathan, and Ernest Souza, eds. 1992. *A companion to epistemology.* London: Blackwell.

Ghiselin, Michael. 1997. *Metaphysics and the origin of species.* Albany: State University of New York Press.

Grene, Marjorie. 1995. *A philosophical testament.* Chicago: Open Court.

Kant, Immanuel. *Critique of pure reason.* Translated by Norman Kemp Smith. London: Macmillan, 1933.

———. *Metaphysical foundations of natural science.* Translated by James Ellington. Indianapolis: Bobbs-Merrill, 1970.

———. *Foundations of the metaphysics of morals.* Translated by Lewis White Beck. Indianapolis: Bobbs-Merrill, 1969.

———. *Lectures on logic.* Translated and edited by J. Michael Young. Cambridge Edition of the Works of Immanuel Kant. Cambridge: Cambridge University Press, 1992.

———. *Prolegomena to any future metaphysics that will be able to come forward as science.* Translated by Gary Hatfield. Cambridge Edition of the Works of Immanuel Kant. Cambridge: Cambridge University Press, 2002.

Lorenz, K. 1941. Kant's Lehre von Apriorischen im Lichte gegenwärtiger Biologie. *Blätter für Deutsche Philosophie* 15:94–125. Translated as "Kant's doctrine of the a priori in the light of contemporary biology," in *Learning, development and culture: Essays in evolutionary epistemology,* ed. H. C. Plotkin, 121–143 (Chichester: John Wiley, 1982).

Quine, W. V. 1969. Epistemology naturalized. In *Ontological relativity and other essays,* 69–90. New York: Columbia University Press.

Richards, Robert J. 1986. A defense of evolutionary ethics. *Biology and Philosophy* 3:265–93.

Ruse, Michael. 1993. Une défense de l'éthique évolutionniste. In *Fondements naturels de l'éthique,* ed. J. P. Changeux, 35–64. Paris: Editions Odile Jacob.

Russell, Bertrand. 1926. *Our knowledge of the external world.* 2nd ed. London: Allen & Unwin.

———. 1978. *Autobiography.* London: Allen & Unwin. Quoted in Cunningham 1996, 75.

Vollmer, Gerhard. 1987. What evolutionary epistemology is not. In *Evolutionary epistemology: A multiple paradigm,* ed. W. Callebaut and R. Pinxten, 203–21. Dordrecht: D. Reidel.

Williams, Michael. 1992. Death of epistemology. In Dancy and Souza 1992, 88–91.

CHAPTER 10

On the Problem of Direction and Goal in Biological Evolution

Dieter Wandschneider

To doubt the fact of biological evolution today would be absurd—especially since, on the basis of the arguments developed by Darwin, a remarkably powerful theory of biological evolution is available. Its present main variant—connected, above all, with the name of Ernst Mayr—is the so-called synthetic theory, synthetic in the sense of including the state of present-day research of various biological sciences. In the following discussion, such phrases as "Darwinian theory" and "Darwinian explanation" always refer to the modern synthetic form of Darwinism.

The rejection of every form of teleological interpretation and the limitation to principally causal patterns of explanation are characteristic of Darwinian arguments. Highly simplified, the popular basic idea can be given as follows: the random occurrence of variations within a species (e.g., caused by gene mutations) leads, in the struggle for survival for limited vital resources, to a selection of the best-adapted individuals, whose survival advantages are passed on and preserved through reproduction. The principle of Darwinian evolution can thus be characterized as the *survival of the best adapted under the conditions of variation, selection, and reproduction,* or what I will call, in short, the (biological) "principle of survival."[1] It is a principle because evolution is by definition nothing other than survival of the best adapted for survival. What is meant by the "best adapted" in a specific situation is, in contrast, a question that can be answered only empirically.[2]

The principle of survival is not to be problematized in the following discussion (this was done in chapter 3); rather, it is accepted as a plausible basis of the Darwinian theory of evolution. Now, what are the philosophical implications of the principle of survival, especially concerning the direction and goal of evolution? What consequences do such explanations have? Interest in these questions arises out of the special position of *Homo sapiens* as the most highly developed animal. More concretely, the questions to be discussed in the following chapters are: (a) Is there a tendency toward higher development in evolution? (b) If so, is man then to be considered the goal of evolution? (c) With regard to this, how is the role of the mind to be valued? (d) What are the anthropological implications of the answers to these questions? and (e) What, then, are the implications for a metaphysics of nature?

THE PROBLEM OF DIRECTION IN EVOLUTION

The process of evolution suggests an effective underlying tendency toward more highly developed organisms. How does the biological principle of survival pertain to this tendency, and what does *higher development* mean? As a rule, biologists do not like such normative attributes; Darwin himself noted, "Never use the words higher or lower."[3] Nonetheless, a development toward higher complexity is obvious. As the philosopher Wolfgang Stegmüller summarizes the theory put forth by Manfred Eigen,[4] "[T]he formation of ever more competitive mutants . . . is really compelled by thermodynamic laws" and "for this reason can even be described as *physically necessary*."[5] *Higher* accordingly means "more capable." However, does it also mean "more capable of survival"? There is a widespread opinion that it does. The philosopher Nicolai Hartmann, for example, considers it obvious that the higher type is "the more effective in the competition of the struggle for existence."[6] *Homo sapiens,* as the last link of evolution, should then be the organism most capable of survival.

At this point doubts arise. Compare the chance of survival of, say, infusoria with that of humans:[7] risk increases with an increase in capability. A glance at inorganic structures makes this even clearer. The Alps are obviously characterized by considerable stability. The proton is absolutely stable (or at least is considered to be almost stable) according to physicists. In short, as the philosopher Hans Jonas has emphasized, stability, the ability to survive, cannot be everything in the process of evolution: "The survival standard itself is inadequate for the evaluation of life. If mere assurance of permanence were the point that mattered, life should not have started out in the first place. It is essentially precarious and corruptible, being an adventure in mortality, and in no possible form as assured of enduring as an inorganic body can be."[8] The stability of inorganic forms obviously cannot be exceeded by that of living species. Yet how can evolution be said to progress if maximal stability is already attainable in the inorganic realm? Is the principle of survival not so relevant for Darwinian arguments after all?

One could object that for Darwinian arguments it is not simply the ability to survive that is decisive but the ability to survive in a particular situation—that is, the *adaptation* of an organism to its environment. Yet in this case too the previously mentioned arguments apply: protons are also optimally "adapted." Are humans better adapted than infusoria? "In my view," according to the biologist Leo von Bertalanffy,

> there is . . . not the slightest glimmer of a scientific proof that evolution, in the sense of a progression from simple to more complex organisms, has anything at all to do with increased adaptation, selectional advantage, or the production of a larger progeny. Adaptation is possible at every level of organization. An amoeba, a worm, an insect or a nonplacental mammal is just as well adapted as a placental mammal; if it weren't, it would have become extinct a long time ago. The equivalence of evolution and adaptation can, therefore, in no way be considered as proven.[9]

"Each of these levels," Hoimar von Ditfurth also says, "is perfect in itself." And, he continues, "The secret of the continual progress of evolution appears to be even greater in view of this fact."[10] So the question is posed more urgently: How can there be progress in evolution at all, and how can it be explained?

As for the first part of the question, Bertalanffy and biologists generally consider it a fact that "a general progress in evolution manifests itself in the direction toward higher organization"[11]—that is, toward more intense differentiation and consequently higher complexity. We proceed from these generally accepted conclusions in what follows.

At this point, however, a problem arises: In terms of Darwinian argument the principle of survival is extraordinarily plausible, but what does it have to do with higher complexity? Because of the uncertainty on this point, the principle of survival is accepted by some biologists, such as Bertalanffy and Adolf Portmann, only with reservations. We therefore need some clarification.

It will be instructive first to consider a form of evolution *not* connected with higher development, a development that is instead horizontal, extending in breadth. It must be remembered that evolution, according to Darwinian theory, presupposes competition, which is, for the most part, for the same territory, food, and so forth, since individuals of the same species are competing for them. Competition does not take place between earthworms and tigers, who have completely different needs and biospheres. Earthworms and tigers can coexist; they are, to use an expression from Leibniz, *compossible* species, which as such exist in different "ecological niches." How does this happen? To use a striking example, one can argue that because land exists, land-living animals finally develop from aquatic animals, with which they no longer compete and with which they are thus compossible. Because air exists, birds, too, develop from aquatic or land-living animals, with which they no longer compete and with which they are compossible as well.

To generalize: competition leads to a selection pressure in the direction of the occupation of niches and biospheres that are still free, and so to "horizontal" development in breadth. That means a negation of competition and, as a result, a multiplication of compossible species. Horizontal evolution leads to diversification.

In this regard, an interesting insight arises: such Darwinian processes of selection are doubtless driven by the principle of survival, as the term *selection pressure* indicates, but selection in this sense does not lead to an increase in the capacity for survival: its result is not more stability but diversification. What takes place is the occupation of still available niches, and that means biospheres already available— in the sense of a horizontal development that extends in breadth.

The other form of evolution has, in contrast, a vertical character: it is a development not in breadth but in "height." Horizontal evolution consists of the occupation of available biospheres, but vertical evolution creates new biospheres. For example, the existence of plants makes the existence of herbivores possible; the existence of herbivores, in turn, makes the existence of carnivores possible. In the first case the botanical world created by evolution provides a food resource, but only for a completely different kind of living being, namely herbivores, and these in turn provide a food resource for carnivores. The level of development reached by evolution itself becomes at this point a "niche," the basis of existence for new forms of life. Each level attained becomes the starting point for a new development. The result of this vertical evolution is a series of levels, a development in height, in the sense of a *self-upgrading of nature,* as I will call it.

The process of self-upgrading necessarily has consequences for the organization of living beings. For example, herbivores must possess completely different functions from plants. They need a chewing apparatus to chew up plant food and a digestive system to utilize it. But that does not suffice; they must first look for food, namely the plant species that are good for them, and that means they must move around; they cannot take root but need limbs for movement that must be coordinated and controlled; this demands a nervous system and brain as a control center. Furthermore, the ability to move around presupposes the ability to orient in the environment, and consequently a sensory organization associated with information processing by the nervous system and brain.

In short, herbivores exhibit fundamental differences in organization from plants: they are necessarily more complexly organized than the latter. Because their existence presupposes that of plants, they need functions that exceed those of plants. Of course, they also lack certain things that plants have, such as chlorophyll and the capability for photosynthesis. But they no longer need to meet their energy requirements as plants do because they can eat plants instead. And they are not really lacking chlorophyll and photosynthesis, for these are necessary only for the growth of plants, which are in turn necessary for the herbivores' nourishment. What has been attained at one level is — at least (as in the example of photosynthesis) — indirectly available at the following level. In this respect, the later level of development in vertical evolution is in fact more complex and in this sense higher. I admit that much more could be said about "higher." Evolution here tends toward higher complexity brought forth by itself. This form of self-upgrading of organic nature explains, therefore, the appearance of not only new but also more highly developed species on the basis of a strict Darwinian argument.

Although the principle of survival drives this "vertical" tendency of evolution toward higher complexity, higher complexity does not produce an increase of fitness for survival in any way. A chimpanzee is not more capable of survival than a tick. What can be said about this in the light of the argument developed?

In the case of horizontal evolution, it has already been shown that selection on the basis of the principle of survival leads, not to an increase in fitness for survival, but rather to a diversification of the species. Similarly, in the case of vertical evolution, the principle of survival drives development toward more complex structures of organization, yet this form of evolution is not connected with an increase in fitness for survival; it is simply based on the fact that selection pressure toward vertical evolution is effective. And selection pressure, though based on the principle of survival, indeed leads to new species, but in no way to species more capable of survival. This is because nature is perfectly satisfied with survival. Survival alone is required by the principle of survival; and that is also true in the case of vertical evolution.

The apparent paradox of a process of upgrading controlled by the principle of survival without an increase of the fitness for survival makes the difficulty facing evolutionary theorists understandable: although they justifiably adhere to the universal validity of the principle of survival, they cannot gain a yardstick from it for the height of development. Thus the connection between the principle of survival and higher development can be characterized by the formula that nature is satisfied with simple survival but not with the survival of the simple. Rather, a vague but "goal-directed" urge toward the complex is effective in it, an urge that drives the process of the self-upgrading of nature. That can be firmly based in the framework of Darwinian arguments

It must be emphasized that we are not describing the actual course of evolution here but rather formulating an ideal-typical statement. In Darwinian theory, it would be expected that an undisturbed, continual evolution of organisms would lead to higher development in the long run: that is, development toward more complex forms of organization and consequently toward acquisition of new biospheres. But the qualification *undisturbed* is important here because the immanent dynamics of evolution can be crucially disturbed by natural catastrophes. Evolution in the Darwinian sense unavoidably includes variation; thus an element of chance influences the course of evolution, which may also regress from the level of organization already attained or lead up various blind alleys and side streets. Nevertheless, the characterized vertical tendency of evolution remains—as a tendency—unaffected by this: there always remains a tendency toward higher development in evolution.

If the direction of evolution is not only horizontal but also—in tendency—vertical, one can further ask whether the tendency toward higher forms of organization is directed toward a goal in the sense of an end of biological evolution. This question is discussed next.

DOES BIOLOGICAL EVOLUTION HAVE A GOAL?

Before we can address the question of a possible goal of biological evolution, we must clarify what "development to higher complexity" means concretely; "higher complexity" is still a very comprehensive expression. The example of the transition from plants to herbivores at least shows that the organization of the latter demands not only limbs for motion and a digestive system but also organization of the senses, a nervous system, and a brain. The animal must look for its food; to do this it must move around, orient itself in its environment, process information, and regulate, coordinate, and control sequences of motion. These demands evidently increase with the level of organization. Orientation, regulation, and control mechanisms are more complex in higher animals than in lower ones. The perceptual capacities of the organism are decisive for this: orientation, regulation, and control are information-related processes and are thus very dependent on perceptual capacities.

Perception at first gives information about the external environment. Animals already necessarily possess knowledge of objects, though this is not the same as "objective knowledge," which is available only at the level of abstract thought. The perception of an animal is subjectively colored. It has, for example, a strongly selective character: the animal perceives only what is existentially relevant for it—in other words, what is beneficial or detrimental to it. Such subjectivity is not at all to be understood as a shortcoming. On the contrary, because it directly concerns the existentially meaningful relationship of the animal-subject to its object, it is especially efficient economically. It is simply the matching of object knowledge to a specific animal subject. Besides, perception's subjective coloring does not alter the fact that we are dealing with a highly remarkable cognitive accomplishment of animals.

Higher animals are more complex and live in more complex environments. The demands concerning orientation, regulation, and control are accordingly higher. Thus it becomes increasingly important for behavior that perception include not only data from external situations but also subjective data like skin contact and muscle tension; even data of an "existential" character, like pain, are "inserted," as it were, into the perception of the environment. Higher evolution has thus led to a crossing over of outer and inner perception and consequently to the capability of sensation.[12] Regarding the sensation of touch, for example, what I sense about a stone I sense about myself at the same time. The same goes for the sensation of pain when I touch a hotplate. This reflexivity of perception, made possible by the crossing over of outer and inner perception, results in quite new possibilities for orientation and fine regulation of movements and, consequently, behavioral efficiency. At the same time a completely new sphere of being has been established: an inner dimension of the organic subject, the world of the psychic—a sphere of self-perception in which the subject meets itself and acquires reflexivity.

Of course, numerous unsolved problems are involved here, which are addressed by a widespread and controversial contemporary discussion on the mind-body problem.[13] Nevertheless, in the present connection, it is essential for the plausibility of my argument only that in the course of biological evolution a development of sensation finally takes place, and thus a crossing over of outer and inner, or existential, perception. The subject continually senses itself in its perception and thus explicitly becomes the focus of its own intended actions and executed actions. Implicitly, this has always been the case because of the organism's instinct for self-preservation. In short, through the evolution of sensation, the animal subject is simultaneously present and thematic to itself in perception. Sensation means continual "self-thematization" of the subject.

In summary, two principal tendencies are important for higher forms of evolution: the development of *cognition* and of *self-thematization*. Both developments make sense within the framework of organic behavior. That they make sense is guaranteed by the fact that they have already passed their test, so to speak, as a result of the evolutionary process. That means that, normally, animals automatically possess adequate cognition and action impulses precisely so that they optimally succeed at subsisting in the realm of life to which selection has fitted them. They can rely on their innate instincts.

Unfortunately, as philosophizing biologists[14] suggest, evolution seems, nevertheless, to have made a mistake with the development of man. Through hypertrophy of the cerebrum, a completely "heady" being came into existence. His brain was no longer the "ratiomorphic apparatus" (as the normal brain of the animal has been called)[15] optimally adapted to a specific realm through natural selection but rather the seat of reason emancipated from all vital bonds, which, as the history of humankind appears to confirm, turns everything upside down. Man is consequently no longer only animal but also mind—as will be shown, a precarious, momentous combination of vitality and reason.[16]

At first, the occurrence of mind had consequences that aided life: with the development of language new forms of cooperation became possible. Consequently, the group hunt, for example, could be organized more efficiently. Techniques for securing subsistence, such as those for making clothing, food, and tools, could be developed. However, with the development of language, thought also arose and with it the dimension of *consciousness*. Consciousness is, however, due to the characteristic tendency toward self-thematization typical for higher animals, principally connected with the capacity for self-awareness. Self-awareness, in turn, is to be understood as the root of the essential consciousness of freedom for human beings.[17] In other words, with the occurrence of *mind* the consciousness of freedom also arose in humans and with it the potential for questioning natural relations, dissociating and possibly freeing oneself from them.

One expression of this is the development of technology. It aims at freeing humans from natural constraints, but at the same time it always raises the possibility of being able to oppose nature.[18] The history of humankind, which can be viewed as a continual process of the perversion of the natural order, also illustrates that this is no illusion. Man, although himself a child of nature, develops unnatural ways of life, causes damage of civilization, employs weapons of destruction, destroys ecological systems, and conceives of ideologies destructive to life. Even more: with the mind "evil" entered into the world—not the "so-called evil" that Konrad Lorenz[19] has expounded, with terrifying harmlessness, simply as "aggression." Certainly aggression exists in animals and, as a phylogenetic heritage, in humans too; however, consciously created evil is a pure human specialty.[20] On the other hand, mind has also brought forth forms of the most highly developed humanity, outshining all that is natural. In short, the rise of human mind is at the same time a process of the separation of man from nature and of the self-authorization of man, negatively as well as positively—the establishment of another, thoroughly artificial nature: *culture.*

In this sense man has been designated (by Herder) as the being freed from nature.[21] Human freedom, as Schelling[22] tells us, is an expression of the fact that mind has stepped out of the center of nature (and thus also has the potential to go mad). In this respect, it even has the ability to oppose God; this would be the metaphysical origin of evil. Similarly, from an anthropological perspective, Helmut Plessner[23] characterizes man as that eccentric being who has freed himself from the "centered" adjustment of the animal to its environment. And for Max Scheler,[24] the ability to say "no," the capacity to transcend every factual situation, belongs constitutively to the nature of mind. In other words, not only is culture an artificial nature, but it transcends nature to become a supernature.

The undeniable consequence is the end of natural selection. This simply results from the fact—hinted at already—that the contrivances of culture have made the mechanisms of natural selection obsolete. In a world in which sickness can effectively be cured, clinics and spas are at people's disposal, artificial limbs are applied, and replacement organs are implanted, the biological principle of survival has been "unhinged" *(ausgehängt).*[25] And that means, too, that natural evolution has come to an end. In this respect, man stands at the end. Is man, then, the goal of evolution?

One could object that the human species changes biologically even today—for example, in muscle structure, susceptibility to sickness, and life span. That cannot be denied. But these changes are manifestations of the "self-domestication" of man and thus consequences of civilization, which as such are not the results of *natural* selection. On the contrary, they are expressions of an evolution that is now taking place under completely different conditions, namely those of *cultural evolution*. This is the continuation of natural evolution only in a temporal sense; in its character, it is completely different from the former. Though the biological principle of survival has been unhinged, an analogous principle is operating on the cultural level with regard to competing strategies, institutions, theories, and so forth.[26] The motors of cultural evolution are intellectual and technological innovations. Linguistic communication, forms of passing down traditions (e.g., through written language), and creative processes play an important role.

A central aspect of this development is the speed with which it takes place. The speed of natural evolution is obviously dependent on the reproductive rate of the organisms. Cultural evolution, on the other hand, has accelerated more and more in its historical course—think only of the development in technology. All these processes initiated by human consciousness are far more rapid than natural selection, which blindly gropes along its way. For this reason alone it would have no chance against cultural evolution, if the disconnection with the biological principle of survival were not already reason enough. To return to the objection mentioned, if humans have also biologically changed in the course of history, then the changes are due not to natural selection but to culture alone.

We can conclude, then, that natural evolution does have an end, and that it is man himself. The idea is not strikingly new. In the Judeo-Christian tradition, man has always been the "crown of creation." Criticisms of this view are also not new. Biologists tend to suspect it of anthropocentrism. From a phylogenetic perspective, man too is only an animal, even if late on the scene. That he has the advantage of reason, in contrast to other animals, does not warrant assigning him a special position. Animals have other advantages, such as perfect instinct, that man lacks.

Besides, biological natural selection has by no means been turned off. Where man does not control, the selection mechanism regulated by the principle of survival continues, for example, (unfortunately) in the case of pathogenic viruses and bacteria, which due to their high rate of reproduction[27] can change very quickly and thereby become resistant to treatment. Thus new evolutionary lines arise continually that, according to the previous considerations, will show a tendency toward development of higher complexity. One could, then, speculate that such continual evolutionary processes outside the sphere of man could lead at some time to the origin of another species of humanoid being and, in the long run, even to a superman. According to this viewpoint, an evolutionary development that would surpass man himself would refute the idea that natural selection ends with man.

What might such a development look like? The trend toward higher complexity concerns, above all, the development of the brain, so, to evoke a brief science fiction scenario, mind might at some point appear in such an extrahuman line of evolution. Once mind appeared, however, natural selection would be finished here too, because it would again be replaced by cultural development, and the principle of survival would thus be unhinged. Indeed, through natural evolution, other, again humanoid, beings could arise perfectly well, but, in any case, not of a kind that would surpass man with regard to the character of his mind. That would be impossible through natural selection, whereas the succeeding cultural evolution of intellectual and also biological traits would lead to the greatest differences.

All in all: With the appearance of mind, natural evolution, which is at the same time a development toward higher complexity, is irrevocably ended, and in this sense man—or (with respect to the structure of his mind) indeed any humanoid being—represents the end of natural evolution both positively and negatively (negatively since, as was mentioned, "evil" came into the world by means of mind). Seen in this way, man is the crown and cross of the creation in one.

MIND AS THE ELEVATION OF NATURE

On one important point I would like to amend what I have stated thus far. If the goal of natural selection is the evolution of mind, and if mind is the negation of naturalness, then is not the goal of natural evolution the negation of naturalness?[28] That would indeed be a strange consequence.

To clarify this, it is important to see what the appearance of mind means concretely. In comparison with animals, it means more richness of being: language, thought, law, art, religion, science, technology, and so on are dimensions principally closed to animals. In this respect it could be said that the negation of naturalness carried out by mind does not mean a loss for man; from the perspective of natural selection theory, however, it seems highly paradoxical.

At this point it is helpful to recall the difference in cognitive abilities between animals and men: animal "cognition" is, as already mentioned, strongly subjectively colored perception. Due to his emancipation from natural bounds, man, in contrast, strives for objective knowledge: that is, a knowledge of the object that no longer depends on subjective-private states but rather does justice to the thing itself. That is the basic ethos of science. The object of natural science is, accordingly, nature as such and not one's subjective experience of nature. No animal is able to abstract from its subjective experience and thematize nature itself.

But what is "nature itself"? Certainly not the actual state of nature in its transient manifestations. For knowledge of nature, only the lawfulness underlying nature can be of interest; accordingly, the object of science is not any single natural object existing here and now but rather the law of nature: that is, a universal of nature that transcends time and space or—to use a classical philosophical term—nature's underlying essence.

The essence of nature is not just any part of nature, so in space and time the essence of nature will be sought in vain; it cannot be found there as one would find a stone. Consequently, if scientific cognition aims to discover the essence of nature, then it is searching for something that does not exist in natural reality because it remains hidden as the essence underlying it. Cognition reveals something that completely determines natural being but that itself does not possess the form of natural reality. In other words, insofar as knowledge of nature comprehends the underlying essence of nature, it accomplishes something that nature cannot. So something is added to the being of nature that is not realized in it—precisely knowledge of the essence of nature.

An instructive example in this regard is again technology. It can be characterized as the venture that frees the possibilities lying in the essence of nature. Indeed, the fact that nature essentially holds possibilities results from its lawlike character. For example, the movement of the earth around the sun is determined by the law of gravitation, but that law's validity is not limited to the earth and sun: it holds for bodies in general. Thus the law contains possibilities that go far beyond this case of the movement of planets. Nature, which at first appears to be a finished order willed by God, presents itself from a technical perspective as a field of unforeseeable possibilities. Think of the taming of fire, the airplane surmounting gravity, or the transistor making worldwide communication possible. Technology adds something to the actually realized being of nature that stems from the dimension of possibility in nature, from the lawfulness essentially underlying it. Technology brings something from the hidden essence of nature to light to create a second, artificial nature.

As already said, this endeavor becomes possible through the knowledge of natural laws, which, in turn, is an accomplishment solely of mind. Animals are not in the position to develop physical theories and translate them into technology, let alone plants and stones: nature possesses—as nature—no knowledge of itself. Knowledge of nature transcends the possibilities of nature. Mind can accomplish something that nature itself cannot, namely to understand nature. Nature only exists; it is not aware of existing.

Now, man himself is a child of nature. Nature thus appears to have brought forth, in humanity, a being that is capable of understanding nature, and in this way it remedies the previously named flaw in nature. In the form of man, evolution has developed the organ of cognition that nature itself lacks. Only the activity of man's cognition can make visible what nature is in its essence. From this perspective, mind appears as the completion and perfection of nature, which, as we have said, also manifests in the completely new possibilities of technology in contrast to those of "biologically developed" nature.

As mind, man has the ability to oppose nature and to pervert and destroy it, yet nature has completed itself only in the human mind. Both results come from the same root: the ability to understand the underlying essence of nature. Animals can perceive only the outer appearance of nature. But man's ability to understand the essence is simultaneously his liberation from natural limitations and the negation of naturalness.

The mind's negation of naturalness has the consequence that man constitutes the end of natural selection. Consequently, I have asked whether the negation of naturalness is to be understood as the goal of natural evolution, a thought that appears paradoxical. Now the case presents itself as follows. Nature, in the form of mind, has in fact brought forth a being that can transcend nature by means of reason and can understand nature in its essence and translate it into technical reality. Viewed in this way, evolution is a gigantic process of the self-clarification of nature that is completed in man. In this sense, nature develops, beyond its mere natural being, an organ of self-cognition that nature itself lacks. The negation of naturalness by means of the mind is thus at the same time the completion and elevation of nature—though in the ambivalent sense that this can always turn into a perversion of the natural order. Everything shows this fundamental ambivalence, the possibility of positive or negative manifestation, and inevitably this is also true for the human mind. Nevertheless, it can also be said that with man nature, as it were, gains a consciousness of itself.

We had asked about the goal of evolution. From the argument developed here, we can conclude that evolution *has* a goal: the human mind. In it nature transcends itself as nature and gains at the same time the potential to reveal the essence underlying nature. One thus underestimates mind if one views it only as something unnatural. It is also basically the *supernatural* and thus the potential for the completion and elevation of nature.

ANTHROPOLOGICAL IMPLICATIONS

A brief comment is in order concerning the consequences of this line of argument for an understanding of human beings and especially of human cognition and action. Since we are dealing here only with the characterization of the essential difference between animals and humans, their extensive common ground, which of course also exists, will not be considered. Such a common ground is manifested in the explorations, by both evolutionary epistemologists and sociobiologists, of what is, in a certain sense, a continuum of development from the animal to the human mind. These extraordinarily interesting and important results for epistemology as well as for ethics will not be treated here, since we are dealing only with the difference between humans and animals.

Our argument has consequences for cognition: (nonhuman) animals are optimally adapted to their environment, so their perception supplies them with automatically adequate cognition within their subjective circle of life. For humans, in contrast, an epistemological problem results from the nature of thought. Because of liberation from natural constraints, thought is indeed an act of will and thus an arbitrary act in principle as well. This means thought can also miss its goal, it can make mistakes. Mistakes in this sense can be viewed as the price of freedom. Animals cannot make mistakes because they have no freedom of thought. Precisely for this reason, they also are not capable of objective knowledge. Only a being that can make mistakes can also possess objective knowledge.

The argument also has consequences for action. The instinctual behavior of animals is adapted to the environment and thus, like their perceptual faculties, is automatically adequate. In contrast, human action, freed from the constraints of instinct, is entirely problematic; it must at first find its purpose and may therefore also miss it. Further, the nature of the human mind makes possible genuine human behavior like lying, deceit, and malice. In short, due to its liberation from instinctual constraints, human behavior is no longer automatically correct. Consequently, moral laws are needed, and humans can negate and break them. So, together with mind, the possibility of guilt has entered into the world. This is, as it were, "original sin," belonging to man from the very beginning, basically only a religious metaphor for man's self-authorization connected with the appearance of mind and transcending the natural order: not only a theological but also an evolutionary fact!

To sum up: the mind frees man from his natural constraints; the possibility of error and guilt is the price of freedom. But this freedom is also the surmounting of naturally given limits. Cognition and freedom on the one hand and error and guilt on the other thus belong intrinsically together, metaphorically and biblically speaking: hell always belongs to heaven, the two cannot be separated. To exist in this ambivalence is man's inevitable destiny.

METAPHYSICAL IMPLICATIONS

In conclusion, we must ask what presuppositions enter into the argument carried out here. Let us first recall the basic ideas developed earlier: I showed that in evolution there is a tendency toward higher development—that is, a self-upgrading of nature—and toward the development of cognition and "self-thematization" and thus finally of mind. The appearance of mind, as was further argued, has as a consequence for man the possibility of freeing himself from the constraints of nature. At the same time, it also makes possible for him to understand what nature is in its essence.

What has been presupposed in this argument? In each case, it is the fact that in the process of nature, precisely in the form of evolution, the appearance of new forms is possible. In other words, using Darwinian arguments, we assume from the very beginning that nature contains possibilities that are still hidden in primitive forms and become visible only in more complex structures. The concept of natural evolution makes sense only under this assumption that nature is more than its factual being in that it contains potentials that increasingly appear in the natural process.

But where does the potential contained in natural being come from? Doubtless from natural laws. Only that which is compatible with them is naturally realizable. This holds for technology as well as for the development of nature itself, especially biological evolution. The basic presupposition for this is a natural being of a lawlike character. Incidentally, the main point of Darwinian theory is that it reasons only according to the law of causality and thus can do without the Aristotelian premise of a nature that is purposive in itself. This causal nature of the Darwinian approach becomes apparent with regard to the biological principle of survival: variation is based on the mutation of genes, for instance, due to radiation; natural selection in its broadest sense on the "interaction" of organic systems with their environment; and reproduction on the function of hereditary mechanisms. Viewed in this way, the lawfulness of nature is in fact the basis of the Darwinian explanation of natural evolution. According to recent theories, it even includes the abiotic formation of life from inorganic material.[29]

The lawlike character of nature cannot, of course, be founded in experience, since the claim to universality connected with the concept of natural law transcends every possible experience. On this point, Hume's criticism must be accepted. In fact, we are dealing here, not with a result of empirical research, but with its premise, which, viewed more closely, represents a metaphysical presupposition underlying all natural sciences.

Indeed, the laws of nature have a completely different character of being than the being of nature. The law of the motion of planets is not itself in motion; the law of electromagnetism is not itself electromagnetic; the law of the earthworm is not itself a worm. The laws of nature are rather like the logic determining the process of nature: they do not exist in time and space like a stone or an earthworm but possess *ideal* character. This ideal, then, is what was described above as the essence of nature underlying the natural being. In short, the concept of natural evolution implies that nature contains potentials that, in turn, stem from the laws of nature and thus from an ideal basis of nature—an inevitable metaphysical premise for all natural sciences.

At this point the question arises of how this metaphysical aspect is to be valued from a philosophical point of view. This is the question of the philosophical concept of nature, which I have discussed in detail elsewhere.[30] This much can be said about it: we are forced to assume something like a logic underlying nature in the form of natural laws, and this logic leads to the evolution of the faculty of cognition, which in turn is capable of directing itself toward nature and of penetrating and understanding it logically. All in all, this points to an (objective-) idealistic ontology of nature. Mind you: *objective* here designates an idealism not of a Berkeleyan but of a Hegelian type, for which there are, I think, good reasons and to which we owe probably the most well-thought-out philosophical concept of nature that occidental philosophy has brought forth.

I have argued here that biological evolution finds its end in the appearance of mind, with the new potential for understanding the laws of nature and translating them into technology. In this manner the biological principle of survival is "unhinged": cultural principles take its place (which may well be analogous to the principle of survival) in the course of a new form of evolution, the cultural evolution of man.

Thus the condition for natural evolution (including abiotic evolution) is the lawlike character of nature. This ultimately leads to the appearance of mind, which, in turn, understands the lawfulness of nature as the ideal essence underlying it and thus adds something to nature that, to be sure, belongs to it but that is, nevertheless, not realized within its own horizon: precisely a knowledge of the lawfulness of nature as the ideal essence underlying it. In other words, the ideal underlying nature initiates an evolution that finally reveals this ideal, or, expressed more trenchantly, the ideal underlying nature prosecutes its own self-revelation by means of evolution. This would have to be, I think, the final answer to the question of the direction and goal of evolution from a metaphysical standpoint.[31] At the same time this result, it seems to me, can be thought of as a restitution of teleology, not in an Aristotelian sense but in a new, modern sense that is no longer incompatible with science.

The metaphysical character of this answer transcends every possible experience. Surprisingly, the arguments developed here have shown this. The answer, however, follows as a consequence of Darwinian arguments and indeed relies only on the presupposition of lawfulness underlying nature—a general premise of all natural science too, as has already been said. But what distinguishes Darwinian arguments from other forms of natural scientific thought is that, thought out to its end, it also reveals the metaphysical implication contained in it: that the ideal essence underlying nature drives an evolution that develops—in the form of the human mind—an organ for cognition of precisely this ideal. The natural scientist makes the immanent logic of nature visible qua knowledge of nature; in the Darwinian view, he himself is a product of this logic that he makes visible and that reveals itself in him at the same time. In this respect, this apparently completely unmetaphysical, causal-analytically oriented Darwinian theory ultimately reveals a metaphysical dimension of nature as a whole[32]—indeed, including the understanding mind itself—an interrelation that suggests an objective-idealist ontology of nature and that is adequately comprehensible, and probably only comprehensible, within this framework.

NOTES

This chapter has been translated from the German by Edward Kummert and edited in association with Timo Klein. Elisabeth Magnus made a thorough, judicious final revision.

1. The core of this idea can already be found in Empedocles, as described by Aristotle in his "Lectures on Physics" (198b–199a): namely, only animal species with sharp teeth suitable for chewing and fighting were able to survive; otherwise they would have become extinct. What is missing in Empedocles's thought is the element of variation and the transmission of acquired features by means of reproduction.

2. Whether the principle of survival amounts to a tautology is the subject of a controversial discussion. See, e.g., L. von Bertalanffy, "Gesetz oder Zufall: Systemtheorie

und Selektion," in *Das neue Menschenbild: Die Revolutionierung der Wissenschaften vom Leben,* ed. A. Koestler and J. R. Smythies (Vienna, 1970), 80; R. Dawkins, *The Selfish Gene* (Oxford, 1976), 92: N. Hartmann, *Philosophie der Natur: Abriß der speziellen Kategorienlehre* (Berlin, 1950), 646; R. Spaemann and R. Löw, *Die Frage Wozu? Geschichte und Wiederentdeckung des teleologischen Denkens* (Munich, 1981), 242; W. Stegmüller, *Hauptströmungen der Gegenwartsphilosophie,* vol. 2 (Stuttgart, 1975), 439; V. Hösle and C. Illies, "Der Darwinismus als Metaphysik," in *Jahrbuch für Philosophie des Forschungsinstituts für Philosophie Hannover 1998,* vol. 9, ed. P. Koslowski and R. Schenk (Vienna, 1997). Hösle and Illies, in "Der Darwinismus," mention a "quasi-tautology" (114); see the clarifying explanations about this too (104). See also V. Hösle, *Moral und Politik: Grundlagen einer Politischen Ethik für das 21. Jahrhundert* (Munich, 1997), 259; V. Hösle and C. Illies, *Darwin* (Freiburg Br., 1999), 98 ff.

3. Quoted in Hösle and Illies, "Der Darwinismus als Metaphysik," 106.

4. See, e.g., M. Eigen, "Wie entsteht Information? Prinzipien der Selbst organisation in der Biologie," *Berichte der Bunsen-Gesellschaft für physikalische Chemie* 80 (1977): 1059–81; instructive, too is the representation of Eigen's theory in Stegmüller, *Hauptströmungen der Gegenwartsphilosophie.*

5. Stegmüller, *Hauptströmungen der Gegenwartsphilosophie,* 441.

6. Hartmann, *Philosophie der Natur,* 652, see also 651, 679.

7. See the report in *Der Spiegel,* May 1, 1981: "Tardigrades, e.g., were heated in a cryptobiotic state up to over 150 degrees centigrade for several minutes and showed as little damage as anorganic matter would. Freezing down to 0.008 degrees above absolute zero did not affect them either. The tiny beings resisted ionizing radiation with an energy of 570,000 röntgens for 24 hours (they would survive an atomic war the best). Even after a stay in a vacuum, as in space, they became mobile again."

8. H. Jonas, *The Phenomenon of Life: Toward a Philosophical Biology* (New York, 1968), 106.

9. Bertalanffy, "Gesetz oder Zufall," 82. Compare also the clueless discussion of the experts on this point at the end of the article.

10. H. von Ditfurth, *Der Geist fiel nicht vom Himmel: Die Evolution unseres Bewusstseins* (Hamburg, 1976), 321.

11. Bertalanffy, "Gesetz oder Zufall," 81.

12. For a more detailed discussion of this point, see D. Wandschneider, "Das Problem der Emergenz von Psychischem: Im Anschluss an Hegels Theorie der Empfindung," in *Jahrbuch für Philosophie des Forschungsinstituts für Philosophie Hannover 1999,* vol. 10, ed. V. Hösle, P. Koslowski, and R. Schenk (Vienna, 1998).

13. See P. Bieri, ed., *Analytische Philosophie des Geistes* (Königstein/Ts., 1981); Hösle and Illies, "Der Darwinismus als Metaphysik," 121 ff., 124 ff.; Hösle and Illies, *Darwin,* 176 ff.

14. For instance, R. Riedl, *Biologie der Erkenntnis: Die stammesgeschichtlichen Grundlagen der Vernunft* (Berlin, 1980), 28 ff., 79, 185 ff.

15. Ibid., 35, following E. Brunswik.

16. See the instructive discussion in V. Hösle, "Sein und Subjektivität: Zur Metaphysik der ökologischen Krise," in *Prima Philosophia* 4 (1991): 519–41. Hösle emphasizes that mind itself has a natural aspect: only because of this it is possible to influence natural being and to destroy it too—which, of course, would lead to an endangering of its own natural existence (520, 535, 537).

17. See D. Wandschneider, "Selbstbewusstsein als sich selbst erfüllender Entwurf," *Zeitschrift für philosophische Forschung* 33 (1979).

18. Cf. Hösle, *Moral und Politik,* 288: "Man is that organic being that completes the tendency of life by his ability to negate life."

19. See K. Lorenz, *Das sogenannte Böse: Zur Naturgeschichte der Aggression,* 9th ed. (Munich, 1981).

20. "If by this we understand a kind of behavior that damages others but offers no advantage for itself and even sometimes disadvantages, then . . . it becomes immediately clear that malice cannot be a result of genetic evolution" (Hösle, *Moral und Politik,* 267); on this, see also D. Wandschneider, "Ethik zwischen Genetik und Metaphysik," *Universitas* 38 (1983): 1139–49.

21. J. G. Herder, *Ideen zur Philosophie der Geschichte der Menschheit,* vol. 1, bk. 4 (Leipzig, 1897).

22. F. W. J. Schelling, *Philosophische Untersuchungen über das Wesen der menschlichen Freiheit und die damit zusammenhängenden Gegenstände* (1809; reprint, Stuttgart, 1964).

23. H. Plessner, *Die Stufen des Organischen und der Mensch: Einleitung in die philosophische Anthropologie* (1928; reprint, Berlin, 1975).

24. M. Scheler, *Die Stellung des Menschen im Kosmos,* 6th ed. (1928; reprint, Bern, 1962).

25. This expression, I think, has been introduced by A. Gehlen.

26. See, e.g., Hösle and Illies, "Der Darwinismus als Metaphysik," 116, 119; Hösle and Illies, *Darwin,* 109, 180.

27. It is said that these species reproduce about once every twenty minutes.

28. Cf. Hegel's assertion that "the goal of nature is to kill itself and to break through its rind of the immediate and sensual, to burn itself as Phoenix in order to go forth rejuvenated out of this superficiality." *Hegel-Werkausgabe,* 20 vols., ed. Eva Moldenhauer and Karl Markus Michel (Frankfurt am Main, 1969 ff.), 9538 add.; "add." refers to the editorial additions to Hegel's *Enzyklopädie der philosophischen Wissenschaften im Grundrisse.* See also the impressive discussion on this subject, above all concerning the problem of human identity, in Hösle, *Moral und Politik,* ch. 4.2.1., especially 293 f.; see also n. 18.

29. See M. Eigen, "Wie entsteht Information?" as well as the presentation of Eigen's theory in Stegmüller, *Hauptströmungen der Gegenwartsphilosophie.*

30. See, e.g., D. Wandschneider, "Die Absolutheit des Logischen und das Sein der Natur. Systematische Überlegungen zum absolut-idealistischen Ansatz Hegels," *Zeitschrift für philosophische Forschung* 39 (1985); D. Wandschneider, "Das Problem der Entäußerung

der Idee zur Natur bei Hegel," in *Hegel-Jahrbuch 1990,* ed. K. Bal and Heinz Kimmerle (Bochum, 1992), 25–33; D. Wandschneider, "From the Separateness of Space to the Ideality of Sensation: Thoughts on the Possibilities of Actualizing Hegel's Philosophy of Nature," *Bulletin of the Hegel Society of Great Britain* 41/42 (2000): 86–103; D. Wandschneider, "Hegel und die Evolution," in *Hegel und die Lebenswissenschaften,* ed. O. Breidbach and D. von Engelhardt (Berlin, 2001).

31. In the light of the argument developed here, the ideal represents itself as fundamental and therefore—in comparison with natural being—as an outstanding mode of being that again refers to an (objective-)idealistic ontology.

32. "In man the whole of being manifests itself" (Hösle, "Sein und Subjektivität," 520).

CHAPTER 11

Objective Idealism and Darwinism

Vittorio Hösle

One of Wilhelm Dilthey's most interesting contributions to philosophy is his famous trichotomic subdivision of all the different philosophies that have developed in the course of the history of philosophy into the main types of naturalism, subjective idealism, and objective idealism. If one accepts this typology, it seems obvious that a worldview that takes Darwinism seriously must belong to the first type, namely to naturalism. Indeed, the opinion is often enough voiced that a naturalist philosophy is the only acceptable one after the triumph of Darwinian biology. When we recognize that humans are themselves the result of natural evolution and that this evolution is, at least to a certain degree, the result of chance, both the Kantian version of subjective idealism, according to which nature is itself determined by a priori categories of the mind, and the objective-idealist tradition, which sees in the world the manifestation of a Reason not reducible to the human mind, seem obsolete. According to naturalism, the basic entities are natural substances or events, and everything that seems to transcend nature must in fact be reduced to those basic natural entities. Particularly dangerous for naturalism is the alleged autonomy of the normative realm, and here indeed Darwinism seems to offer a reduction not easily accessible to earlier forms of naturalism. What the tradition had called "the transcendentals"—the most relevant of them are, with regard to a possible autonomy of the normative realm, the true, the good, and the beautiful—seem now to be explainable in a naturalist way. According to evolutionary epistemology, truth is nothing else than a survival guaranteeing matching of the representations, conscious or not, of the organism with the external world. According to evolutionary ethics, the basic principle of behavior is the maximization of one's inclusive fitness, and all morally praised forms of "altruism" are in fact epiphenomena of genetic egoism. According to evolutionary aesthetics, finally, beauty is nothing else than a function of sexual attractiveness. The opinion that Darwinism is compatible only with a naturalist interpretation of reality is strengthened by the fact that Darwinism is usually regarded as radically opposite to Platonism, the paramount representative of objective idealism in the Western tradition.

At least it belongs to the topoi of the history of ideas that the recognition of the evolution of species destroyed one of the pillars of Platonism, given that in Platonism the biological species had enjoyed the status of paradigmatic *eide*. Darwin's scientific adversaries Richard Owen and Louis Agassiz regarded themselves as Platonists,[1] so it seems to make perfect sense when we read:

> In the first place it [the Darwinian revolution] dethroned Reason, for the order of the universe could be accounted for by unadulterated mechanism. . . . It did away with Platonism, including, we should emphasize, its attendant mystical attitude toward numbers. . . . At the same time, and by implication, the fundamental sequence of priority was reversed. It is individuals that differ from one another, and individuals that struggle for life. Platonic Ideas, essences, types, forms . . . turned out to be hypostatized abstractions. To understand evolution, we need to know what varies, not just what remains constant. . . . A Platonist would forsake our world for a transcendent realm of ideas. A Darwinian seeks to be at home in the land of his birth.[2]

The aim of the following discussion is to question this opposition and show that it is logically possible, and even a wise choice, to be both a Darwinian and an objective idealist (and thus a Platonist in a *very* broad sense of the word). This may surprise some, so I will begin by sketching some basic tenets of Platonism and of later forms of objective idealism and by granting the partial truth I see in statements that refer mainly to the revolution in the concept of species (like the one quoted above from Ghiselin) (section I). But even Plato and certainly later forms of objective idealism distinguished between different types of *eide,* and while one can imagine forms of objective idealism that are not committed to specific ontological assumptions about the status of biological species, the central tenet of objective idealism will always be the impossibility of interpreting the ideas of the true, the good, and the beautiful in a naturalistic way—namely by reducing them to states of affairs within the empirical world. Not only do I think that the arguments in favor of this tenet remain cogent and are not confuted at all by Darwinism, I want to argue that the most reasonable philosophic interpretation of Darwinism will even support this tenet. Thus I will discuss the relation of Darwinism and objective idealism with regard to the idea of truth (section II), the idea of the good (section III), and the idea of the beautiful (section IV)—the first two very quickly, since I rely on work done elsewhere. Since objective idealism has always granted great importance to teleological thinking and since it has been again and again integrated with theism (even if there are atheistic forms of objective idealism and there are forms of theism in contradiction with objective idealism, such as voluntarist theism), I will end with some programmatic reflections on the relations between Darwinism and a teleological and theistic worldview (section V). Of course, I share the opinion that Darwinism is one of the greatest and most original scientific theories ever devised by the human mind and that it has profoundly changed not only our biological views but in general the way we interpret the world. I think, however, that we honor this theory better if we recognize its limited ontological relevance and do not make a First Philosophy out of it. It speaks for Darwinism that it is, as Charles Darwin himself knew,[3] compatible with a host of metaphysical positions and that it does not entail naturalism; for otherwise the classical arguments against naturalism would also strike at Darwinism.

I

Since objective idealism encompasses historically distant and systematically different philosophical positions, it is not easy to find a common denominator for all its forms—the differences between Plato and Whitehead, for example, are considerable. In a certain sense it would be a "Platonic" problem itself to distinguish between the idea of objective idealism and its various manifestations—a task far too ambitious for this chapter. I will start with some reflections on the historical Plato, since he is the father of objective idealism and since a study of the origin of a tradition is always helpful in order to understand both its potential and its problems. Then I will show why some of the historical Plato's tenets are indeed confuted by the Darwinian revolution, and finally I will sketch a broader concept of objective idealism compatible with it.

The core of Plato's written philosophy is certainly his doctrine of ideas. To present this complex doctrine is very difficult, partly because it tries to answer—for our modern categories—too many questions that differ too much from one another. Thus there is little doubt that the Platonic theory of ideas is rooted in ethical considerations—both the Socratic origin of the Platonic philosophy and the nature of the first Platonic dialogues prove this. Furthermore, concerns related to the philosophy of mathematics, a discipline developed by the Pythagoreans, are crucial for it (even if Plato distinguishes mathematical entities from ideas). At the same time the doctrine of ideas is a general theory of universals, linked to Plato's surprising epistemological position, which seems to deny the possibility of *episteme,* knowledge par excellence, with regard to perceptible objects. To make matters even more difficult to grasp, Plato possibly denies—at least according to Aristotle[4]—that there are ideas of artifacts, though the ideas of plants and animals play a central role in his ontology. In fact, there is little doubt that the efforts to establish a biological taxonomy so amply developed by Aristotle began, even if in a rudimentary form, in the Academy, where species and genera were distinguished and ordered in a hierarchical way by comparing similarities and differences.[5] Even if the ideas enjoy ontological priority over perceptible objects, Plato does not deny value to the sensible world; on the contrary, the *Timaeus* is a great attempt to propose a theory of nature as the place in which the ideas are mirrored—that is, to understand the physical world as structured by ideal principles. Arthur Lovejoy has pointed out that this duplication of the ideas in the physical world is an important corrective to the otherworldliness characteristic of Plato: "The Intellectual World was declared to be deficient without the sensible." This conviction led Plato to the formulation of what Lovejoy calls the principle of plenitude, the idea "that the extent and abundance of the creation must be as great as the possibility of existence and commensurate with the productive capacity of a 'perfect' and inexhaustible Source, and that the world is the better, the more things it contains."[6] It is easy to understand that this principle entailed the principle of continuity: "If there is between two given natural species a theoretically possible intermediate type, that type must be realized—and so on ad infinitum; otherwise, there would be gaps in the universe, the creation would not be as 'full' as it might be."[7] One of the serious hermeneutical problems related to Plato's doctrine of ideas is, of course, whether Plato thought that ideas had causal power with regard to perceptible objects. His use of the concept of *aitia* in relation to ideas does not entail this, since *aitia* means far more than the modern concept of cause and also encompasses reasons. Aristotle's famous doctrine of the four *aitiai* is a further development of the two factors distinguished in the *Timaeus* (the *causae materialis* and *efficiens* of *to anagkaion,* the *causae formalis* and *finalis* of *to theion*). Now, as it is obvious that the belief in a final cause in Aristotle never replaces the search for efficient causes[8]—the different types of causes complement each other, they do not compete with each other—Plato does not want to deny that there are mechanical causes of biological processes. However, both he and Aristotle probably reject the modern view that the efficient causes are sufficient to determine the factual course of the world. Independent of this special problem, Plato and Aristotle share the opinion that the understanding of efficient causes, as propagated by Democritus, does not answer all the legitimate questions, such as those of the essence and the telos (a concept that in Aristotle, unlike his Christian successors, does not entail any design in a divine mind). Even if Aristotle considers as efficient cause of an organism another organism of the same species, both thinkers believe that the reproductive behavior of organisms is an expression of the ontological priority of the species—that is, the essence of the individual.[9] While Aristotle teaches the constancy of species, Plato at the end of the *Timaeus* (90 e ff.) sketches an inverted evolution from man through woman to the lowest animals, and even if the mythical character of the work clearly does not allow us to take this conception at its face value, it is probably fair to say that Plato was, like Empedocles, open toward evolutionary ideas, even if he wanted to insist, as in the case of political history in the *Republic,* books 8 and 9, that the telos is what counts when one tries to understand a development. With Aristotle, Plato believes that there is a natural hierarchy in nature, culminating in man.[10] What is incompatible in this approach with the Darwinian revolution? Certainly it would not be fair to say that Plato and Aristotle were not interested in causal mechanisms for organic phenomena. Even if their knowledge is very limited, both recognize the search for these mechanisms as legitimate. However, Plato does not dispose of a satisfying theory of empirical research—here he is inferior to Aristotle, even if he is superior to him in being able to justify the application of mathematics in natural science. (The mathematization of biology in the twentieth century is certainly more Platonic than Aristotelian by nature.) There would be nothing wrong in Plato's theory of ideas either if it only meant that science deals with universal concepts (which can hardly be regarded as merely subjective constructions as long as we think that some set of concepts does more justice to the world than an alternative one) and that the search for similarities and differences, existing independently of us, is an important part of taxonomic work (after Darwin it has to be balanced with genealogical concerns, but it cannot be replaced by them if one aims at something more than cladification). As Plato's rudimentary evolutionary ideas show, the mere fact of evolution is not in contradiction with his theory of ideas, which does not entail that species taxa are eternal. Nor must the idea of the scale of nature necessarily be given up by a Darwinian—Charles Darwin uses the term himself,[11] even if he is less sure about the criteria for "higher" than the tradition[12] (but let us not forget that Aristotle, too, considers several criteria) and even if a certain interpretation of Darwinism, which can sometimes, but not always, be found in Darwin himself, would deny that such normative criteria have any objective validity. But this, again, is a matter of the philosophical interpretation of Darwin's theory, not of the theory itself.

Nevertheless, there are at least three differences between Plato's and Darwin's concept of species. First, one of Darwin's central points is the antiessentialist denial of an absolute difference between species and variation. "From these remarks it will be seen that I look at the term species, as one arbitrarily given for the sake of convenience to a set of individuals closely resembling each other, and that it does not essentially differ from the term variety, which is given to less distinct and more fluctuating forms."[13] Still, Darwin occasionally defends the concept of type: "We could not, as I have said, define the several groups; but we could pick out types, or forms, representing most of the characters of each group, whether large or small, and thus give a general idea of the value of the differences between them."[14] Sometimes Darwin's reflections remind one of Ludwig Wittgenstein's concept of *Familienähnlichkeit* (family resemblance), since he insists that species that have nothing in common may be connected through intermediary links. "Nothing can be easier than to define a number of characters common to all birds; but in the case of crustaceans, such definition has hitherto been found impossible. There are crustaceans at the opposite ends of the series, which have hardly a character in common; yet the species at both ends, from being plainly allied to others, and these to others, and so onwards, can be recognized as unequivocally belonging to this, and to no other class of the Articulata."[15] Despite this difference from traditional essentialism, one must understand that one of the starting points of Darwin's theory is the acceptance of the principle of continuity, *Natura non facit saltum,*[16] which, as we have seen, is shared by Plato and the tradition founded by him. One must furthermore recognize that the later development of biology has found it difficult to do without a more precise concept of species than Darwin himself was satisfied with; Ernst Mayr, for example, criticizes the nominalist conception that "a person (not nature) makes species by grouping individuals under a name. . . . Nothing brought this point home to me more forcefully than the fact that the Stone Age primitive natives in the mountains of New Guinea discriminate and name exactly the same species that are distinguished by the naturalists of the West. It requires a vast ignorance of both living organisms and human behavior to adopt the nominalist species concept."[17] The species concept favored by Mayr and most modern biologists—not present, however, in *The Origin of Species,* even if hinted at in Darwin's early notebooks[18]—is the biological species concept (which does not apply to asexual organisms). Michael Ghiselin, for example, defines biological species as "populations within which there is, but between which there is not, sufficient cohesive capacity to preclude indefinite divergence"[19] and insists that species are individuals.[20] Of course, Mayr is right that only species taxa can be individuals and that they must be distinguished from the species concept and the species category.[21] Ghiselin's definition is a good Platonic attempt to grasp the essence of a concept (in this case of the concept of species in general, not of a single species), and the concept itself is hardly an individual (in the usual sense of the word). Still, one can agree that what counts in the biological species concept are not morphological traits but a certain type of behavior in individuals, and this is indeed the second difference from Plato. The reason why the biological species concept is called "biological" is that it gives the biological cause for the formation of species—the isolating mechanisms limit the procreation of hybrids, which usually are not a very good investment by the parents. In any case, the biological species concept makes belonging or not belonging to the same species a *function* of a behavioral trait—the possibility or impossibility of interbreeding—while the tradition would certainly have liked to regard the possibility or impossibility of interbreeding as the *consequence* of more "essential" traits. (I will not pursue here the question whether we would be as satisfied with the biological species definition if we happened to live in a world in which there was no similarity at all between the different members of an interbreeding population and their offspring, or in which we could never find any plausible cause for the impossibility of interbreeding in more fundamental features of the organisms such as their DNA.) Thus we approach the third and most important difference between the Platonic and the Darwinian approach, that for Plato (and Aristotle) the *eide* are something given, the patterns structuring reality, but do not have themselves causes or reasons. For Darwin, on the other hand, there is a causal explanation for why there are these and no other species taxa—the operation of natural selection. Henri Bergson has distinguished ancient and modern science by insisting that the basic concepts of the former are the *eide,* the forms, whereas the basic concepts of the latter are the laws of nature and that modern science has to be defined mainly by its aspiration to take time as an independent variable.[22] The modernity of Darwinism consists exactly in this: Darwin claims to have found a natural law (the principle of natural selection) that explains together with antecedent conditions (and other natural laws) the existence of species taxa, the emergence of which is a function of time.[23] By doing so, Darwin gives a causal explanation of the finality to be found in organisms, and this is certainly an idea Aristotle would have abhorred: one need only read his criticism of Empedocles' theory of evolution, of all the ancient theories the one that comes closest to Darwin—a criticism that significantly enough presupposes that there are no general natural laws.[24] Christian Illies and I have tried to show elsewhere that the metaphysics most appropriate to Darwin's enterprise is the Spinozist,[25] and there is little doubt that Spinoza's metaphysics represents a radical break with Platonism. However, this does not mean that a metaphysics centered in the laws of nature cannot be made compatible with an eidetic approach: Leibniz's philosophy is the first such attempt.

One of the most vexing problems of Plato's philosophy is that Plato accepts a peculiar ontological status not only for logical-mathematical and ethical-aesthetical ideas but also for all universals (among them, even entities with a negative value), and even if he clearly distinguishes between these four groups,[26] he does not seem to have at his disposal an elaborate epistemological theory about the different ways we grasp these diverse types of ideas. While there are good arguments for the theory that the knowledge of the idea of the good is purely a priori, the same cannot be said of the idea of the horse, and it is not easy to defend an ontological priority of the ideas if perceptible objects enjoy an inevitable epistemological priority—the suspicion that a hypostatization has occurred is justified. (Nevertheless, a Platonist need not yet give up; he could still argue for the priority of universals by interpreting particulars as bundles of universals.) Furthermore, even if Plato may be right that there is a certain normativity in every concept, it is indispensable to distinguish between descriptive concepts (ideal types in Weber's sense, which can be formed also of immoral institutions) and normative concepts proper, and doubts remain whether Plato was able to do this. In any case, he assumes that a theory of being is incomplete if it acknowledges only physical and mental objects or events; if, therefore, nature is understood as the sum of all physical and mental objects/ events,[27] Platonism, like all forms of objective idealism, stands in opposition to naturalism. I explicitly included mental alongside physical objects and events, for Cartesian dualism also fails to render justice to the central idea of objective idealism: Platonic ideas can be reduced to mental objects or events as little as to physical ones. In the twentieth century, Gottlob Frege's and Karl Popper's three world theories come close to Platonism's central idea, even if Popper's third world, differently than Frege's third realm, combines in an awkward way two very different types of entities, only one of which can be regarded as an heir of the Platonic ideas; the other is linked by Popper[28] to Hegel's objective spirit and encompasses, for example, social institutions, which may or may not be reducible to a combination of physical and mental events but certainly belong to the real, spatiotemporal world, while numbers and values do not. To be more precise (and to return to the above-mentioned criticism that Plato does not do justice to the distinction between the descriptive and the normative dimension), the latter is true only of values in one sense of the word. In fact, the word *value* is, like many other conceptually linked words, profoundly homonymous—there are values that are a part of the social world, of the "objective spirit," and can be described by sociologists (and sociobiologists) as those normative representations that guide the behavior of a group. In this sense one can speak of the values of a gang, of a totalitarian state, and so on. If, however, the desire arises to distinguish between those values that are perhaps successful but morally abominable and those to which one can subscribe with a good conscience, and if one thinks that the choice of values is more than a subjective preference, then, so objective idealists argue, one must inevitably assume that there are some ideal values that ought to be the standards of our—intellectual, moral, aesthetic—behavior, even if they cannot be found in the empirical world, since they are never completely implemented in it. Differently from Kantians, objective idealists, however, are not satisfied with an absolute dualism between descriptive and normative realm, between Is and Ought. Even if they deny the possibility of grounding the Ought in the Is, they think that the Is, nature, participates in some way in the Ought and is a place for its partial implementation. In the following I want to test whether this objective-idealist worldview can be made consistent with the impressive array of facts discovered by evolutionary biology. Of course, this is not yet a positive justification of objective idealism, which is not part of my task today. It is far more modest: I want only to show that nothing in Darwinism confutes an objective-idealist stance with regard to the ideas of the true, the good, and the beautiful. A non-naturalistic and nevertheless objectivistic epistemology, ethics, and aesthetics can profit from evolutionary epistemology, ethics, and aesthetics but cannot be replaced by them.

II

Evolutionary epistemology goes back to Darwin's notebooks and was quite developed in the second half of the nineteenth century, but its most elaborate articulations date from the 1970s: I mention the works of Karl Popper, Konrad Lorenz, Donald Campbell, Gerhard Vollmer, and Rupert Riedl.[29] Even if there are remarkable differences between these authors, the basic idea is quite simple: the process of knowledge begins, not with human efforts, but already with the first organisms. In order to survive, every organism has to gather information about its environment. The genome develops hypotheses about it, which are then confronted by reality and subjected to a process of selection. The survival of an organism is a sign that its hypotheses are not completely wrong—otherwise, it would disappear quickly from the stream of evolution. Of course, the overwhelming majority of these hypotheses are not conscious ones (some are, one could say, present in the form of its organs); and evolutionary epistemology insists that also in humans—whose cognitive difference from the other animals is not denied—a vast array of cognitive processes do not occur in a conscious way and that some of them are innate. The philosophically most ambitious claim of evolutionary epistemology is that it is able to give a "scientific" explanation of Kant's concept of synthetic a priori knowledge. Our concept of causality or our intuition of a three-dimensional space, for example, is explained as being de facto a posteriori, based, however, not on our own experience but on that of our ancestors: the purported ontogenetic a priori proves to be a phylogenetic a posteriori. This justifies its validity not apodictically but partially—if innate structures did not mirror anything in reality, they could hardly have been selected.

Without any doubt evolutionary epistemology has increased our knowledge of cognitive processes. Every complete epistemology has to recognize that conscious acts are only a small part of our cognition and that cognitive efforts characterize even the most primitive organisms. But all these insights belong to the descriptive part of epistemology, as do also, for example, the results of cognitive psychology and of the sociology of knowledge. They do not contribute anything, or at least not much, to answering the normative question: why certain assumptions about reality are in fact knowledge. A sociologist of knowledge can explain why certain beliefs that we do not regard as true have enjoyed high respect for centuries, but the question whether they are true cannot be answered by sociological means, for the social success of a theory is compatible with blatant falsity. But is the biological success of a hypothesis implicit in the morphology or the behavior of an organism not an argument on behalf of its truth? Does evolutionary epistemology not at least show this by the argument quoted above? Here, too, several restrictions have to be made. One can grant that an organism could hardly survive for a long time if its assumptions did not correspond at all to basic facts in its environment—which is not a very far-reaching concession; but since the environment of the organism is limited, there is, first, no reason to assume that its hypotheses would render justice to any traits of reality outside its environment. Indeed, some of our own innate convictions have proven cumbersome or misleading: the fact that we cannot imagine a three-dimensional non-Euclidean space is certainly not an argument against a physical theory teaching the existence of such a space, even if our ancestors could survive quite well without caring for such a space in the mesocosmos they inhabited. Second, the following argument is certainly circular, namely that we should trust, for example, our innate belief in causality because it is the result of an evolutionary process, which operates in a selecting manner also on our ratiomorphic apparatus. For this argument already presupposes the theory of evolution, which is itself a causal theory; and we do not gain any new insight by finding out that a causal theory, together with other assumptions, entails the validity of causality.

Third, even where the belief selected is not an immediate presupposition of the theory that has to justify it, such an argument obviously presupposes the validity of evolutionary theory. Now there may be good arguments to trust the validity of scientific theories, but we need such arguments, and evolutionary epistemology certainly cannot offer such arguments because it already presupposes this validity. It is particularly absurd if evolutionary epistemology claims to be a replacement of Kant's doctrine of synthetic a priori knowledge. Innate knowledge and synthetic a priori knowledge have nothing to do with each other; the first category belongs to descriptive, the second to normative epistemology. Evolutionary epistemology has shown that not all the content of our consciousness can stem from the experience of the individual organism; it is not a contribution to the issue of whether experience can justify the content of this consciousness as knowledge. The question of whether a cognitive attitude is innate or acquired has nothing to do with its truth claim, while Kant assumes that there are necessarily true synthetic propositions a priori. His *Critique of Pure Reason* is the grand attempt to justify the validity of modern science within a framework that rejects empiricism with regard to both concepts and judgments, and even those who do not share Kant's whole theory—for example, his doctrine of things-in-themselves—should recognize that his arguments against an empiricist epistemology are considerable, the problem of induction being one of them.

Still, it is clear that every epistemology needs a theory of experience as encounter with reality, and the merit of evolutionary epistemology is to have insisted that experience began long before humans and that we can learn—if we have a satisfactory theory of experience and of those moments of knowledge that cannot be reduced to it—from the experiences, successful or not, of other organisms. But still more can be said. Darwin's idea of natural selection is not limited to biology (and can therefore be called metaphysical); for it applies to all entities that vary, multiply, and compete for scarce resources, such as human cultures or theories, which are developed further, compete for attention and recognition, and are ontologically multiplied by finding access to more and more minds. In fact, we must distinguish from evolutionary epistemology an evolutionary theory of science, as proposed by Karl Popper and Stephen Toulmin,[30] which is Darwinian but not biologistic. Even abstract metaphysical theories compete with each other—one tries to explain more phenomena than the other, to avoid problems the other has incurred, et cetera, so the principle of competition applies to every epistemic effort, not only to those of organisms and not only to those of empirical theories. The principles of competition and selection seem to be transcendental principles of epistemology—and an objective idealist need not be disturbed at all to see that they operate on a primitive level in the history of life. Truth cannot be reduced to scientific truth, and science cannot be reduced to experience, but experience and science are attempts to approximate truth by a process of trial and error, and that such a process began already three billion years ago shows only that the idea of truth, to be determined by the competition and selection of hypotheses, has operated in natural history for quite a long time.

III

According to Popper, one aspect of progress is that humans are able to let their theories die instead of themselves.[31] But is this late result of evolution not irrelevant compared with the amount of brutality one can observe in the natural world? And is there any more convincing confutation of an objective-idealist interpretation of nature than the theory of the selfish gene advanced by sociobiologists like Edward O. Wilson and Richard Dawkins?[32] I want to abstract here from possible objections based on more holistic models and focus on the central structure of the theory. The arguments developed by modern sociobiology against the traditional theory that organisms' behavior may contradict self-interest in the interest of the species are indeed strong. Not only empirical evidence but a simple thought experiment speaks against this assumption. Even if a group whose members had a tendency to sacrifice themselves for it might have advantages compared with other groups lacking such a tendency, its members would be disadvantaged with regard to those new members of it who would accept the sacrifice of the others but who would not return an analogous favor when it was their turn. It is a simple tautology that genes that program a behavior conducive to a maximization of the replication of these genes are selected at the expense of genes that are altruistic in the sense that they help other organisms even when this means a decrease in their own offspring. (It is essential to keep this definition of *altruism* in mind.) But how to deal with the forms of altruism that characterize the animal world? According to sociobiology, they must be reduced to altruism for kin and reciprocal altruism. Altruism for kin—for example, the self-sacrifice of a mother for her offspring—can be a form of genetic egoism, since it may increase inclusive fitness: if a mother dies but thereby saves three children, her altruistic gene is not lost but still persists on average one and a half times in the gene pool. If the ratio of gain in inclusive fitness to loss in inclusive fitness exceeds the reciprocal of the average coefficient of relationship, then such a behavior may evolve because it is in the selfish gene's interest. Analogously, reciprocal altruism can occur if, and only if, the diminution of fitness linked to the altruistic act is more than compensated for by the advantages the organism receives when it is the beneficiary of the altruistic act. This, however, entails that organisms learn to react against those who benefit from others but are not willing to reciprocate. To survive in the process of evolution, a strategy must be evolutionarily stable, to use a term introduced by J. Maynard Smith and based on game theory: it must not be possible for alternative strategies to replace it. A strategy to benefit everybody, including those who do not reciprocate, is evolutionarily unstable because it will inevitably be replaced by the strategy to profit from others, even under conditions where this latter strategy may lead, by destroying reciprocal altruism, to the extinction of the population. For this fact alone will not save the evolutionarily unstable strategy.

The picture of nature proposed by sociobiology has often been interpreted as profoundly depressing, since it seems to regard egoism as the only possible root of organic behavior. An ethics based on such a doctrine would be appalling, and in fact no reasonable person who has heard about the naturalistic fallacy will jump from sociobiological facts to normative conclusions.[33] If there is a moral law, it is grounded in something else. Still, I do believe that ethics can learn from sociobiology and that by doing so it may inspire a different view of nature. Since many concrete moral norms are the conclusion of a mixed syllogism, containing both a purely normative and a descriptive premise, knowledge about human nature can be, first, useful in determining our duties, if only in order to know against what forces one has to fight. Even if human behavior will never be reduced to genetics alone,[34] the sociobiological explanation of human nepotism and of the differences in the sexual morality of males and females is quite plausible—an explanation that, of course, should not be identified with a justification. It is also correct to say that an altruism going beyond altruism for kin and reciprocal altruism cannot have a basis in our genes. This does not at all exclude a basis for altruism in human culture, particularly if one understands that the "selfishness" of the genes is compatible with the altruism of the organism: the bird mother who sacrifices herself has to overcome her instinct for self-preservation; and even if it is not the species as such for which she immolates herself, it is still something more universal than herself and transcending herself, namely her genes. Since this tendency is innate, though limited to kin, it is intelligible that human culture could produce the idea of an altruism no longer limited to one's relatives.

But the importance of sociobiology for ethics goes farther—due to the concept of evolutionarily stable strategy. In fact, this concept is not at all limited to biology; rather, it has a general importance in normative ethics. But have I just not warned against the naturalistic fallacy? We have already seen that Darwinism is more than a biological theory, since one can formulate a Darwinian model of cultural evolution.[35] Now, if we grant that a moral norm should have a lasting impact on reality—that is, that the Ought ought to be—it follows immediately that a moral strategy must not be evolutionarily unstable. It is reasonable that a religious doctrine usually contains as one of its precepts that it be propagated, and it makes perfect sense that an altruist who in a situation of conflict, due to scarcity of resources, has to choose between two different recipients of her charitable activities selects that person who is more likely to continue with altruistic activities. This is, again, a morally sound decision, for in the long term the person who so chooses will achieve far more than the person whose altruism is spread without selection and may well foster egoists, thus undermining the principle it itself represents. Of course, I do not claim that every evolutionarily stable and not self-destructive strategy is moral; I am speaking, not about a sufficient, but only about a necessary, condition of the morality of strategies. And it goes without saying that on the level of human culture the restriction to kin cannot be justified, for virtuous behavior is determined only to a very limited degree by genes. Before the development of more complex intellectual capacities, however, there was hardly an alternative to limiting altruism to kin; for only here was there a sufficient probability that one's own altruistic behavior would not be lost in the game of evolution.

When we understand this, a new evaluation of the results of sociobiology becomes possible. The standard interpretation of it is that egoism is the dominating force, altruism being nothing else than an epiphenomenon linked to selfish genes.

In truth, the feature of replicating itself, so characteristic of life, is a fascinating combination of egoism—replication of *oneself*—and self-transcendence: for it remains true that organisms are able to sacrifice themselves for kin and that the series of offspring, not one's own self, is what counts in the course of evolution— a point correctly seen by the ancient philosophy of life, even if it was linked to a misleading metaphysics of *eide*. Even the limitation of altruism to kin, where it is not reciprocal, makes perfect sense, for only thus can the organism increase the probability that altruism will continue in the next generation. One can agree with sociobiology that the organism sacrificing itself behaves as if it cares for its genes— as long as one adds that it cares for genes programming altruism. In human culture, based on teaching, models, and so on, altruism must no longer be limited to kin—but here too an altruistic strategy should try to become evolutionarily stable. The saint will not behave charitably for the reason that he desires to draw personal advantage from his actions being reciprocated; he must inevitably wish, however, that the altruistic principle be adopted by his beneficiaries. An objective idealist will therefore not deny the results of sociobiology, but he will interpret them differently: He will see in the biological limits to altruism the only way that the idea of the good may be implemented in the prehuman organic world and conceive of it as of a preparation for a higher form, as it becomes possible and real in human culture.

IV

Organisms relate cognitively to their environment and care for themselves and their offspring; these forms of behavior link them to the ideas of the true and of the good.[36] What about the idea of the beautiful, which traditional metaphysics has often regarded as the third of the transcendentals? I have to ignore here the difficulties that any conceptual articulation of the idea of the beautiful encounters, difficulties far greater than those related to the other two ideas, for it seems impossible to develop this idea without relating it to sense experience. My task is, again, to show that, if there is a satisfying account of the idea of the beautiful, Darwin's theory not only does not contradict it but fits very well with it. I said intentionally "Darwin's theory" and not "Darwinism," for, while modern Darwinism has developed strongly since Darwin himself,[37] I am referring to an idea that was particularly dear to Darwin himself but was ignored for many decades after his death and is still somehow controversial in modern biology. I have in mind, of course, his idea of sexual selection. It is an astonishing fact that Darwin's most provocative book, *The Descent of Man and Selection in Relation to Sex,* seems to fall apart in two quite different sections, the first part dealing with man's natural origin, the second and larger part dealing with this other form of selection (which had been already named alongside natural selection in the fourth chapter of *The Origin of Species*). The connection with the first part lies in the problem of the origin of the differences between human races, which according to Darwin are partly caused by sexual selection (he does not yet have at his disposal the theory of genetic drift); but the reader has the feeling that Darwin believed in a more profound link, as has been proposed very recently by Geoffrey Miller in *The Mating Mind: How Sexual Choice Shaped the Evolution of Human Nature.*[38] Darwin's theory did not convince his contemporaries—particularly Alfred R. Wallace, the co-discoverer of the principle of natural selection, who rejected it. In his autobiography, Wallace mentions four scientific differences between himself and Darwin, the first being that he did not think that the moral and intellectual qualities of man could be explained by natural selection, the second that he regarded what Darwin believed to be effects of sexual selection—for example, the splendid colors of certain animals—as effects of natural selection. It is clear that the two points are connected: Wallace did not share Darwin's opinion that animals choose their mates on the basis of aesthetics because for him the gap between animal and human intelligence was far greater than for Darwin. While in his first point of difference he ascribed less to the power of natural selection than Darwin did, his refusal to consider sexual selection as an independent force made him a more stubborn defender of the principle of natural selection: "Some of my critics declare that I am more Darwinian than Darwin himself, and in this, I admit, they are not far wrong."[39] But one could argue that the willingness to consider alternatives to his own theory is a sign of the greater originality and versatility of Darwin's mind. Still, his theory of sexual selection found few friends until quite recently—Thomas Hunt Morgan criticized it, and even Ronald A. Fisher's remarkable attempt to reconstruct it in the context of his genetic theory of natural selection met with strong objections from Julian Huxley.

The starting point for Darwin was the difficulties that the theory of natural selection encounters when confronted with organs that do not help the animal survive and sometimes are even an impediment to this purpose, such as bright colors that attract predators or a long tail that prevents a bird from flying. His answer is that these structures increase the chance of leaving offspring because they are regarded as beautiful by the animals' mates, particularly the females, to whom he ascribes the bulk of sexual choice in the animal realm (not, however, among humans).[40] The argument presupposes that not every male gets an equal chance to reproduce, a hypothesis particularly plausible in polygamic species. According to Darwin, the beauty of the colors and patterns we can observe and of the instrumental and vocal music we can hear in the animal world, the complexity of the love antics and dances of so many birds and of the architecture of the bowerbirds, are all the result of the aesthetic sense guiding sexual choice. Darwin admits that there can be other causes for beauty in organisms: "Hardly any colour is finer than that of arterial blood; but there is no reason to suppose that the colour of the blood is in itself any advantage; and though it adds to the beauty of the maiden's cheek, no one will pretend that it has been acquired for this purpose."[41] But in many cases beauty is the result of sexual selection:

> The case of the male Argus pheasant is eminently interesting, because it affords good evidence that the most refined beauty may serve as a charm for the female, and for no other purpose. We must conclude that this is the case, as the primary wing-feathers are never displayed, and the ball-and-socket ornaments are not exhibited in full perfection, except when the male assumes the attitude of courtship. . . . Many will declare that it is utterly incredible that a female bird should be able to appreciate fine shading and exquisite patterns. It is undoubtedly a marvelous fact that she should possess this almost human degree of taste, though perhaps she admires the general effect rather than each separate detail. He who thinks that he can safely gauge the discrimination and taste of the lower animals, may deny that the female Argus pheasant can appreciate such refined beauty; but he will then be compelled to admit that the extraordinary attitudes assumed by the male during the act of courtship, by which the wonderful beauty of his plumage is fully displayed, are purposeless; and this is a conclusion which I for one will never admit.[42]

In the last chapter Darwin returns to the Argus pheasant and insists that both the beauty of the male and the taste of the female have evolved gradually,

> the aesthetic capacity of the females having been advanced through exercise or habit in the same manner as our own taste is gradually improved. . . . Everyone who admits the principle of evolution, and yet feels great difficulty in admitting that female mammals, birds, reptiles, and fish, could have acquired the high standard of taste which is implied by the beauty of the males, and which generally coincides with our own standard, should reflect that in each member

of the vertebrate series the nerve-cells of the brain are the direct offshoots of those possessed by the common progenitor of the whole group. . . . He who admits the principle of sexual selection will be led to the remarkable conclusion that the cerebral system not only regulates most of the existing functions of the body, but has indirectly influenced the progressive development of various bodily structures and of certain mental qualities. Courage, pugnacity, perseverance, strength and size of body, weapons of all kinds, musical organs, both vocal and instrumental, bright colours, stripes and marks, and ornamental appendages, have all been indirectly gained by the one sex or the other, through the influence of love and jealousy, through the appreciation of the beautiful in sound, colour or form, and through the exertion of a choice; and these powers of the mind manifestly depend on the development of the cerebral system.[43]

It is important that Darwin asserts in this passage, as elsewhere, that there is a fundamental harmony in aesthetic taste among at least all of the vertebrates. Even if he simultaneously recognizes different aesthetic standards among different species and even among different human races,[44] his aesthetic ideas are clearly nonrelativistic: beauty is for him not a species-relative concept. An obvious sign of this is when he speaks of an increase of beauty in the evolution of birds[45]—a statement that presupposes both a species-transcendent criterion of beauty and, because of Darwin's theory of sexual selection, also progress in the aesthetic taste of the birds themselves. Not so obvious is whether Darwin thinks (as Schopenhauer did)[46] that organisms regard mates as beautiful because they excite them sexually, and as sexually exciting because they promise many and healthy descendants, or whether, on the contrary, he wants to suggest that a mate excites sexually because he or she is beautiful. He seems to assume that in most cases beauty and strength go together,[47] even if in such a case sexual selection would be reducible to natural selection.

What is the *status quaestionis* on this issue?[48] First, there is a consensus that natural and sexual selection, even if they may drive in different directions, have to be regarded as two aspects of one general force, which is itself called "natural selection"; for what counts in evolution is the number of the genes of an organism passed on to the next generation. One has to survive in order to have offspring, but if an organism lives very long without leaving offspring, its genes are lost. On the other hand, the descendants must themselves be able to survive in order to leave offspring, so that fitness has to be measured by both survivorship and fertility. Second, it is insisted on that sexual selection is not necessarily linked to sexual dimorphism, Darwin's starting point, even if he does occasionally recognize that sexual monomorphism, too, could be the result of sexual selection. Third, within sexual selection, intrasexual and intersexual (epigamic) selection are distinguished—a differentiation not completely alien to Darwin, even if he did not introduce names for the two forms.[49] (A peculiar subform of epigamic selection is cryptic female choice—that is, female choices, affecting male reproductive success, that occur after the male has succeeded in coupling his genitalia with those of a female.)[50] The existence of the first form, namely competition for mates between members of the same sex, was always uncontroversial and was also accepted by Wallace; the more challenging claim regards the existence of epigamic selection (the operation of which was in any case overrated by Darwin, who neglected alternative explanations in terms of natural selection). Research in the last decades has proven beyond any reasonable doubt that epigamic selection exists, since there are mating preferences even at the level of insects.[51] But while Darwin regarded these preferences as inexplicable facts—"No doubt the perceptive powers of man and the lower animals are so constituted that brilliant colours and certain forms, as well as harmonious and rhythmical sounds, give pleasure and are called beautiful; but why this should be so, we know no more than why certain bodily sensations are agreeable and others disagreeable"[52]—beginning with Ronald A. Fisher, people have pursued the hypothesis that there is an evolution in mating preferences too. Of course, not all mating preferences have a genetic basis—imprinting is an important cause for sexual preferences. But Fisher's famous theory tried to show how there could be a genetically determined co-evolution of sexual preferences and sexually preferred traits. For if a hen prefers a peculiar plumage in her mate, she will tend to inherit to her offspring, if male, this plumage and, if female, this preference, and this genetic correlation between sexual traits and sexual preferences will usher in a "runaway process."

> The two characteristics affected by such a process, namely plumage development in the male, and sexual preference for such developments in the female, must thus advance together, and so long as the process is unchecked by severe counterselection, will advance with ever-increasing speed. In the total absence of such checks, it is easy to see that the speed of development will be proportional to the development already attained, which will therefore increase with time exponentially, or in geometric progression. There is thus in any bionomic situation, in which sexual selection is capable of conferring a great reproductive advantage, the potentiality of a runaway process, which, however

small the beginnings from which it arose, must, unless checked, produce great effects, and in the later stages with great rapidity.[53]

Of course, Fisher assumes that there are checks and that the sexual ornaments we observe are the result of a sudden spurt of change stopped by the principle of natural selection. Fisherian models for the evolution of sexual preferences (which under certain conditions may lead to an equilibrium state instead of an exponential process) have not remained the only ones; the main alternative are the "good-genes models." According to these often speculative models, there is a hidden connection between sexually preferred traits and those favored by natural selection (in the narrower sense of the term): strong colors may correlate with parasite resistance and good health. One could also regard the so-called "handicap models" as almost a subform of the good-genes models: in an influential essay,[54] Amotz Zahavi, Darwinism's Veblen, developed in 1975 the theory that an organism displaying cumbersome ornaments signals to its mates a particular fitness—he must be particularly strong if he managed to survive with them. The theory met with serious problems when it was mathematized, but newer models by A. N. Pomiankowski, Y. Iwasa, and L. Sheridan render probable that it works, and Geoffrey Miller's book tries to explain a good deal in human culture on its basis. Today many biologists assume that the two (or three) models for the evolution of mating preferences are compatible with each other.[55] Still, many questions remain unanswered. Anne Houde ends her splendid book with a list of such questions, two of which are particularly relevant for my topic. Her first question is "What is the sensory basis for female mating preferences?" and her sixth is "Do female mating preferences always lead to sexual selection on male traits?"[56] To begin with the latter, Houde considers the possibility that females could express their mating preferences depending on the presence or absence of predators or other factors belonging to natural selection in the narrower sense of the word. In any case, it remains an astonishing fact that mating preferences that run contrary to the demands of the principle of natural selection are remarkably constant; natural selection does not cause them to disappear as quickly as one would expect.[57] To explain this, the theory of sensory bias may be helpful: animals may prefer beautifully colored mates to those with a better camouflage because this preference is correlated (to use a concept central for Darwin's theory) with a better perception of certain other objects in the environment, favored by natural selection. "The possibility that selection on sensory systems in contexts other than male choice (e.g., food finding) can affect the evolution of mating preferences also needs to be investigated."[58]

What does all this entail for the issue of Darwinism's compatibility with objective idealism? Even if it is inevitable that Darwinism will try to reduce sexual selection to a form of natural selection in the broad sense of the term, it is an essential point of Darwin's doctrine, first, that animals choose their mates and, second, that aesthetic criteria play a role in this process, even when they run against the demands of natural selection. Whatever the causes for the aesthetic taste of animals may be, it can hardly be doubted that animals have one[59] and that it is not always possible to reduce the sense of beauty to an anticipation of health and strength. If the handicap principle should prove to work also in the prehuman world—I have little doubt that it works in human societies, where engaging in risky behavior enhances prestige and is furthermore an expression of human nature's superiority with regard to life—then we would already have in nonhuman life a negation of what is immediately useful for life, a negation that, however, could be selected only because it proved attractive to possible mates. Of course, no reasonable person denies that even in humans the attractiveness of certain features of the other sex is related to biological factors, such as youth; but Rensch is right when he insists that one should call such sexual signals "attractive," not "beautiful."[60] It is obvious that human art transcends the boundaries of such attractiveness and even consciously denies them, and it is plausible to assume with Darwin that, perhaps on the basis of sensory bias, animals develop, sometimes in an unusually rapid process due to exponential growth, clear-cut preferences for certain more striking and more dangerous colors (instead of safer grayness) and that this purely aesthetic preference is not a function of sexual attractiveness but on the contrary limits what can prove sexually attractive. The idea of the beautiful thus proves to be more fundamental than mating behavior, and animals', mainly females', sense of beauty is one of the causes both of the evolution of life and of the splendor we find in the organic realm—a splendor that Darwin never tired of recognizing with an awe almost religious.

V

We have seen that it is possible to interpret the evolution of life as the slow and gradual process of the implementation in the material world of the ideas of the true, the good, and the beautiful. But even if objective idealism may not be refuted by Darwinism, is it not obvious that a teleological and theistic worldview has been disproved? I want to end with some programmatic and quite assertive statements on this issue.

Darwin (who in his autobiography claimed to deserve to be called a theist, even if he later added some agnostic remarks)[61] never thought that his discovery had destroyed the cosmological proof of God (the ontological and the moral he did not mention); he asserted only that "[t]he old argument of design in nature, as given by Paley, which formerly seemed to me so conclusive, fails, now that the law of natural selection has been discovered."[62] Even this claim is very modest; for it is related specifically to Paley's version. Is it really true? One can agree that the theory of natural selection replaced special creationism, but two important restrictions have to be made. First, the physico-theological proof had already been subjected to a deleterious criticism in the eighteenth century by Hume in the *Dialogues concerning Natural Religion* and by Kant in the *Critique of Pure Reason* (Hume somehow anticipates Darwin's insight,[63] but this thought is not crucial for his argument and is completely missing in Kant). Darwin's theory, therefore, was not necessary from a philosophical point of view, even if he clearly helped public opinion in Britain to reject Paley's theology. But many decades before Darwin it had become clear that the existence of an omnipotent and omniscient being could never be induced from nature as we know it. Only in connection with the other proofs could the argument from design be of any value.

It is possible to reformulate the argument from design after the Darwinian revolution in such a way that it merges with the cosmological proof. Even if God is no longer the secondary cause of special contrivances, one can still regard him as the First Cause of a universe to which all the organisms we know of belong. Darwinism does not deny the existence of wonderfully adapted organisms;[64] it merely states an efficient cause for their adaptation—a cause that operates only if there are other appropriate natural laws; for of course natural selection as such does not have to lead to higher organisms; given different natural laws, evolution could have stopped at the level of bacteria. Already in the early notebooks Darwin had written that we profane God's most magnificent laws when we do not think them capable of producing every effect of every kind that surrounds us (including human religion) without separate acts of God.[65] There is no reason to doubt Darwin's sincere religious feeling in this passage.[66] In a certain sense Darwin and Wallace have done nothing else than discover the metaprinciple that is at the root of all concrete adaptations—natural selection as a very rational mechanism. As Wallace's comparison of it with "the centrifugal governor of the steam engine, which checks and corrects any irregularities almost before they become evident"[67] clearly shows, the argument from design can be restated on a higher level of abstraction. If, according to Leibniz,[68] who in the famous letter to Varignon seems to come close to evolutionary ideas in biology, two criteria for the best possible world to be created by the God of theism are the richness of the effects and the parsimony of the causes, then the Darwinian theory, which offers such a simple explanation for the organic plenitude we witness, fits such a world quite well.

But are simplicity of causes and richness of effects enough? Is not the brutal struggle for existence incompatible with the moral predicates of God? I cannot enter the theodicy debate here but must limit myself to the following remark: the amount of suffering characteristic of the animal world was by no means discovered by Darwin (even if he insisted more on intraspecific than on interspecific competition); what Darwin did was to offer an explanation for it—the Malthusian fecundity of organisms—and to discover positive consequences of it. "When we reflect on this struggle, we may console ourselves with the full belief, that the war of nature is not incessant, that no fear is felt, that death is generally prompt, and that the vigorous, the healthy, and the happy survive and multiply."[69] All this eases the task of theodicy (I do not assert that it solves it). In any case, the fact that God is no longer the secondary cause for specific animal behavior has considerable religious advantages, for it limits God's responsibility for the undeniable cruelties that nature exhibits. "Finally, it may not be a logical deduction, but to my imagination it is far more satisfactory to look at such instincts as the young cuckoo ejecting its foster-brothers,—ants making slaves,—the larvae of ichneumonidae feeding within the live bodies of caterpillars,—not as specially endowed or created instincts, but as small consequences of one general law, leading to the advancement of all organic beings."[70] Darwinism leaves room for variations and for the choices of mating animals and breeders as secondary causes of evolution. God is responsible for the world as a whole, not for every detail in it.

But is not one of the central beliefs of religion, the central role of humans in the universe, utterly incompatible with the chance process driving evolution? The answer is very simple: Darwin himself was, like the rationalists of the seventeenth century, a rigorous determinist.[71] He assumed, therefore, that given the laws of nature and antecedent conditions, all events occurred necessarily, including the evolution toward humans. A Darwinist who shares Darwin's determinism can therefore accept the idea that the evolution of moral organisms belongs to God's plan for the universe (even if he wisely will leave it open whether *Homo sapiens* is the only possible instantiation of moral organisms). But if theism is not refuted by Darwinism, is there anything in Darwinism that theism can learn from? I do think so—namely the idea that there is an astonishing continuum between prehuman organisms and humans and that it well befits us, not only because of the ecological interdependence that connects all living beings, to be more caring toward our relatives. But it will not be a naturalist interpretation of Darwinism that teaches us this lesson—for there is no ethic grounded in naturalism—it will be an objective-idealist one. To return to the quotation from Ghiselin at the beginning, it is much easier "to be at home in the land of [our] birth" if we see with Plato a sun that gives us light, orientation, and warmth.[72]

NOTES

I thank Peter Martens for having corrected my English.

1. See M. Ruse, *The Darwinian Revolution,* 2nd ed. (Chicago, 1999), 122 ff., 237.

2. Michael T. Ghiselin, *The Economy of Nature and the Evolution of Sex* (Berkeley, Calif., 1974), 23 f. Compare similar reflections in S. J. Gould, *Full House* (New York, 1996), 40 ff.

3. As G. Altner wittily put it, "Darwin war kein Darwinist!" G. Altner, "Darwin, seine Theorie und ihr Zustandekommen," in *Der Darwinismus: Die Geschichte einer Theorie,* ed. G. Altner (Darmstadt, 1981), 5–8, 7.

4. See, e.g., *Metaphyics* 991b4 ff., 1070a17 ff. The philological problems relating to the reconstruction of Plato's doctrine cannot be discussed here. See V. Hösle, *Wahrheit und Geschichte: Studien zur Struktur der Philosophiegeschichte unter paradigmatischer Analyse der Entwicklung von Parmenides bis Platon* (Stuttgart, 1984), and "Platonism and Its Interpretations," in *Eriugena, Berkeley, and the Idealist Tradition,* ed. S. Gersh and D. Moran (Notre Dame, Ind., forthcoming).

5. Cf. *Statesman,* 263c ff., and Speusippus's *Homoia* (frag. 123–45 Isnardi Parente).

6. A. Lovejoy, *The Great Chain of Being* (Cambridge, Mass., 1936), 52.

7. Ibid., 58.

8. *De partibus animalium* 658b2 ff., 672a13 ff., 673a32 ff., 677b21 ff., 685b15 f., 694b5 f. and *De generatione animalium* 739b28 ff., 743b16 ff., 755a21 ff., 776b31 ff., 788b29 ff. prove the complementarity of both types of "causes." It is important to remark that finality is seen by Aristotle only within the organism, not between different species (the only exception is *De partibus animalium* 696b27 ff.). Recent studies on Aristotle's biology are collected in W. Kullmann and S. Föllinger, *Aristotelische Biologie: Intentionen, Methoden, Ergebnisse* (Stuttgart, 1997).

9. *Symposium* 208a f. Cf. *De generatione animalium* 731b 31 ff. and *De anima* 415a25 ff.

10. On the conception of the *scala naturae* in Aristotle, see, e.g., *Historia animalium* 502a16 ff., 532b29 ff., 588a16 ff.; *De partibus animalium* 655b37 ff., 681a9 ff.; and *De generatione animalium* 732a3 ff., b26 ff.

11. C. Darwin, *The Origin of Species* (Harmondsworth, 1968), 126, 187.

12. Ibid., 421. On this issue, see V. Hösle and C. Illies, *Darwin* (Freiburg, 1999), 89 ff. *Higher,* of course, cannot be defined in terms of "better adapted"—bacteria are very well adapted.

13. Ibid., 108. See also 456: "we shall at least be freed from the vain search for the undiscovered and undiscoverable essence of the term species."

14. Ibid., 414.

15. Ibid., 403.

16. Ibid., 223.

17. Ernst Mayr, *This Is Biology* (Cambridge, Mass., 1997), 131.

18. C. Darwin, *Charles Darwin's Notebooks, 1836–1844* (Ithaca, N.Y., 1987), 289 (C 161).

19. M. Ghiselin, *Metaphysics and the Origin of Species* (Albany, N.Y., 1997), 305.

20. On Darwin's predecessors with regard to a historical, non-"Platonic" conception of species, see P. Sloan, "Buffon, German Biology, and the Historical Interpretation of Biological Species," *British Journal for the History of Science* 12 (1979): 109–53.

21. Mayr, *This Is Biology*, 133.

22. H. Bergson, *L'évolution créatrice* (Paris, 1969), 332 f., 335.

23. The title of A. R. Wallace's essay in the *Annals and Magazine of Natural History* of 1855 is revealing: "On the Law Which Has Regulated the Introduction of New Species." In this essay, however, a causal explanation of the phenomena observed to be structured according to a law is still missing; Wallace found it only in 1858.

24. See *Physics* 198b10 ff.

25. V. Hösle and C. Illies, "Der Darwinismus als Metaphysik," *Jahrbuch für Philosophie des Forschungsinstituts für Philosophie Hannover* 9 (1998): 97–127; now in V. Hösle, *Die Philosophie und die Wissenschaften* (Munich, 1999), 46–73.

26. *Parmenides* 130b ff.

27. This is not the case with all concepts of nature one can find in the tradition; one need think only of Johannes Scotus Eriugena's *De divisione naturae*.

28. K. Popper, *Objective Knowledge* (Oxford, 1972), 106 ff., particularly 125 f.

29. This section summarizes V. Hösle, "Tragweite und Grenzen der evolutionären Erkenntnistheorie," *Zeitschrift für allgemeine Wissenschaftstheorie* 19 (1988): 348–77, now in Hösle, *Die Philosophie und die Wissenschaften*, 74–103. See Popper, *Objective Knowledge;* K. Lorenz, *Die Rückseite des Spiegels* (Munich, 1973); D. Campbell, "Evolutionary Epistemology," in *The Philosophy of Karl Popper*, ed. P. A. Schilpp (La Salle, Ill., 1974), 413–63; G. Vollmer, *Evolutionäre Erkenntnistheorie* (Stuttgart, 1975); R. Riedl, *Biologie der Erkenntnis* (Berlin, 1979).

30. S. Toulmin, *Human Understanding*, vol. 1 (Oxford, 1972). A recent attempt of a Darwinian theory of memes was proposed by S. Blackmore, *The Meme Machine* (Oxford, 1999).

31. This section summarizes a chapter of V. Hösle, *Moral und Politik* (Munich, 1997), 258–74. This book has appeared in English as *Morality and Politics* (Notre Dame, 2004); see 197–210.

32. E. O. Wilson, *Sociobiology* (Cambridge., Mass., 1975); R. Dawkins, *The Selfish Gene* (Oxford, 1976).

33. See, e.g., B. Gräfrath, *Evolutionäre Ethik?* (Berlin, 1997).

34. See, e.g., W. R. Clark and M. Grunstein, *Are We Hardwired? The Role of Genes in Human Behavior* (Oxford, 2000).

35. See the impressive comparison of biological and cultural evolution in B. Rensch, *Das universale Weltbild,* 2nd ed. (Darmstadt, 1991), 140 ff.

36. This section draws partly on Hösle and Illies, *Darwin,* 126–39, but adds new material; G. Miller's book *The Mating Mind: How Sexual Choice Shaped the Evolution of Human Nature* (New York, 2000), with whose ideas the chapter has some affinities, did not yet exist at that time.

37. On the evolution of Darwinism, see D. J. Depew and B. H. Weber, *Darwinism Evolving* (Cambridge, Mass., 1995), and M. Weber, *Die Architektur der Synthese* (Berlin, 1998).

38. Miller, *The Mating Mind.* See also Ghiselin, *Economy of Nature,* xi.

39. A. R. Wallace, *My Life* (London, 1905), 2:22 and in general 2:16 ff. (in the 2nd ed. [1908], 237 and 236 f.). Against sexual selection, see also his *Darwinism* (London, 1891), 274 ff.

40. C. Darwin, *The Descent of Man, and Selection in Relation to Sex* (London, 1871), photoreproduction (Princeton, N.J., 1981), 2:371.

41. Ibid., 1:323.

42. Ibid., 2:92 f.

43. Ibid., 2:402 f.

44. Ibid., 1:64, 2:67, 281, 310, 353.

45. Ibid., 2:223. A species-relativist interpretation of the passage would have to assume that Darwin meant an increase of the birds' beauty only according to our human standards, but there is no hint for this being Darwin's intention.

46. A. Schopenhauer, *Die Welt als Wille und Vorstellung,* bk. 2, ch. 44.

47. Ibid., 1:262; 2:400.

48. On the following, see E. Mayr, "Sexual Selection and Natural Selection," in *Sexual Selection and the Descent of Man, 1871–1971,* ed. B. Campbell (Chicago, 1972), 87–104, as well as J. L. Gould and C. Grant Gould, *Sexual Selection* (New York, 1989).

49. Darwin, *Descent of Man,* 1:279, 2:398.

50. See W. G. Eberhard, *Female Control: Sexual Selection by Cryptic Female Choice* (Princeton, N.J., 1996).

51. See D. Mainardi, *La scelta sessuale* (Turin, 1975), 27 ff.

52. Darwin, *Descent of Man,* 2:353.

53. R. A. Fisher, *The Genetical Theory of Natural Selection* (Oxford, 1930), 137. Fisher had defended Darwin's theory of sexual selection for the first time in an essay of 1915.

54. A. Zahavi, "Mate Selection: A Selection for a Handicap," *Journal for Theoretical Biology* 53 (1975): 205–14. See the more recent book by A. Zahavi and A. Zahavi, *The Handicap Principle* (Oxford, 1997).

55. See A. E. Houde, *Sex, Color, and Mate Choice in Guppies* (Princeton, N.J., 1997), from which I have taken much information.

56. Ibid., 160 f. See also 144 ff. on a mismatch between preferences and male traits.

57. See the examples given by Mayr, "Sexual Selection," 101f.

58. Houde, *Sex, Color,* 160.

59. See Rensch, *Das universale Weltbild,* 163 f., on his own experiments with apes, monkeys, and birds.

60. Ibid., 172 f.: "Wir bezeichnen solche Menschen dann als schön. Dieser zunächst rein ästhetische Begriff ist aber eigentlich nicht ganz berechtigt. Gewiß mögen auch rein ästhetische Komponenten wie Harmonie der Proportionen von Körper und Gesicht oder der Kontrast von Haarfarbe und Gesichtsfarbe zur Erregung des Wohlgefallens beitragen, aber das eigentliche Wesen der Wirkungsweise ist mit der Bezeichnung 'schön' nicht erfaßt. Zutreffender wäre es, nur von 'reizvoll' zu sprechen, denn es handelt sich um sexuelle 'Signalreize.'" See also Thomas Mann—whose reflections on life have inspired no less a figure than Manfred Eigen—on the difference between beauty incarnated in the female and the female use of beauty. Manfred Eigen, *Stufen zum Leben* (Munich, 1987); Thomas Mann, *Joseph und seine Brüder,* vol. 2, *Joseph in Ägypten* (Frankfurt, 1978), 864.

61. C. Darwin, *The Autobiography of Charles Darwin,* ed. N. Barlow (New York, 1993), 93 f.

62. Ibid., 87.

63. D. Hume, *Dialogues concerning Natural Religion* (Indianapolis, 1947), pt. 8, 185.

64. The comparison of organisms with machines is unavoidable, even if one recognizes that their causes are different. See M. French, *Invention and Evolution: Design in Nature and Engineering* (Cambridge, 1988).

65. Darwin, *Charles Darwin's Notebooks,* 553 (M 136).

66. On Darwin's religious ideas, see N. C. Gillespie, *Charles Darwin and the Problem of Creation* (Chicago, 1979). An analogous protest against the pettiness of the special creationists' God can be found in Wallace; see H. L. McKinney, *Wallace and Natural Selection* (New Haven, Conn., 1972), 45.

67. A. R. Wallace, "On the Tendency of Varieties to Depart Indefinitely from the Original Type," (1859), in *A. R. Wallace: An Anthology of His Shorter Writings,* ed. C. H. Smith (Oxford, 1991), 293–300, 300.

68. G. W. Leibniz, *Principes de la nature et de la grace, fondés en raison,* § 10.

69. Darwin, *Origin of Species,* 129.

70. Ibid., 263. See also 447. Something analogous can be said about sexual selection, for not all organs selected are really attractive (cf. Darwin, *Descent of Man,* 2:239). G. Miller misses the point completely when he writes: "This psychologizing of evolution was Darwin's greatest heresy. It was one thing for a generalized Nature to replace God as the creative force. It was much more radical to replace an omniscient Creator with the pebble-sized brains of lower animals lusting after one another. Sexual selection was not only atheism, but indecent atheism." Miller, *The Mating Mind,* 46.

71. Darwin, *Origin of Species,* "a cause for each must exist" (203); "what we must in our ignorance call an accident" (241).

72. See K. R. Miller, *Finding Darwin's God* (New York, 1999).

PART III

What Is the Epistemological
Relevance of Darwinism?

Not only has Darwinism challenged our concept of nature and being; it has, as the early Darwin already understood, epistemological consequences. Discussion of the achievements and shortcomings of evolutionary episte-mology, which had already begun at the end of the nineteenth century, experienced a certain peak thirty years ago, but a consensus was not reached. The first of the essays in this section connects with ontological issues, as it starts by addressing the problem of the metaphysical status of species so profoundly altered by Darwin. Are species classes, or are they individuals? If the biological species concept is the correct one, how does it relate to the morphological one? Is the purported coin-cidence of the two an exception or the rule? If species are individuals, one could explain why there are not laws for single species but only for classes of populations. How do these laws relate to the historical dimension so peculiar to the theory of evolution? And what does the complex biological-historical genesis of the human metaphysical capability entail with regard to the status of the eternal verities meta-physics is supposed to cherish? Would it not be a crude genetic fallacy to deny the latter status because of the former? And could not the inclusion of possibilities and counterfactuals in metaphysics allow an explanation of those human moral ca-pacities that transcend genetic egoism? These are some of the questions Michael Ghiselin discusses.

Gerhard Vollmer pursues the later questions in detail, focusing on the form of evolutionary epistemology he helped to articulate in the 1970s. What is evo-lutionary naturalism? Can realism be supported by selectionism? Is the main argument of evolutionary epistemology based on a vicious circle, insofar as the theory of evolution already presupposes the realism that it is supposed to support? How can cognition and language be compared from an evolutionary point of view?

One of the major philosophical problems connected with Darwinism is, of course, the role of chance in evolution. But what does *chance* mean? Is it an epistemological or an ontological concept? What is the meaning of probability? How does it change within the contexts of a realist and an instrumentalist interpretation of science respectively? Marcel Weber discusses various attempts to argue for an ontological indeterminism within Darwinism and shows how problematic they are. Also, for him, the ontological structure of Darwinism has important epistemological consequences, Darwinism being the proper theory for finite beings.

CHAPTER 12

The New Evolutionary Ontology and Its Implications for Epistemology

Michael T. Ghiselin

Darwinism as presented in *The Origin of Species* was clearly out of line with what was supposed to be proper metaphysics and metaphysically correct epistemology. Therefore either something had to be fundamentally wrong with Darwinism, or conventional metaphysics needed more epicycles. Over a hundred years passed before it was realized that the metaphysics of evolutionary biology could do without the epicycles altogether, holding out the prospect for an epistemology that is more pleasing to the mind.

Much as Copernicus revolutionized astronomy by moving the sun to the center of the solar system, Darwin revolutionized biology by recognizing the central position of the individual in our understanding of the living world (Ghiselin 1971). At the level of the organism, that meant recognizing the significance of individuals competing reproductively with one another. But what role, if any, species play in the evolutionary process was far from clear at the time. There were all sorts of befuddling ambiguities, such as *species* being both the singular and the plural form of the noun. A proper theory of what a species is began to emerge only in the 1930s. Others were involved, but Ernst Mayr (1940) gets, and I believe deserves, much of the credit. However, the dispute as to whether species are real, which goes back to Darwin and his predecessors, was not resolved at that time. The main barrier to finding the solution was a fallacy of question framing that was all too easy, given the kind of metaphysics that was generally presupposed. Species are classes. For a nominalist, individuals are real, but classes are not, species are classes, therefore species are not real. For a realist, both classes and individuals are real, species are classes, therefore species are real. It fell to me, then a very young postdoctoral fellow under the sponsorship of Mayr, to propose the radical solution (Ghiselin 1966, 1974b). Irrespective of whether one is a nominalist or a realist, individuals are real, species are individuals, therefore species are real.

To anybody who knows the appropriate definitions of both *species* and *individual* there is no question that this inference is correct. It follows deductively and thus is analytic and a priori, though one cannot deduce from it whether there are any instances of the class of species or whether *Homo sapiens* is one of them. Since it is part of the "deductive core" of the synthetic theory, one might be tempted to treat it as synthetic a priori, but I for one would not go that far. Be that as it may, my very suggestion that species are individuals ran flagrantly counter to the expectations of the logical empiricism that dominated academe at the time. My manuscript (Ghiselin 1966) was refereed by the philosopher David Hull, who flatly rejected the individuality thesis; however, a long correspondence resulted. Ultimately Hull changed his mind and became its most effective advocate (Hull 1975, 1976, 1978, 1988). What changed Hull's mind will be explained in due course, and his contribution will be emphasized here.

My mature reflections on the individuality thesis have been embodied in an inevitably difficult book that attempts to place everything in perspective (Ghiselin 1997). Here we need only summarize important points germane to the present discussion. In the first place, classes are kinds, and individuals are instances of kinds. The problem with species and some other individuals is that they are composite wholes with parts that are also individuals. A university is such a whole, one that has professors as components. The University of Notre Dame is not a class of professors. Rather, it is an organization made up of professors (and other things). A second important point is that classes may have instances but individuals do not. It would be absurd to say that Professor Phillip Sloan was a University of Notre Dame. For precisely the same reason it would be wrong to say that he was "a" *Homo sapiens,* sloppy usage notwithstanding. Such terms as *human being* and *American* are what I have called componential sortals, meaning in these cases organism-level components of a species and of a nation respectively.

Classes are spatiotemporally unrestricted, whereas individuals, though they may move around a bit, have definite positions in space and time. They have a beginning and an end. The University of Notre Dame is again a good example. It is located in Indiana and was founded in 1842. But where is University? That kind of question makes no sense. University is an abstraction, whereas the University of Notre Dame is a concrete, particular thing, like Professor Sloan. Classes are abstract, individuals are concrete. Some people have problems conceptualizing social organizations, the branches of a phylogenetic tree, wars, thoughts, and other "intangible" entities as concrete. Hence they are apt to consider them "abstract," and the resultant folk metaphysics has been a serious barrier to clear thinking about such matters. The mere fact that Don Quixote never actually metabolized has been taken by one of my hostile critics as evidence that there is something wrong with my metaphysical system. I am sorry, but *Don Quixote* is a novel, written by an individual human being at a particular time and place, and the characters in it are parts of that intellectual construct (Ghiselin 1980).

Classes can have defining properties, but individuals can only be described. Definition of species, and also other taxa such as individual genera and families, is purely ostensive, and there is no property such that having it or lacking it can disqualify it as an element of that taxon. Whether or not one believes in essences, a biological species or an organism does not have one, although the class of biological species and the class of organisms are reasonable candidates, since they do have definitions. I agree with John Stuart Mill and many other authors that attributing the essence of a class to the instances thereof is an error, however common it may be. Species have the defining properties of species, but that does not mean that they themselves have defining properties.

Finally, and in some ways most important, the laws of nature refer to classes, not individuals. This was what caused David Hull to change his mind. He read a book by Smart (1963), in which it was asserted that there are no laws for *Homo sapiens*. Of course there are no laws for *Homo sapiens,* but there are no laws for the planet Saturn either, and the simple reason is that both *Homo sapiens* and Saturn are individuals. Smart had drawn the conclusion that biology is not a science because, supposedly, it has no laws. According to logical empiricism science consists of laws and nothing whatsoever else. That created problems when it occurred to him that astronomy was generally considered a science. He said that physicists test their hypotheses by observing astronomical individuals. As if biologists did not do likewise when counting the flies in a bottle! Anything but falsify the metaphysical claims upon which his definition of "science" depends, or the definition itself.

That biological species are individuals is an implicit premise in the synthetic theory of evolution. One cannot deny the truth of the thesis yet affirm the truth of the theory without flagrant self-contradiction. It is part of the "deductive core" of that theory. As such one might treat it as synthetic a priori, though again I think we need not go so far. Yet the resistance to the thesis often comes from those who claim to be evolutionists, and one wonders why they should object to it. One hesitates to resort to ad hominem arguments for the simple reason that they tend to be abusive, with obvious problems. Yet the motives driving the discussion are often quite admirable. Philosophers, like everybody else, enjoy trying to shoot down the views of their colleagues and seeing if they can find an alternative. The more important the topic, the greater the motivation. In many cases it would seem that the opponents of the individuality thesis have been trying to salvage the kind of metaphysics that it calls into question. In the first place there are those who remain committed to the logical empiricist metaphysics and try to salvage it one way or another. In the second place there are those who realize that the individuality thesis suggests that species and other supraorganismal wholes may play an important role in evolutionary, and other, processes and for one reason or another are unhappy about that. In particular there are "reductionists" who would have us trivialize not just species but even organisms in favor of their components.

The salvage efforts have all failed miserably. The best that people seem to be able to accomplish of late has been to beg the question by redefining the terminology. Every effort at finding a law of nature for a particular species having failed, we are asked to believe that there are lawlike statements about that species. But in what sense are they supposed to be lawlike? Evidently such persons have what systematists call "conservative characters" in mind. Rather like bad habits that somebody tries in vain to eradicate, some properties of evolving individuals are highly resistant to change, so it is hardly surprising that the hairiness of mammals might be confused with a law of nature. We have to redefine *natural kind* if we are to allow such "kinds" to be spatiotemporally restricted. Or somebody will say that species have "historical essences," thereby making essentiality a matter of contingent historical fact rather than something that is logically, physically, or metaphysically necessary. Or they will find an excuse for saying that ostensive definitions of proper names provide defining properties of those names.

It is perhaps too hyperbolic to paraphrase Dobzhansky and say that nothing in the philosophy of biology makes sense apart from the fact that species are individuals. In a positive light, however, we can say that the individuality thesis allows us to make sense out of all sorts of matters that earlier ontology was impotent to deal with. Understanding what kind of thing we are studying allows us to ask the proper questions, and asking the proper questions is often sufficient for getting the right answer. When dealing with such matters it is important to bear in mind that the ontological considerations take priority over the epistemological ones. It is a serious mistake to say that the ontological status of an entity depends upon one's point of view. But it is a very common mistake, and one that manifests itself in repeated efforts to treat classification as if it were something arbitrary, subjective, or phenomenal. Likewise, the effort to establish a new ontology has sometimes been dismissed, or at least downgraded, as if it were nothing more than an exercise in language analysis, perhaps one of merely parochial interest to a few biologists. The notion of the logical positivists that they could slay the dragon of metaphysics with a wave of the linguistic wand was one of philosophy's all-time great metaphysical delusions. I cannot urge too strongly that genuine, substantive ontological questions are at issue. And they are profound.

Let us see how the individuality thesis helps us solve important metaphysical problems. For example, suppose we ask the sort of question already alluded to above. Are there laws of nature in biology? If so, where should we seek them? For the present discussion we may sidestep such issues as whether there are laws of nature in any science. The individuality thesis tells us that there are no laws for individuals. Therefore, trying to find laws for any species or lineage is a waste of time. Mahner and Bunge (1997) claim that species are classes and that there are laws of nature for them. To achieve that end they identify species, not with the individuals that evolve (such as the species of which all human beings are components), but with something else, which is supposedly a class, the members of which are those individuals that are parts of the individuals that evolve. It is like saying that the University of Notre Dame should not be identified with the entity that is designated by its proper noun in legal contracts and on its diplomas but rather with something that does not even have a name and obviously is not worth talking about. Mahner and Bunge assert that there must be laws of nature for such "classes," but they have failed to give an example of one.

On the other hand, the legitimate laws of nature refer to classes of individuals, though as generally understood these days, such classes have to be natural kinds. There would seem to be nothing extraordinary about natural kinds in the daily life of a scientist. The periodic table of the elements, with its halogens and the included kinds of halogens, chlorine, bromine and iodine, is a straightforward example of a classification in which the groups are natural kinds. Given that by definition natural kinds are classes for which there are laws of nature, the requirement that laws be about natural kinds seems a virtual truism. Simply stated, the individuality thesis tells us that if we are to find the laws of biology, those laws are about kinds of species and other kinds of individuals. The (class of) species functions as a natural kind, even though species (being individuals) do not function as natural kinds, in evolutionary theory at any rate. So do other kinds of populations such as the geographical race and the deme. So the place to look for laws of nature in evolutionary biology is in classes of populations. Such laws are not far to seek. For example, as the effective population size goes up, the frequency of fixation of alleles through sampling error goes down. This is, of course, a statistical law. What are we to make of that? Among other things, an appropriate epistemology for evolutionary biology very likely will have much in common with those other branches of science in which statistical laws play an important role.

Clarifying what should be obvious to an evolutionary biologist but not necessarily to a philosopher, I of course am concerned with species as the class is defined by the biological species concept. The biological species functions as a natural kind. That is why I have been able to single out what I call "Mayr's Law" (Ghiselin 1989), according to which under ordinary circumstances there is no speciation without an antecedent period of extrinsic isolation. In the synthetic theory biological species and their approximate equivalents function as individuals, and so do parts of species that are sometimes called "species" under other criteria. But to my knowledge at least there are no laws of nature for so-called morphological species. Even if there are, it seems unlikely that they are the same laws of nature as those for the biological. Ruse (1987) has taken the purported "coincidence" of morphological and biological species as evidence that the species is a natural kind, and perhaps species are natural kinds as well. The supposed coincidence does not exist. To be sure, sometimes they do coincide, to the point that the one might be used as a crude surrogate for the other, but often they do not. In the case of cryptic species the use of a morphological species concept sometimes produces an underestimate with respect to the real thing of something like a full order of magnitude (Knowlton 2000). "Chronospecies" give an overestimate. Agamospecies—that is, "species" without sex—are, so far as the biological species concept goes, a contradiction in terms.

Aristotle said that knowledge is "of" universals, not particulars. He had a legitimate point, but the individuality thesis forces us to recognize that there are problems with such views as well. The taxonomic hierarchy is a powerful intellectual instrument, partly because it subsumes the more particular under the more general. But that subsumption amounts to locating individuals within larger individuals. It is a whole made up of parts, a genealogical nexus, and an epitome of evolutionary history. The epistemological import of this fact is that biological knowledge is historical knowledge and has to be pursued with the appropriate techniques. To be sure, the laws of nature play an important role in biological research, but explanatory historical narrative is its ultimate goal and the criterion upon which the legitimacy of knowledge claims is to be evaluated. I hardly need emphasize how strongly this view of science conflicts with the conventional wisdom.

Consequently it is generally true that in solving biological problems, the appropriate question is necessarily historical. If we are to understand why organisms (and other individuals such as families) have the properties that they do, then right thinking means asking how organisms (and other individuals) have competed reproductively with one another. The answer that is given consequently must take the form of an explanatory historical narrative. If that narrative is to be more than just educated guesswork, it has to be justified on the basis of evidence germane to the truth of that narrative. Such evidence of course includes principles and laws of nature, as well as evidence with respect to temporal succession and other particulars. The explanatory narrative is a theory asserting that evolution has happened in a particular way and as a consequence of the causes that are hypothesized to have been responsible for it happening that way (and not otherwise). The idiographic evidence may be phylogenetic trees that tell us the succession of changes within lineages, but it may also be biogeographical or stratigraphic. Nomothetic evidence may play a role in such inference as well, and I will give a sketchy example that also provides an illustration of biological laws of nature.

There has been much confused discussion about putatively nonfunctional characters. Part of that confusion has been dispelled by pointing out that "character" is equivocal and that the functionality or lack of it of a part and of an attribute are very different issues (Ghiselin 1969, 1984). The connection between this insight and the individuality thesis lies in the recognition that the world is populated by wholes and parts and that these play a very different role in evolution and evolutionary thinking than that of the attributes predicated of them. The epistemological problem of deciding whether parts and attributes are functional is a traditional one among biologists. Whether the parts and attributes of organisms are functional has been a serious bone of contention among various factions both scientific and philosophical. Often claims of nonfunctionality were based upon mere ignorance. But how do we identify a nonfunctional part? One method that is hard to endorse is the "method of residues," which suggests that we find out what is functional and treat what is left over as otherwise. There do exist, however, legitimate canons of evidence. The narrative history may tell us that there have been shifts in selection pressures—for example, when there has been a movement from one habitat to another. The putative nonfunctional organ should be one in which the previous selection pressure no longer operates. Furthermore, the theory of natural selection predicts that whenever selection ceases to operate at the genomic level, evolutionary change is inevitable and will proceed in the direction of loss of organization. Were that not the case, the second law of thermodynamics would have to be false. Accordingly, rudimentary parts show signs of disorganization—for example, often becoming bilaterally asymmetrical. One wing of a kiwi can be strikingly larger than the other. It might be argued whether we are dealing with physical or biological laws here. Although change in the direction of increasing entropy is something physical, the kind of organization in question is something that we find only in organisms and their productions.

To generalize, all scientific claims with respect to adaptive significance entail a historical narrative explanation, whether explicit or implicit, and have to be evaluated in terms of whether the claims are true or false. "Panglossian adaptationism" is illegitimate precisely because it fails to treat the putative explanation as a testable scientific hypothesis and then test it. Daniel Dennett is a good example of an apologist for Panglossian adaptationism who obviously does not understand such matters. He wants to treat adaptation from an ahistorical point of view, as if it could be inferred from general principles or laws. Not surprisingly he has dismissed the individuality thesis, calling it wrong without providing any justification beyond his personal opinion (Dennett 1995).

Panglossian adaptationism is a particular case of teleology, but what does *teleology* mean? At a bare minimum it means the kind of metaphysical delusion that was unmasked by the Darwinian revolution. Unfortunately the word has also meant some other things, and equivocations provide excuses for refurbishing the kind of metaphysics that Darwin discredited. There is a complex interplay between ontology and epistemology here. It is fairly obvious that an ontology that takes occult causes such as an anthropomorphic deity for granted is going to lead to often ludicrous misconceptions. That natural selection cannot be substituted for such occult causes is clear enough, provided that one's ontology is reasonably sophisticated. Even so, one has to be careful. Ayala (1999), an outstanding evolutionary biologist, nonetheless makes the usual mistake when he says that organisms maximize their reproductive success; he should have said that they do that which maximized their ancestors' reproductive success. Of course, things like foresight and understanding do exist and need not be treated as something occult. Nonetheless, we can mishandle them for the same basic reasons, treating human beings and other animals "anthropomorphically." Teleology is best regarded as a fallacious way of thinking, a matter of getting the wrong answer because one asks the wrong question. Getting the ontology more or less right does not mean that one can just turn over and go back to sleep, for the epistemological problems keep disturbing our dreams.

The individuality of biological taxa and the consequent historicity of the sciences that deal with them suggest that a much larger range of entities traditionally conceived of as classes might upon closer examination turn out to be individuals. For epistemology this means finding out how to identify them as such and finding appropriate ways of understanding them. Rejecting organismality as equivalent to individuality was just the first step. It is also crucial that we recognize the existence of individuals that do not fall under the category of substance. Such cultural individuals as languages provide a relatively straightforward example, especially because of the long-standing recognition of their strong analogies with species and because genealogy is the basis of higher-level classification in linguistics. The fact that some philosophers deny what has been the consensus among professional historical linguists for some two hundred years is good evidence for the intellectual bankruptcy of that kind of philosophy. On a more positive note, treating intellectual history from an individualistic point of view, perhaps (but not necessarily) along lines suggested by Hull (1988, 1992), tells us that the objects studied can change indefinitely, remaining the same thing even though hard to recognize as such (much as the English of *Beowulf* is the same language as that of *Huckleberry Finn*). It indicates that we should be looking for continuities, not "paradigm shifts" or other analogues of special creation or macromuation, without downplaying the significance of new ways of thinking or other genuine innovations. Rather, we should be applying such principles as *Funktionswechsel,* as suggested by Darwin's great follower Anton Dohrn (1875; translation in Ghiselin 1994).

The animals that invaded the land so as to give rise to the terrestrial fauna did not "leap" out of the water in making that transition. Rather, they underwent a large number of changes, some of which were adaptive in relation to earlier conditions of existence and others of which came into being during the transition and subsequently. In vertebrates fins were transformed into ambulatory organs. In pulmonate gastropods the cavity that once surrounded the gills became a lung. In a very important sense, the homologues remain the same, but of course that is not the whole story. In the context of a new environment and a remodeled organism, the corresponding parts take on very different (physiological) significance. So too with the evolution of science when there is a shift in the problem situation. The concepts and the terminology come to be quite different from what they were before. The concept of homology is a very good example. A phenomenal term that was defined on the basis of (occult) neo-Platonic metaphysics became a theoretical term that was defined on the basis of evolutionary history. Because of historical circumstances the explication of *homology* and its contrast term *analogy* became egregiously muddled, and the resulting controversies gave rise to a vast literature, almost all of it misguided. The situation was exacerbated by the persistence of the kind of metaphysics that Darwin had rendered obsolete. From the outset there had been a problem insofar as *analogy* had mistakenly been defined in terms of common function. That mistake reinforced another one: a failure to distinguish between similarity and correspondence. Thanks to the individuality thesis, the basic problems have now been solved. Homologies are correspondences between parts of parts of individuals. Analogies are correspondences between parts of members of classes.

The particulars of the example just given are interesting, but that is not the basic message. Theories evolve, and they come to us burdened with the vestiges of what has gone before. They are what they are largely because of historical contingencies, the paleoenvironments in which they previously flourished. Often what have long been accepted as brute facts turn out to be inferences based on theories that ought to be questioned, or mere pedagogical traditions. Given that historicity, and its consequences, it becomes crucial that the scientist undertake a critical examination of previous stages in the evolution of knowledge. And that includes the metaphysical underpinnings and other important chapters in the history of philosophy. It may be necessary to reconstruct the whole edifice upon a new foundation.

We may even need to revise our conception of what philosophy is all about. I have gone so far as to suggest that metaphysics should be redefined as the most fundamental of the natural sciences. A couple of my hostile critics have responded with outraged indignation. One of them insists that metaphysics has to be a matter of belief systems, I suppose along the lines of what is sold in "metaphysical book stores." Another says that metaphysicians do not do their research in laboratories. Having done a lot of research in the field and the library, the laboratory seems to me not quite what we need as an "essence" for science. Furthermore, I at least have derived a great deal of metaphysical insight as a consequence of working in the laboratory. I remember when my colleagues and I were collecting the ribosomal RNA sequence data that led to a major breakthrough in the phylogenetics of the Metazoa (Field et al. 1988). Early one morning, as I was getting the ice and other paraphernalia together for the coming day's work, it occurred to me that here we were, figuring out what had gone on hundreds of millions of years in the past, and I marveled at the power of the human intellect in transcending its apparent limitations. Indeed it was the claims of a faction of biologists who were hostile to the goals of phylogenetics that first got me seriously interested in epistemology and metaphysics.

How should we define *philosophy?* These days philosophy seems to have very little to do with wisdom, or the love thereof, but that situation may prove ephemeral. Is it one of the humanities? One might give very different answers to that question, depending upon the kind of philosophy of taxonomy that gets invoked. Are we to have a classification that, as Plato put it, cuts nature at its joints? Or is academe to be a sort of Procrustean bed, or dormitory, that determines what parts of reality are excised? What are called epistemic values often enter in when decisions are made as to how we will classify. Some of that seems to me quite legitimate. In systematic zoology and botany there is often a trade-off between theoretical relevance and ease of identification. The academic world is organized as it is because of contingencies with respect to how teaching and research have been conducted. It is maintained in the present condition largely because of academic politics and turf battles that, if perhaps inevitable, are not always edifying. Epistemological taste seems to be taking priority over ontological reality.

But, to borrow an expression from John Dewey, what kind of reconstruction in philosophy would seem to be in order? Surely more than just throwing up a buttress or two and replacing a few of the gargoyles. In some thoughtful commentary in the present volume, Vittorio Hösle has suggested that my views might be reconciled with those of Plato. That seems perfectly reasonable at the level of the eternal verities: goodness, truth, and beauty. I certainly believe that what is true or false has nothing whatsoever to do with what anybody knows or opines. That kind of realism seems to me the fundamental precondition for the very existence of knowledge, and I have often fallen back upon the principle of contradiction when under skeptical attack. Pyrrhonism, including that which is manifest in much of contemporary social relativism, is inherently self-contradictory. But where are the eternal verities located, and how do we come to know them? Do they come from within, from without, or perhaps from beyond? Treating knowledge as if it were obtained through a kind of memory, even racial memory, makes its acquisition all too passive, neglecting the role of the active powers of the organism. Although our capacity to deal with the eternal verities may be historically rooted in crass expediency, any effort thereby to degrade the verities themselves crudely exemplifies the genetic fallacy.

Idealism tends to be dualistic, leading to fallacies of false disjunction. It treats the knowing subject as if it were something distinct from, perhaps surrounded by, the rest of the universe rather than an integral part of it. My own position has been characterized as monistic, but it is a peculiar kind of monism. It affirms the unity of the cosmos and the unity of knowledge. At the same time it also affirms the diversity of beings and the need for the enrichment of the understanding. The "pluralism" to which it stands opposed is perhaps better characterized as "syncretism," by which is meant the juxtaposition of incompatible elements with resulting intellectual schizophrenia. Those authors who have maintained that species "are" sets are a fine example. The logical apparatus that serves as the metaphysical basis for their epistemology gets treated as if it told us something about ontology. Given such metaphysics it makes perfectly good sense to believe that there exist no larger connections among the entities that are relegated to an ontological Never-Never Land by attaching a name to them that has no ontological import. As to mind-body dualism, such a dualism need not conflict with the kind of monism that I have in mind, provided that they are not incompatible. A dual-aspect theory is monistic in that sense, whereas one of mind-body identity seems downright self-contradictory.

In evolutionary biology dispositional properties are very important. The animals on which I specialize are often poisonous, or at least distasteful, and such dispositions result in predators' leaving them more or less alone. I have used the phenomenon of toxicity to counter objections to the biological species concept with its notion of "potential" interbreeding. Given the immense amount of contingency that characterizes organic evolution, it makes sense for those who study it scientifically to be concerned with possibilities that may or may not be realized. But possibilities, dispositions, and counterfactuals are also important in the conduct of our everyday lives, including the ethical aspects. Animals with foresight and understanding are able to consider a variety of alternative ways of running their lives. On that basis I justify the notion that morals can be more than just crass expediency carried to its logical conclusion. Reasonably intelligent persons can observe what happens when they and those about them act from unmitigated selfishness and greed. They can imagine what life would be like were things otherwise. They can find ways of mitigating the situation, such as instituting just laws and agreeing to a system of rewards and punishments for those who are subject to them. I have suggested that the institution of society may bring new selection pressures into play. Indeed, the selection that is responsible may be of a different mode, namely unconscious artificial selection (Ghiselin 1974a). Under such circumstances, perhaps, Darwinism might account not just for self-sacrifice irrespective of motive but also for what is still called "altruism" by normal human beings. The world seems to be profoundly confused about the metaphysics of agency, and it is perhaps in this area that we metaphysicians can contribute most to the welfare of humanity.

REFERENCES

Ayala, Francisco J. 1999. Adaptation and novelty: Teleological explanations in evolutionary biology. *History and Philosophy of the Life Sciences* 21:3–33.

Dennett, Daniel C. 1995. *Darwin's dangerous idea: Evolution and the meanings of life.* New York: Simon & Schuster.

Field, K. G., G. J. Olsen, D. J. Lane, S. Giovannoni, M. T. Ghiselin, E. C. Raff, N. R. Pace, and R. A. Raff. 1988. Molecular phylogeny of the animal kingdom. *Science* 239:748–53.

Ghiselin, Michael T. 1966. On psychologism in the logic of taxonomic: Controversies. *Systematic Zoology* 15:207–15.

———. 1969. The principles and concepts of systematic biology. In *Systematic biology: Proceedings of an international conference,* ed. C. G. Sibley, 45–55. Washington, D.C.: National Academy of Sciences.

———. 1971. The individual in the Darwinian revolution. *New Literary History* 3:113–34.

———. 1974a. *The economy of nature and the evolution of sex.* Berkeley: University of California Press.

———. 1974b. A radical solution to the species problem. *Systematic Zoology* 23:536–44; 1975.

———. 1980. Natural kinds and literary accomplishments. *Michigan Quarterly Review* 19:73–88.

———. 1984. "Definition," "character," and other equivocal terms. *Systematic Zoology* 33:104–10.

———. 1989. Individuality, history and laws of nature in biology. In *What the philosophy of biology is,* ed. M. Ruse, 53–66. Dordrecht: Kluwer Academic Publishers.

———. 1994. The origin of vertebrates and the principle of succession of functions: Genealogical sketches by Anton Dohrn. 1875: An English translation from the German, introduction and bibliography. *History and Philosophy of the Life Sciences* 16, no. 1: 5–98.

———. 1997. *Metaphysics and the origin of species.* Albany: State University of New York Press.

Hull, David L. 1975. Central subjects and historical narratives. *History and Theory* 14:253–74.

———. 1976. Are species really individuals? *Systematic Zoology* 25:174–91.

———. 1978. A matter of individuality. *Philosophy of Science* 45:335–60.

———. 1988. *Science as a process: An evolutionary account of the social and conceptual development of science.* Chicago: University of Chicago Press.

————. 1992. The evolution of conceptual systems in science. *World Futures* 34:67—82.

Knowlton, Nancy. 2000. Molecular genetic analysis of species boundaries in the sea. *Hydro-biologia* 420:73—90.

Mahner, Martin, and Mario Bunge. 1997. *Foundations of biophilosophy.* Berlin: Springer-Verlag.

Mayr, Ernst. 1940. Speciation phenomena in birds. *American Naturalist* 74:249—78.

Ruse, M. 1987. Species: Natural kinds, individuals, or what? *British Journal for the Philosophy of Science* 38:225—42.

Smart, J. J. C. 1963. *Philosophy and scientific realism.* London: Routledge & Kegan Paul.

CHAPTER 13

How Is It That We Can Know This World?
New Arguments in Evolutionary Epistemology

Gerhard Vollmer

1. METAPHYSICAL IMPLICATIONS OF DARWINISM?

Most implications of Darwinism are antimetaphysical. This is one reason why Darwin's theory was fought against and why it is still regarded as a scientific revolution. To be sure, Darwin's intention was not to become a revolutionary but rather to solve interesting scientific problems. But his intention did not help him. In that respect, he may be likened to Nicolas Copernicus, Martin Luther, or Max Planck. Indeed, Copernicus is characterized by Arthur Koestler as "the timid canon," and Planck by Helge Kragh as "the reluctant revolutionary."[1] In a similar vein, Carl Friedrich von Weizsäcker, writing about Werner Heisenberg, submits "that only the conservative can be a revolutionary."[2]

Metaphysical or not, there is no doubt that Darwin's theory has philosophical consequences. Some of these consequences refer to epistemology. Darwin himself was well aware of this. Already in 1838, he wrote in his Notebook M: "Plato . . . says in his *Phaedo* that our *'necessary ideas'* arise from the preexistence of the soul, are not derived from experience.—read monkeys for preexistence."[3] Even from this short note it is plain that Darwin replaces a metaphysical concept, preexistence, with a biological one, monkeys as man's predecessors. Whether our ideas remain "necessary" in that move is another interesting problem.

Cognition takes place in our heads. From the signals that we receive from our sense organs our brain builds a picture of the world up to a whole worldview. We construe the world as three-dimensional, ordered and directed in time, regular, even structured by laws of nature, and causally connected. We draw conclusions and proceed from past experiences to expectations of the future. With some of our constructions we are successful, with others we fail.

The principles by which we construct this world picture are not dictated by our sense organs or exclusively by external stimuli. How did they come into our heads? This question is answered by *evolutionary epistemology*. In section 2 we recapitulate its main theses. In section 3 we characterize it as a naturalistic position. In section 4 we answer some typical objections. Later sections are devoted to more recent arguments, concerning language (sections 5 to 7), realism (sections 8 and 9), and the theory of selection (sections 10 and 11). Let's start with the theses.

2. MAIN THESES OF EVOLUTIONARY EPISTEMOLOGY

Thinking and cognition are achievements of the human brain, and this brain originated through organic evolution. Our cognitive structures fit the world (at least partially) because phylogenetically they were formed in adaptation to the real world and because ontogenetically they have to grapple with the environment. The biologist George Gaylord Simpson (1902–84) makes this point crudely but graphically: "The monkey who did not have a realistic perception of the tree branch he jumped for was soon a dead monkey—and therefore did not become one of our ancestors."[4] Hence the fact that our spatial perception is relatively good we owe to our predecessors living in trees with prehensile organs. In this way we may also explain other cognitive achievements.

But why, then, are our cognitive faculties not even better? Again the answer is simple enough: biological adaptation is never ideal, nor is our cognition. There is no evolutionary premium on perfection, only on effectiveness. What is decisive for evolutionary success is not pure quality but a defendable cost-benefit relation. It is not essential to find the best possible solution, only to be better than the competitors. Here we must think not only of interspecific competition but of intraspecific competition. Thus evolutionary epistemology explains not only the achievements of our brain but also its failures.

That section of the real world to which man is adapted in perception, experience, and action we call the *mesocosm*. It is a world of medium dimensions: medium distances and time periods, small velocities and forces, low complexity. Our intuition[5] (what Egon Brunswik calls our "ratiomorphic apparatus") is adapted to this world of medium dimensions. Here our intuition is useful, here our spontaneous judgments are reliable, here we feel at home.

Whereas perception and experience are mesocosmically impregnated, scientific cognition may transcend the mesocosm, extending in three directions: to the very small, the very large, and the very complicated.[6] As we know, intuition fails regularly there. Nobody can visualize the conditions of the quantum realm, relativity theory, or deterministic chaos.

Yet we must constantly deal with complicated systems. To do so we need working tools and thinking tools, instruction and training. The most important thinking tool is language. Other ladders leading beyond the mesocosm are algorithms, calculi, mathematics, and computers.

3. EVOLUTIONARY NATURALISM

Evolutionary epistemology constitutively rests upon organic evolution. This has given evolutionary epistemology its name. The adjective *evolutionary* does not mean that all epistemological problems can or should be solved by reference to the evolution of the universe, organisms, man, or knowledge. However, it documents the claim that the evolutionary origin of our cognitive faculties plays an important role in epistemology, both explanatory and critical.

Thus we explicitly deny Ludwig Wittgenstein's claim in his *Tractatus* that "Darwin's theory has no more to do with philosophy than any other hypothesis in natural science."[7] To be sure, we must support our denial by arguments, and this is done best by *showing* how evolution pertains to philosophy. It may solve old philosophical problems, pose (and even solve) new problems, or shed new light on problems. Such a triple claim is made by evolutionary epistemology.

It remains unclear, however, how general a concept of evolution is meant here. Do we talk about organic evolution, the evolution of organisms, or do we also talk about the evolution of knowledge, maybe even of science? This ambiguity has been quite confusing. Here we are talking about the biological evolution of cognitive faculties. Insofar as we are investigating the development of science using evolutionary concepts in general, we prefer to call this enterprise the *evolutionary philosophy of science*.[8]

The general orientation of evolutionary epistemology is naturalistic. What does that mean? There is talk about naturalism in several areas: theology, philosophy of science, ethics, art. In the present context, we take it to be a conception of natural philosophy and anthropology claiming that there are no secrets anywhere in the world. Hence it is distinguished by two traits: by its *claim to universality* and by the *limitation of tools* admitted for describing and explaining the world.[9]

Philosophical naturalism is both conception and program. As a program it has at least four parts:

1. It calls for and charts a *cosmic view,* a "worldview."
2. It assigns to *man* a definite place in the universe (which turns out to be rather modest after all).
3. It covers *all* human capacities: language, knowledge, scientific investigation, moral action, aesthetic judgment, even religious faith.
4. Under these premises it calls for and develops in particular

    a) a naturalistic anthropology
    b) a naturalistic epistemology
    c) a naturalistic methodology of research
    d) a naturalistic ethics
    e) a naturalistic esthetics

With respect to epistemological questions, W. V. O. Quine has formulated such a naturalistic program, and evolutionary epistemology attempts to carry it out. Occasionally, Quine himself drew on evolutionary arguments: "Natural selection, then, could explain why innate standards of resemblance have been according us and other animals better than random chances in anticipating the course of nature" and "Creatures inveterately wrong in their inductions have a pathetic but praiseworthy tendency to die before reproducing their kind."[10] Quite analogously some philosophers have tried to develop an evolutionary ethics. Generally, we may speak of an *evolutionary naturalism.*[11]

Finally, evolutionary epistemology is *realistically* oriented. More precisely, it defends a *hypothetical realism,* characterized by the following:

1. *Ontological realism:* There is a real world independent (for its existence) of our consciousness, lawfully structured, and quasi-continuous.
2. *Epistemological realism:* This world is partially knowable and understandable by perception, thinking, and an intersubjective science.
3. *Fallibilism:* Our knowledge about this world is hypothetical and always preliminary.

Numerous objections have been raised against evolutionary epistemology.[12] We look at three of them.

4. THREE OBJECTIONS

Does the Concept of Truth Make Sense?

Hypothetical realism makes use of the correspondence theory of truth. By this theory, a proposition is true if what it says corresponds with external reality. But how do we get to know this reality, hence truth? There is no independent access to reality unless for God. We humans cannot take this divine perspective, cannot know the world in itself, and therefore cannot assess truth in the sense of correspondence theory. So goes the objection.

As far as this objection is justified it is directed against all kinds of epistemological realism (except perhaps the internal realism proposed by Hilary Putnam some years ago, which strictly speaking is no realism at all). In fact, we are no gods. But this is not necessary either. The correspondence theory of truth does not supply a criterion of truth, only a definition of it. As epistemologists had to realize after 2,500 years of fruitless search and growing doubts, there are no satisfiable sufficient criteria for factual truth. What we have are necessary criteria like consistency, corroboration, coherence, or consensus, as exhibited by the different theories of truth. For the definition of truth all these theories rely, in the last analysis, on the correspondence concept. (Where this is not the case, the concept of truth is, strictly speaking, superfluous.)

We might object that a God's-eye perspective is an undue idealization. However, no theory of truth can do without such idealizations. Internal realism, for instance, regards as true what at the end of all research will be ascertained about the world. What if not this is an idealization? Thus realism and correspondence theory answer this last objection with a *tu quoque* argument: Yes, there is an idealization, but all other theories of truth use comparable devices.

Is the Fit of Our Cognitive Structures Ascertainable without a Vicious Circle?

Would we not have to know reality *independent* of our cognitive structures? This objection goes beyond the former because what is at stake now is not a definition of truth but our knowledge about reality. Here evolutionary epistemology makes a more ambitious claim.

Let's take an example: physicists and physiologists disclose to us that our eyes are sensitive in precisely that section of the electromagnetic spectrum where—thanks to the optical window of the terrestrial atmosphere—radiation from the sun can pass through the air and reach the surface of the earth. How could they come to know this? Of course, even physics had to start in the mesocosm, but it has definitely left it since. In so doing, physics has objectified its methods as much as its results and its theories. It does not talk about colors anymore but about frequencies, wavelengths, and energies. For the characterization and detection of radiation it does *not* depend on the eye. And it finds electromagnetic radiation of all wavelengths. True, even terms like *wavelength* and *sensitivity* could still be anthropomorphic. However, there is no rational doubt that, first, not all of what in principle could exist does really exist; second, with our senses we can process only a section of what there is; and third, there is a very good fit between (what we call) daylight and the properties of our eye. We can detect this fit without being realists and without an explanation at hand. It is the fit thus established that is interpreted and explained by evolutionary epistemology, not as pure chance, nor as the work of a creator, but as the result of an adaptive process.

Still, one could object that what physicists describe is not the real world but at best a projection, possibly a garbled one, and in the worst case nothing but a wild construction. Certainly we cannot strictly prove the truth, the correctness, the adequacy of our theories. But what on earth can be proved strictly? We cannot even disprove the solipsist claiming or in fact being convinced that there is nothing besides his present consciousness.

But where proofs are missing, good reasons might still be available. For ontological realism (there is a real world independent of our consciousness) and for epistemological realism (this world can be known at least in part and approximately) there are good arguments. For the suggestion, however, that scientific knowledge is nothing but a wild construction there are *no* good arguments. And it is utterly implausible that we should have adapted to constructions that have been worked out by scientists during the last centuries.

Some constructivists think that organisms are adapted, not to an external world, but to survival. This is unbiological thinking. If there are no selective demands by the environment, there can be no traits facilitating survival or traits threatening it; in that case any solution will do. But then the concept of adaptation makes no sense at all.

Let's keep to it: we may sensibly talk of fits and adaptations, we may argue for them, but we cannot prove them. And evolutionary epistemology is happy enough to explain these fits—that is, our cognitive achievements as well as our failures.

How Can Cognitive Structures Be Adapted to an Environment That Would Have to Be Known to the Organism before the Organism Might Adapt to It?

If this objection were sound there would be no eyes! For how could eyes be adapted to terrestrial illumination if they were necessary *before* any light could be processed? But eyes did originate several times, independently, and, if we follow evolutionary biologists, at least forty times. And most of them are perfectly well adapted to light. How could they originate? The answer is simple: eyes originated as all things originate in evolution, namely by trial-and-error elimination, by blind variation and selective retention, by undirected mutations and gene recombinations and preferential reproduction of superior solutions. Nowadays the evolution of the vertebrate eye—and that includes the human eye—can be reconstructed quite well. Similar considerations apply to all other sense organs, to all senses, to all perceptual achievements. There is no reason why they should not be applied to higher cognitive functions, as far as these are genetically conditioned.

For evolutionary origins and explanations, however, it is not essential that an organ be perfect. The intermediate steps are evaluated selectively; they must increase fitness, but it is not necessary that the later function be present and effective from the very beginning. Changes of function are possible and rather common. Here, a trait is built for a function that will be replaced later by another function. From fins arose arms and legs; feathers served at first not for flying but for gliding, catching prey, and keeping warm; the middle-ear ossicles are former jawbones. Since the change of function is not saltatory, it is indispensable that a trait have two or more functions at the same time.[13]

For the eye such intermediate steps and multiple functions are well known because there are so many types of eyes. In other cases we must content ourselves with a scenario—that is, with a sequence of steps for how the appearance of a trait *might* have happened. And sometimes intermediate stages and double functions have yet to be found.

5. LANGUAGE ABILITY: A HELPFUL ANALOGY

With respect to empirical testing, evolutionary epistemology faces two characteristic difficulties. First, it asserts a strong genetic component for cognitive faculties. In this it sides with the classical nativists (who were mostly rationalists, as were Descartes or Leibniz). For, if at birth the brain is a tabula rasa, as John Locke and other strict empiricists have it, if there is no strong innate component, then it remains a mystery how we can ever achieve knowledge. Evolution and genetics could not then be held responsible.

But specifying this supposed innate component is not an easy task. For how do we find out the cognitive inventory of newborn babies? They cannot talk, so we must rely on observations of behavior. But even the behavioral spectrum of newborns is rather limited. And things that a newborn does not master from the outset but only days, months, or years later may always be claimed to be acquired individually, hence to be due not to phylogeny but to ontogeny. Many experiments being, on principle, informative are banned for moral reasons. Thus nobody will intentionally prevent a baby from experiencing color or music merely to find out how it will develop without these stimuli. Therefore, it is very difficult to find conclusive evidence for the genetic preconditions of cognitive achievements.

Now, ethologists find help in comparing species. Traits occurring in many species, especially if the latter are closely related, are supposed, in the sense of a legitimate working hypothesis, to be innate. Thus it is informative to investigate the cognitive achievements of our kin, the great apes. But they talk even less than human babies, and innate components are again difficult to spot.

Happily there is a fertile analogy to cognition: language. True enough, investigating language acquisition faces obstacles similar to those of investigating cognition. As with cognition, it does not suffice to look at the result—that is, at the different linguistic products or different languages. What is at stake is the *ability* to speak—that is, to learn a language, to use and to form it. Why do we speak? Why can humans do something that no other animal can do? Is this due to a biological, a genetic, hence phylogenetic component? What does it look like, and how did it arise?

If our language faculty has its origin in organic evolution—and for a naturalist there is no doubt about that—then there must have been intermediate steps in this development. Unfortunately such intermediate steps are neither recent nor evidenced by fossils. But the comparison of languages gives at least some cues to innate elements. That is why it is enlightening that the evidence suggesting an innate component of language ability in humans has been strengthened in recent years. It comes from research on Creole languages and on deaf-mutes.

6. CREOLE LANGUAGES AS AN ARGUMENT FOR THE EXISTENCE OF INNATE STRUCTURES

Sometimes members of a linguistic group live in an environment where other languages are spoken; they may be merchants, refugees, slaves, or inhabitants of a

colony. In such cases they develop, in order to communicate, typical hybrid languages, so-called *pidgin* languages. (The word *pidgin* comes from the Chinese pro-
nunciation of the English *business*. The term is used for all such mixed languages.)
Pidgin languages are quite simple in their vocabulary and even more in their grammar. They do not count as full-blown languages.

Very often the children of such immigrants develop their own languages,
called *Creole languages*. Originally the term *Creole* was used to describe the descendants of white Romance-language immigrants in all of South America (white
Creoles) or the descendants of black slaves in Brazil (black Creoles). Nowadays, the term applies to all languages developed by immigrants of the second
generation, whether on islands or in coastal areas of Middle America, West
Africa, the Indian Ocean, or the Pacific region. Creole languages are complex
languages whose vocabulary is drawn from totally different "mother tongues,"
mostly of colonists. Thus Jamaica-Creole rests on English, Guyana-Creole on
Dutch, Haiti-Creole on French, Crioulo in West Africa on Portuguese. In their
grammar, however, they are quite autonomous, dependent neither on the original language of the immigrants nor on the language of the "host country" (with
respect to slavery, the expression *host country* seems inappropriate) or of the
colonizers.

Since the colonies are quite separated and have no exchange, or nearly none,
the Creole languages must have developed independently of each other. Yet in
recent years linguists have discovered that these Creole languages are surprisingly
similar in their structure—that is, in morphology and grammar.[14] How can this be
explained?

Many traits common to all humans are explained as being genetically conditioned. If there is an innate language faculty, as claimed for a long time by rationalists and nativists and more recently by Noam Chomsky and Steven Pinker,[15]
then there should be features common to all natural languages. The search for
such linguistic universals has not been extremely successful; it has uncovered only
very abstract principles. Creole languages, however, share many very concrete
traits.

Not conclusive but at least suggestive, therefore, is the conjecture that these
shared traits are due to a biological-genetic component. Precisely this is the claim
of the leading researcher on Creole languages, Derek Bickerton. According to him,
the innate language component can develop freely only if it is not suppressed by
corrections from outside, and just this is the case with Creoles: their immigrated
parents have not mastered the local language yet, and the children usually do not get
a formal education. Therefore, the structural similarity of Creole languages of

independent origin is evidence for the existence and influence of a strong genetic component in language ability.

7. INFANT GRAMMAR AND THE LANGUAGE OF DEAF-MUTES

This conjecture is supported by more recent findings. Children do not master their mother tongue immediately but start with characteristic *mistakes;* they use some kind of an "infant grammar," violating the respective "correct" grammar in many ways—for example, with respect to double negation and interrogative forms. According to Daniel Slobin, these infant grammars are strikingly alike. What is more, they have very much in common with Creole languages. This suggests that infant grammar is partly innate. In most cases this innate grammar is overcome by the native tongue learned from outside—except with Creoles.

As much as these findings confirm Chomsky's thesis that there is an innate language acquisition device, they contradict another of Chomsky's conjectures. According to Chomsky all natural languages should fit into the innate linguistic structure. According to Bickerton and Slobin, however, they do *not* fit this innate structure in every respect; that is why children make typical mistakes, and that is why this infant grammar could be detected.[16]

New investigations with deaf-mute children point in the same direction. To communicate with each other these children develop an extensive system of signs and gestures. American psychologists have analyzed and compared such sign languages of deaf-mute children from America and Taiwan. They found that although these children had never met before, they gesticulated in stunningly similar ways that they could not have learned from their parents.[17] This is a kind of involuntary Kaspar Hauser experiment: since the deaf-mute children grow up without linguistic stimulation from outside, they have to develop such structures themselves. Here again, a biological-genetic explanation suggests itself.

Suppose such explanations are correct: What does that mean for our cognitive abilities? Language and cognition, though not identical, are closely intertwined: without language there is no higher cognition, and without cognition language does not make much sense. For Chomsky, language is therefore a kind of probe giving us insight in the organization of mental processes. Thus the evolution of language faculty must have gone along with an evolution of cognitive faculties. If the one part is plausible, so is the other. No wonder, then, that Chomsky's disciple Steven Pinker wrote not only a book on language but also one on thinking.[18]

8. THE SUCCESS OF THEORIES AS AN ARGUMENT FOR REALISM

In arguing from success we use success as evidence for the quality of a premise. Science is successful as far as it achieves its goals. And a scientific theory is called successful if it promotes our goals. Such successes corroborate the premises made by the respective theory. One fundamental premise of natural science, possibly of all empirical science, is *realism*. How can we argue for realism?

Often enough, it is the success of science that counts as the best argument in favor of realism. According to the early Hilary Putnam, "The typical realist argument against Idealism is that it makes the success of science a *miracle*."[19] In fact, the realist can explain the success of science, whereas the antirealist cannot. For if quarks and quasars really exist, then it is no wonder that theories claiming or presupposing their existence are successful. If, however, these objects do not exist at all, how is it that with these theories we make correct predictions and solve so many more problems?

But even the success of science is of course no *proof* of realism. And vice versa: the fact that idealism, positivism, instrumentalism, or constructivism cannot explain something does not refute them. Still we may say that realism *explains more*. In theories of empirical science explanatory power is an important trait by which theories are judged. (Other traits are noncircularity, internal and external consistency, testability, and test success.)

True, neither realism nor its counterparts are theories of empirical science. They rather help us to do science and to interpret our results. But if we want to judge metatheories, metaphysical positions, methodological attitudes, and heuristic rules as well, we need criteria on this level, and then explanatory power plays again an important role. And by this criterion realism fares much better. Some philosophers even think that realism is historically *testable,* but they disagree on the question whether it has stood up to this test.

There is an important objection: Couldn't there be several ways to do justice to the same experiences? Could there not be empirically equivalent theories contradicting each other in their basic premises? Could we not—following radical constructivism or conventionalism—even work with arbitrary theories?

It is not easy to name concrete examples for empirically equivalent theories contradicting each other. Even so, such theories are thinkable. As a matter of principle, for a finite number of experiences an infinite number of empirically adequate theories may be constructed. Thus, from the success of a theory, we cannot derive its truth. Nor may we infer the truth of realism from its success. Is there a better argument for realism?

9. THE FAILURE OF THEORIES AS A BETTER ARGUMENT FOR REALISM

Failure is the opposite of success. A theory fails if something runs counter to what the theory makes us expect. This applies on the theoretical level—for instance, with predictions—as well as on the practical level—let's say with bridges or tools. What we mean when ascribing success or failure to a theory evidently does not depend on our answer to the question of realism. This independence applies not only to the meaning of the concepts "success" and "failure" but also to the assessment of whether a prediction is confirmed or whether a tool works. There is no danger, then, that the realist will see successes where the antirealist sees none. Nor will the realist want to explain something where the antirealist does not see any problem or any need for an explanation.

Historically, there are more wrecked theories than successful ones. We are not aware of this because we care so little about wrecked theories. And we don't care because, in normal education, there is no time to teach, analyze, and criticize refuted theories.

But what makes so many theories fail? The antirealist has no answer to this question. He may describe the failure: he may say that the set of acknowledged observational statements has turned out to be inconsistent or that his tool did not meet his expectations. But these rewordings do not explain anything. They just say in what sense the theory has failed; they elucidate the failure accepted before. But they don't answer the actual question, they don't explain the failure.

For the realist, the answer is easy enough: A theory fails because it is wrong: the world is not as the theory submits. But to be different the world must not only exist but also have a specific structure that we may hit or miss.

Thus realism explains not only the success but also the failure of theories. Even so, there is an asymmetry: for success there are more explanations, even nonrealistic ones. But not for failure. The failure of theories is therefore a much better argument for realism, presumably the best one.

10. EXTINCTION AS AN ARGUMENT FOR THE EFFECTIVENESS OF NATURAL SELECTION

Natural selection is differential reproduction due to varying fitness. According to evolutionary epistemology, cognitive abilities raise fitness; therefore, selection works for better cognition, at least in cases where such improvements are useful, available, and not too expensive. As far as our cognitive ability is (taken to be) reliable — that is, in the mesocosm — we may explain this reliability by the effect of selection.

The fact that humans survived evolution under competition makes plausible the reverse conjecture, namely that our cognition cannot be too bad. This inverted argument is not altogether compelling; above all, it is not sufficient to exhibit our cognitive faculty as unfailing or to specify some bit of knowledge as certain. By this argumentative step we may justify our (limited) trust in our cognitive apparatus.

Obviously evolutionary epistemology makes constitutive use of the theory of evolution, primarily the principle of natural selection. Without natural selection, both argumentative possibilities mentioned above would escape us. For evolutionary epistemology it is therefore relevant whether this important factor of evolution is effective.

What testifies to the effectiveness of natural selection? Usually the multiplicity of species counts as the best argument. Wasn't it the different finches on the Galapagos Islands that aroused in Darwin the idea of natural selection? And if we are told that there exist on earth at least five million, possibly even twenty million, different kinds of organisms[20] (not to count bacteria or viruses), all occupying their own ecological niches, then we are even more easily convinced of the effectiveness of natural selection.

But again there is an objection: Could there not be several, even many, ways to adapt to the same environmental conditions? Could not totally different species occupy the same ecological niche? Is it then not natural selection but mere chance and respective histories that determine which species are formed and populate the earth?

There are in fact arguments supporting this interpretation. We have concrete cases where similar ecological niches are occupied by completely different species: the niche of the great pasture animals is occupied in the savannas of Africa by hoofed animals, in Australia by kangaroos. According to the neutral theory of evolution, developed since 1968 by Motoo Kimura, many genetic changes follow pure chance processes. From this slow and uniform "genetic clock" we can even determine the age of a species — that is, the time elapsed since it branched off its next relatives. Is organic evolution a mere chance process with natural selection playing a minor role or none at all?

Again, there is a better argument for the effectiveness of natural selection: the extinction of species. For that, we must recall how many species already have died out. Evolutionary biologists take the number of extinct species to be at least one hundred times that of the existing ones. Ernst Mayr even guesses that 99.9 percent of all evolutionary lines are extinct, so that the recent ones are surpassed in number a thousand times. Why did so many species die out?

As with individuals it happens occasionally that species go extinct more or less accidentally, by a flood or by the impact of a meteorite. As with individuals, we might talk here of situational death. It would be absurd, however, to classify all extinctions under situational death. In contrast to individual aging and dying, there is, as far as we know, no preprogrammed species extinction. Thus we must look for external causes in most cases. Hence we might also ask: What makes organisms, populations, and species fail?

For selectionists the answer is simple: populations and higher taxonomic units die either because they can no longer adapt to environmental conditions, primarily when these change relatively fast, or because they are displaced by fitter organisms, possibly by superior members of the same species. Both cases instantiate mechanisms of natural selection.

And how do antiselectionists, such as neutralists, explain species extinction? Not at all. The reason is not that they have problems with the concept of extinction. Even antiselectionists can state that species become extinct and feel a need to explain it. However, they cannot offer a plausible explanation. The theory of natural selection has more explanatory power than any antiselectionist theory, such as the neutral theory.

And now we may repeat the last section of the previous chapter nearly word for word: selection theory explains not only the success but also the failure of species. Again, there is a pronounced asymmetry: there are more explanations for success than for failure. The failure of species is therefore a much better argument for the theory of selection, presumably the best one.

11. BOTH ARGUMENTS SUPPORT EACH OTHER

It should now be conspicuous why we switched so abruptly from the epistemological problem of realism to a problem of evolutionary biology. By that move we uncovered a far-reaching analogy that we could follow right into its verbal formulations. It is tempting to use this analogy as a further argument. Then both conceptions, the realist one and the selectionist one, could support each other. The two arguments do not, however, depend on each other.

A further good analogy with mutual support is furnished by the phenomenon of convergence. In the development of science we find a phenomenon we could call convergence of research. There are several kinds of convergence: convergence of measurements, measuring methods, and theories. How do they come about? Again, the antirealist has no answer, whereas the realist has a ready reply: research converges because there are real structures that we may uncover and do indeed uncover slowly. This is what the realist rates as scientific progress. Here again the superior explanatory power of realism is remarkable.

Now, convergence is also observed in evolutionary biology. There it refers to similar traits that originated independently, such as the streamlined design of the ichthyosaur, shark, tunny, and dolphin. Here the external conditions, especially the need to advance fast in water, have promoted this trait. The effectiveness of natural selection is especially conspicuous here. Again the analogy between the two lines of thought is striking. That the word *convergence* is used in both cases is not essential, but it makes the analogy all the more suggestive.

It is tempting to apply the concept of convergence not only to body traits but to cognitive achievements. Thus we might say that our different sense organs supply us with a convergent view of the world—for example, when an apple is seen, felt, and tasted. The signals from the sense organs are different but are combined to build an undivided object in perception. Similar considerations apply to higher cognitive achievements. This kind of convergence may be interpreted in favor of realism. For only if there are unified outside objects does it pay to reconstruct such objects in our imagination.

Thus it is obvious how much evolutionary epistemology is connected with other conceptions without being displaced by them: with realism, with naturalism, with evolutionary theory, with the development of science, with an evolutionary philosophy of science. This concatenation cannot be further delineated here.

NOTES

1. A. Koestler, *The Sleepwalkers* (London: Hutchinson, 1959), pt. 3; H. Kragh, "Max Planck: The Reluctant Revolutionary," *Physics World* 13 (Dec. 2000): 31 ff.

2. C. F. von Weizsäcker and B. L. van der Waerden, *Werner Heisenberg* (Munich: Hanser, 1977), 8.

3. I owe this note to M. T. Ghiselin, "Darwin and Evolutionary Psychology," *Science* 179 (1973): 965.

4. G. G. Simpson, "Biology and the Nature of Science," *Science* 139 (1963): 84.

5. E. Brunswik, "'Ratiomorphic' Models of Perception and Thinking," *Acta Psychologica* 11 (1955): 108.

6. See the recent book by Roger Penrose, *The Large, the Small and the Human Mind* (Cambridge: Cambridge University Press, 1997).

7. L. Wittgenstein, *Tractatus Logico-Philosophicus,* trans. C. K. Ogden (London: Routledge and Kegan Paul, 1922), proposition 4.1122.

8. G. Vollmer, "What Evolutionary Epistemology Is Not," in *Evolutionary Epistemology: A Multiparadigm Program,* ed. W. Callebaut and R. Pinxten (Dordrecht: Reidel, 1987), 203–21.

9. G. Vollmer, "What Is Naturalism?" *Logos,* n.s., 1 (Jan. 1994): 200–19.

10. W. V. O. Quine, "Epistemology Naturalized," in *Ontological Relativity and Other Essays* (New York: Columbia University Press, 1969), 90, and "Natural Kinds," in Quine, *Ontological Relativity,* 126.

11. This is even a book title: M. Ruse, *Evolutionary Naturalism* (London: Routledge, 1995). To be fair we should mention that Roy Wood Sellars has already sketched such a naturalism in his *Evolutionary Naturalism* (Chicago: Open Court, 1922).

12. For a more extensive discussion of objections to evolutionary epistemology, see G. Vollmer, *Was können wir wissen?* vol. 1, *Die Natur der Erkenntnis* (Stuttgart: Hirzel, 1985), esp. 217–327.

13. For the concepts of multiple functions and of change of function, see G. Vollmer, "Die Unvollständigkeit der Evolutionstheorie," in Vollmer, *Was können wir wissen?* vol. 2, *Die Erkenntnis der Natur* (Stuttgart: Hirzel, 1986), 1–38.

14. D. Bickerton, "Creole Languages," *Scientific American* 249 (July 1983): 108–15, and "The Language Bioprogram Hypothesis," *Behavioral and Brain Sciences* 7 (Feb. 1984): 173–88.

15. N. Chomsky, *Cartesian Linguistics: A Chapter in the History of Rationalist Thought* (New York: Harper & Row, 1966); S. Pinker, *The Language Instinct* (New York: Morrow, 1994).

16. This is a good example of the cognitive value of failures: the fact that we may learn especially much about a system if it does *not* work.

17. For the gestures of deaf-mutes, see S. Goldwin-Meadow and C. Mylander, "Spontaneous Sign Systems Created by Deaf Children in Two Cultures," *Nature* 391, no. 6664 (1998): 279–80.

18. S. Pinker, *How the Mind Works* (New York: Norton, 1997).

19. H. Putnam, "What Is 'Realism'?" *Proceedings of the Aristotelian Society* 76 (1976): 177.

20. R. W. Kaplan, "On the Numbers of Extant, Extinct, and Possible Species of Organisms," *Biologisches Zentralblatt* 104 (1985): 647–53.

Darwinism as a Theory for Finite Beings

Marcel Weber

1. INTRODUCTION

In the *Origin of Species,* Darwin (1859, 131) wrote: "I have hitherto spoken as if the variations—so common and multiform in organic beings under domestication, and in a lesser degree in those in a state of nature—had been due to chance. This, of course, is a wholly incorrect expression, but it serves to acknowledge plainly our ignorance of the causes of each particular variation." In these lines, Darwin can be interpreted as expressing a subjective view of chance—most likely a consequence of a metaphysical doctrine that was held by many nineteenth-century scientists. This doctrine, of course, is determinism and was most succinctly formulated by Pierre Laplace. It states that our failure to predict or explain a natural event cannot be attributed to probabilistic causes or objective chance. Instead, the reasons for such failures must be sought in our ignorance of the real causes determining the event. In other words, for the determinist, chance is merely an appearance or an illusion that arises because of the finite nature of human reason.

Was Darwin right about this? Or should we revise our thinking about chance in evolution in light of the more advanced, quantitative models of neo-Darwinian theory, which make substantial use of statistical reasoning and the concept of probability? Is determinism still a viable metaphysical doctrine about biological reality after the quantum revolution in physics, or do we have to abandon it in favor of an objective indeterminism? In light of such reflections, what is the relevant interpretation of probability in evolutionary theory? Do biologists use the concept of probability because they are finite cognitive agents or because the evolutionary process is fundamentally probabilistic? In this chapter, I will show that we do not yet fully understand the nature of chance in evolution.

In the next section, I will review the different evolutionary contexts in which chance and probability appear. In section 3, I take a brief look at chance and probability in physics, since I believe that there are lessons to be learnt from the philosophy of physics. In sections 4 and 5 I critically discuss an instrumentalist and a scientific realist account of the statistical nature of evolutionary theory. I show that they are both unsatisfactory. In section 6, I show that there are viable alternatives to these accounts. I sketch one such alternative, which is compatible with both determinism and realism and which tries to do justice to the scientists' own thinking on this problem. In section 7, I turn to the question of whether evolutionary processes are deterministic. An examination of the arguments presented by indeterminists leads me to agnosticism on this question. Finally, in section 8, I show that there are limits to our understanding of the metaphysical foundations of Darwinism.

2. CHANCE AND PROBABILITY IN EVOLUTION

In the passage from the *Origin* that I have quoted above, Darwin was concerned with the causes of genetic variation. Modern geneticists, of course, know much more about the causes of genetic variation than Darwin did. Molecular geneticists have discovered a variety of mechanisms that can cause changes in an organism's genetic material, also known as mutations. It is a central tenet of neo-Darwinism that mutations are random in the sense that their probability of occurring is causally independent of their effect on an organism's fitness. In other words, mutations that are beneficial or detrimental for an organism's survival and reproduction are not made more likely by virtue of their being beneficial or detrimental. The mechanisms that cause mutations are "blind" to adaptive value. I believe that this is the correct sense in which mutations can be said to be random. However, this account of chance variation leaves an important question unanswered: What exactly do we mean by *probability* or *likelihood* when we say that mutations are not made more probable or more likely by virtue of their adaptive value? Does this probability express an irreducibly stochastic disposition of the kind postulated in quantum mechanics? Since the neo-Darwinian definition of random mutation is also compatible with strict determinism, questions like these are far from trivial.

I am going to argue that we cannot give a definitive answer to this question, even though molecular geneticists know a great deal about the mechanisms causing mutations, because we do not know, at present, whether evolutionary processes are deterministic. Thus we do not know exactly what the nature of chance is in the mechanisms generating genetic variation. For most of this essay, I want to leave the issue of random mutation aside. For the concepts of chance and probability arise in different contexts in neo-Darwinian theory, and it is these other contexts with which I will be mainly concerned here. The main issues I will address are the following. First, what is the relevant meaning of *chance* and *probability* in different evolutionary models? Second, why do evolutionary biologists use probabilistic concepts? Is it due to the nature of the evolutionary process or the finite nature of human reason?

Probabilistic concepts occur in different contexts in contemporary models of evolutionary change.[1] First, they appear in the theory of natural selection. Unless a population of organisms is infinitely large—which is physically impossible—the outcome of natural selection is not fully determined by the fitness values of all the individuals in the population. In finite populations, there is always a nonzero probability that the fittest will not survive (Beatty 1984). Furthermore, if we focus our attention on individual organisms, we find that the individual's actual reproductive success is not fully determined by its fitness value. This fitness value provides only a statistical expectation: for example, a probability distribution that the organism has 0, 1, 2, 3, . . . or *n* surviving offspring (Beatty and Finsen 1989; Mills and Beatty 1979). Second, probabilities occur in the theory of random genetic drift. Genetic drift is defined as an evolutionary process in which the change of gene frequencies is independent of any gene's contribution to fitness. Drift has also been described as a random sampling process, where *random* means that population sampling is indiscriminate with respect to phenotypic properties (Beatty 1984, 1987). Third, probabilities are used in certain models of macroevolutionary change. One can model macroevolution as a stochastic branching process in order to explain certain long-term phylogenetic patterns (Sober 1988). Fourth and finally, probabilities arise in evolutionary models that consider the effect of new mutations on a population.

As this brief overview demonstrates, modern evolutionary theory is profoundly statistical. Statistical concepts are used by evolutionary biologists not just for data analysis but as an integral part of most of the theoretical models they advance to explain evolutionary change. As in other scientific disciplines that deal with statistical theories, the use of probabilistic concepts raises a number of difficult foundational, epistemological, and metaphysical issues. To make a general observation that should be uncontroversial, the reasons why scientists resort to statistical theories may differ considerably depending on the theories' subject domain. This is evident if we take a brief look at modern physics.

3. CHANCE AND PROBABILITY IN PHYSICS

Probably the most important physical theories that use statistical reasoning are the theories of statistical mechanics and quantum mechanics. During the 1960s, it became clear that these theories differ fundamentally with respect to the foundational issues surrounding chance and probability.[2] Classical statistical mechanics assumes that the systems it models are governed by fully deterministic laws of motion, namely Newton's laws. Theories of statistical mechanics are statistical because their subject domain consists of systems composed of an extremely large number of particles: for example, a macroscopic container filled with gas molecules that interact with each other as well as with the walls of the container. It would be impossible to solve the equations of motion for each of these particles in order to explain the bulk properties of this physical system. Consequently, physicists abstract from the trajectories of each individual gas molecule and investigate probability distributions: for example, the probability that a system composed of N particles will be found in a certain region of an abstract state space of $6N$ dimensions. According to the standard interpretation, such probability assignments generalize over hypothetical, infinitely large sets of systems — so-called ensembles — that differ only in initial conditions and are otherwise identical. Thus, in statistical mechanics, we find objective probability statements even though the systems investigated are fully deterministic. And even though probability is objective, in statistical mechanics the reason for scientists' use of probabilistic thinking is the finite nature of human reason.

The case is very different in quantum mechanics. There, probabilistic reasoning is used because states of quantum systems exist that do not fully determine the outcome of measurements. Even though unobserved quantum systems are governed by deterministic laws — namely the time-dependent Schrödinger equation — as soon as a measurement is performed, the dynamics of the system changes radically and it starts to behave unpredictably. However, this unpredictability is fundamentally different from the unpredictability found in classical statistical mechanics, for, as a matter of principle, it cannot be overcome by any amount of additional information and computing power. Quantum mechanics is statistical because of the indeterministic nature of the measurement process, not because of the finite nature of human reason. Accordingly, in quantum mechanics, probability statements express irreducibly stochastic propensities of quantum systems.

As these two examples demonstrate, concepts like chance and probability can differ even within a single scientific discipline. If we turn to evolutionary theory—as I will do in a moment—there is thus little reason to expect that the concepts of chance and probability are the same as in physics, or even that they are the same in different biological theories.

4. AN INSTRUMENTALIST VIEW

In the recent literature in philosophy of biology we find two different accounts of the statistical nature of evolutionary theory. The first account has been developed independently by Alex Rosenberg (1994) and Barbara Horan (1994). They hold that all the processes that are relevant for evolutionary change are deterministic. Even though genuinely indeterministic events exist at the microphysical level, this indeterminism vanishes asymptotically as we move from the micro to the macro level. In other words, indeterminism may play a role at the level of chemical bonds or below, but it plays no role in the macroscopic world of biological organisms. This raises the question of why evolutionary theory uses statistical reasoning and probabilistic concepts.

Rosenberg (1994, 71–73) develops his answer to this question in the context of the theory of genetic drift. For this purpose, he invents the following fictional example. Consider a population of giraffes that, for unknown reasons, shifts away from its adaptive peak in neck length. A team of conservation biologists attributes this change to random genetic drift: in other words, to the chance survival of genotypes that have shorter necks. Rosenberg then assumes that the real reason for the change in neck lengths is illegal poaching. In other words, long-necked giraffes are killed by poachers on a regular basis. The conservation biologists are unaware of this poaching activity, so they attribute the evolutionary change observed to random genetic drift. Rosenberg concludes that the theory of genetic drift, in this fictional example, fails to give us the true explanation of evolutionary change. It is merely a way of expressing the biologists' ignorance concerning the real causes of evolutionary change. Rosenberg then generalizes from this example and claims that the same is true for all cases where biologists invoke genetic drift to explain evolutionary change. In all the alleged cases of drift known to biologists, there is a hidden cause or a set of hidden causes that are responsible for the observed changes in gene frequencies. In his fictional example, the hidden cause is given by the poachers. In other cases of drift, the hidden causes may be something else. Rosenberg's claim is that there are always hidden causes when biologists see genetic drift. Thus he concludes that the theory of drift is merely a "useful fiction."

Rosenberg's argument does not stop here, for he eventually arrives at far-reaching conclusions. He concludes that the probabilities that feature in evolutionary theory are subjective (61). According to this view, the probability, say, that a certain allele of a gene is fixed in a small population by chance merely expresses a human agent's degree of belief in the statement that the allele is fixed. As a consequence, the theory of random genetic drift says nothing about the real causes of evolutionary change. It only says something about what humans can rationally expect to happen in a small population of organisms in the absence of any additional causal information. Thus Rosenberg is an instrumentalist about the theory of genetic drift.

Roberta Millstein (1996) has shown that Rosenberg's example that is supposed to demonstrate the instrumentalist nature of drift theory is fundamentally flawed. The way he has set up his example, it really shows selection rather than drift. What is actually going on is that the poachers select for short-necked giraffes. The example is ill chosen for making any claims about genetic drift because it simply does not qualify as a case of random drift. Furthermore, it is unlikely that biologists would invoke genetic drift in a case where there seemed to be a systematic bias in the survival rate toward short-necked giraffes. They would instead look for a cause for this bias in survival rate. Thus Rosenberg's fictional example fails to support his subjective account of probability and the resulting instrumentalist view of genetic drift theory. Of course, this does not yet mean that such an account is incorrect. It could still be that Rosenberg is right and that the theory of genetic drift fails to give us true explanations of certain cases of evolutionary change. In the next section, I will offer additional criticism of this view. But first I would like to critically review a similar account of the statistical character of evolutionary theory, the one given by Barbara Horan.

Like Rosenberg, Horan (1994) is a determinist with respect to evolutionary processes. Like him, she has also argued that the probabilities that appear in evolutionary models are subjective or "epistemic" and that consequently only an instrumentalist account of evolutionary theory is justified. However, her argument is sufficiently different from Rosenberg's to merit special consideration.

Horan wants to resolve the following potential problem for instrumentalist accounts of evolutionary theory. If evolutionary models are merely instruments telling rational agents what evolutionary outcomes they should expect, how can evolutionary theory provide causal explanations of evolutionary change? She attempts to resolve this problem by rejecting the premise that evolutionary models are causal. Horan argues that the models of population genetics—widely seen as the core of neo-Darwinian theory—fail to give us causal explanations. The gist of her argument is that population genetic equations do not relate causes to their effects; they merely relate effects to one another. Take, for example, models for selection at a single genetic locus. Such models feature a theoretical quantity called the selection coefficient, which specifies the proportion of alleles that survive to the next generation. Horan argues that these coefficients are defined by their effects, namely the rate of transmission of alleles into the next generation.

I think that this argument proves too much, for it could be used to show that Newtonian mechanics fails to give us causal explanations. In some interpretations of classical mechanics, force is defined by acceleration—that is, by its effect—so Horan's argument could be applied *mutatis mutandis* to Newtonian mechanics. Thus the argument, if taken at face value, would undermine not only the causal nature of evolutionary theory but the causal nature of all dynamical theories. Surely this is against Horan's intentions, for her argument seems to be directed at evolutionary theory specifically. Of course, she could argue that there is a substantial difference between Newtonian mechanics and population genetic models in that force is not *defined* by acceleration. The relationship between force and acceleration could be contingent or empirical. But this defense invites the question of why the relationship between selection coefficients and gene frequency changes cannot be viewed as empirical. On the standard view of population genetic models, selection coefficients represent the fitness of the genotypes involved. What stops us from saying that fitness differences are a cause of gene frequency changes? If viewed this way, the relationship between selection coefficients and gene frequency changes is not definitional, it is causal. Thus population genetics emerges as a causal theory after all. Of course, Horan could argue that it is inappropriate to view fitness as a causal disposition. But such an argument is missing from her account. Her attack on the causal nature of population genetic theory thus reduces to the old "tautology objection" to the empirical status of selection theory, an objection that has been rejected by several authors (e.g., Mills and Beatty 1979; Hodge 1987).

So far, we have seen that Rosenberg's and Horan's arguments for a subjective view of probability and a resulting instrumentalist view of evolutionary theory are deficient. It must be emphasized again that this does not imply that their views are incorrect; it means only that they have failed to produce positive arguments in their support. It is therefore worth examining whether their position could be strengthened.

It is important to realize that Rosenberg's and Horan's arguments fail even if determinism is true. Let me come back to Rosenberg's attempt to show that the theory of genetic drift is merely a "useful fiction." Is it possible to reject this conclusion without rejecting the deterministic premise on which it rests? I suggest that this is possible, and we don't even have to construct any new arguments to show it. In his influential book *The Nature of Selection,* Elliott Sober (1984, 126) has argued that even if evolutionary processes are assumed to be deterministic, there are reasons to believe that the statistical models of evolutionary theory provide genuine causal explanations. To show this, Sober imagines a Laplacian demon who can calculate the fate of any population of organisms. Sober then asks whether such a Laplacian supercalculator would have any need for statistical models of evolutionary change. The answer is yes, according to Sober, for the following reasons. Statistical evolutionary models, such as the models for random genetic drift, abstract from the specific causal details that determine survival and reproduction in particular populations and arrive at significant generalizations that hold for a large class of very different populations that have nothing else in common. These significant generalizations define natural kinds, which are invisible in the Laplacian demon's convoluted calculations. The knowledge of these higher-level natural kinds — formulated in the language of probability — contributes to our understanding, so the statistical generalizations are genuinely explanatory. Laplace's demon, on his part, simply wouldn't see the forest for the trees.

Rosenberg, of course, is aware of this argument. However, he can reject it by denying that the generalizations of evolutionary theory define natural kinds. Sober's significant generalizations, according to Rosenberg (1994, 76), appear significant only to cognitively limited beings such as humans. Sober's only argument for the reality of these kinds seems to be that if we don't allow such higher-level natural kinds, then other special sciences such as psychology are doomed as well. For example, the psychological concept of intentional states, according to Sober, is analogous to the concept of probability in evolutionary theory in that it provides us with natural kinds that are invisible at the micro level. Rosenberg finds this unconvincing because the existence of natural kinds in the realm of psychology is far from established. And even if it were established, Sober would have only an argument from analogy for the reality of natural kinds of evolutionary processes.

The question is where the burden of proof lies in this argumentative standoff. Does Sober have to show that the statistical generalizations of evolutionary theory define genuine natural kinds, or does Rosenberg have to show that they fail to do so? I think this is largely a matter of one's greater metaphysical predilections. For those of us who do believe that special sciences such as biology or psychology are in the business of discovering natural kinds and that they succeed at least sometimes, Sober's arguments carry some weight. For someone like Rosenberg, for whom the epistemological status of special sciences is more precarious, the argument appears weak.

I find myself on Sober's side in this question: I would not want to confine natural kinds to the micro-physical level. At the end of the day, Rosenberg's and Horan's instrumentalist accounts of evolutionary theory rest on nothing but an extremely strong and problematic reductionist assumption, namely that only theories that treat phenomena at the most fundamental level can be true. If this assumption is rejected, determinism about biological processes does not imply instrumentalism about evolutionary theory (Weber 2001).

5. A REALIST VIEW

Robert Brandon and Scott Carson (1996) reject the view that biological processes are deterministic. In their account of the statistical character of evolutionary theory, the source of probability is not the finite nature of human reason but objective chance events in the development of individual organisms. They have offered two independent arguments in support of this contention. I will review these arguments in section 7. Right now, I want to point out some internal difficulties in their account of the statistical nature of evolutionary theory.

The starting point of Brandon and Carson's analysis is their belief that the processes and relations postulated in evolutionary theory are real. Thus, in contrast to Rosenberg and Horan, Brandon and Carson are realists with respect to evolutionary theory. Their justification for their belief is the fact that evolutionary theory is "one of the most highly confirmed theories in the history of science" (316). Thus they think that a high degree of empirical confirmation of a scientific theory provides reasons to believe that the processes and relations postulated in the theory are real and that contemporary evolutionary theory qualifies as a highly confirmed theory. I am not going to argue this point, since this position is certainly respectable. What I want to take issue with are the conclusions they draw from these starting assumptions.

Brandon and Carson want to reject the view defended by Rosenberg and Horan that the probabilities that appear in evolutionary models are subjective probabilities or degrees of belief. They undermine this account by rejecting the central premise from which Rosenberg and Horan started out, namely that biological processes are deterministic. If evolutionary theory provides us with true descriptions of reality, they argue, it *must* be assumed that the concept of probability represents stochastic dispositions of individual organisms. For example, fitness is such a stochastic disposition. Fitness does not uniquely determine how many offspring a biological individual will have, and this is because indeterminism keeps the organism's physical properties from fixing exactly how many offspring the organism will produce. Thus Brandon and Carson prefer a propensity interpretation of fitness (first developed by Mills and Beatty (1979). Probabilistic propensities are thought of as real properties of individual organisms, so realism about evolutionary theory is saved from the instrumentalists' attack.

My first objection to this line of reasoning is as follows. Brandon and Carson jump to the conclusion that the realist about evolutionary theory must accept a propensity interpretation of probability too quickly. Although they are right that realism requires an objective interpretation of probability, the propensity interpretation is not the only objective interpretation conceivable. There are other objective interpretations—for example, the limiting frequency interpretation of probability. Frequency interpretations are neutral with respect to whether the events for which probability is measured are deterministic. There are also interpretations of probability that are not neutral with respect to this question. Attempts have been made to introduce a concept of probability that applies to deterministic systems (e.g., von Plato 1982). Popper's (1959) original formulation of the propensity interpretation suggests that it is also applicable to deterministic systems, even though, for Popper, the main motivation for introducing the propensity interpretation was quantum mechanics. Let us also remember that statistical mechanics, which treats fully deterministic systems, conceives of probabilities as objective features of ensembles (see section 3). Thus Brandon and Carson are mistaken in thinking that objective probability implies objective chance. Statistical mechanics especially is living proof that this is not so, for—as I have pointed out earlier—it combines objective probability with subjective chance.

Nothing in what I have said so far is an argument against Brandon and Carson's view that evolutionary probabilities do in fact represent genuine stochastic propensities. Now I want to show that this view is problematic too.

To see this, consider the following thought experiment. A colony of genetically identical plants is found, in a given year, to produce different numbers of seeds. A biologist samples the numbers of seeds produced by each individual plant and calculates the probability distribution that a plant of this genotype will produce 0, 1, 2, 3, . . . , or *n* seeds. For most theoretical purposes, the arithmetic mean of this distribution,

$$\frac{1}{n}\sum_{i=1}^{n} i p_i$$

(where p_i is the probability that an individual plant produces *i* seeds), will provide an estimate of the genotype's fitness.[3] Now let the variation in seed number be the result of some objective chance events during the plants' development. Under this assumption, the probabilities p_i could be interpreted as representing an irreducibly stochastic disposition of plants of that genotype to produce *i* seeds. In other words, the probabilities could then be read as expressing genuine probabilistic propensities. It is Brandon and Carson's view that *all* applications of the concept of probability in evolutionary theory are to be interpreted in this manner.

But now let us change our hypothetical example only slightly. Let us assume that the variation in seed production is not the result of objective chance events. Instead, there is a set of hidden variables that determines the number of seeds produced. This could be any set of causal variables that influence the growth of the plant and that we do not know. In this case, we could not read the probabilities p_i as irreducibly stochastic propensities. The stochasticity observed is now a mere illusion created by our ignorance of the hidden variables. However, the probabilities p_i would not be rendered subjective or "merely epistemic" in this scenario—contrary to what Brandon and Carson as well as Rosenberg and Horan would have us believe. It is still possible to interpret the probabilities p_i as expressing a property that plants of this genotype actually possess, a property that is determined by the plant's physical properties and their environment and that is invariant with respect to the different values of the hidden variables.

It is clear that both of my hypothetical scenarios—the one with and the one without the hidden variables—are compatible with current evolutionary theory. In fact, in most theoretical contexts, the scenario chosen will not matter at all for using the probability values in question to predict or explain what will happen to these plants, for example, if they have to compete with plants of a different genotype. However, I believe that my two scenarios demonstrate the inadequacy of Brandon and Carson's account of the statistical nature of evolutionary theory. The reason is that their account can treat only the first scenario—the one without the hidden variables. But surely, the second scenario—the one with the hidden variables—also occurs in nature. I do not know whether objective chance events exist in the biological domain or how frequent they are *if* they exist. I will come back to this

TABLE 14.1. Summary of Different Positions with Respect to the Statistical Character of Evolutionary Theory

	Rosenberg, Horan	Brandon & Carson
Determinism	Yes	No
Eliminability	Yes	No
Source of Probability	Cognitive limitation	Objective chance
Interpretation of Probability	Subjective	Propensity
Realism	No	Yes

question in section 7. All I want to claim right now is that it is a safe bet that a substantial proportion of biological variation that we observe is a result of some hidden variables. If this is true, then Brandon and Carson's account of chance and probability in evolution has to be rejected because it sees objective chance as the *only* source of probability. Their account does not apply to cases where there are evolutionarily relevant hidden variables in biological processes, yet the existence of such variables is highly likely regardless of whether the biological domain is strictly deterministic (Weber 2001).

I therefore conclude that Brandon and Carson's account of the statistical character of evolutionary theory fails to give us an adequate account of evolutionary processes as we know them. Since the same is true about Rosenberg's and Horan's accounts, we urgently need an alternative. This is what I turn to now.

6. ARE THERE ALTERNATIVES?

Table 14.1 summarizes the two accounts of the statistical nature of evolutionary theory that I have examined so far.

This table contains various epistemological and metaphysical claims that do not necessarily imply each other. For example, I have argued—contra Rosenberg and Horan—that determinism about biological processes does not imply instrumentalism about evolutionary theory and—contra Brandon and Carson—that a scientific realist is not committed to indeterminism or a propensity interpretation of probability. This suggests that, by playing around with this table, we can easily generate alternative accounts of the statistical character of evolutionary theory and check them for coherence. In theory, we could try all possible combinations of po-

TABLE 14.2. A New Position with Respect to the Statistical Character of Evolutionary Theory

	Rosenberg, Horan	Brandon & Carson	Weber
Determinism	Yes	No	Mostly
Eliminability	Yes	No	Don't care
Source of Probability	Cognitive limitation	Objective chance	Hidden variables
Interpretation of Probability	Subjective	Propensity	?
Realism	No	Yes	Yes

sitions with regard to the different claims included in the table. Some of these combinations will clearly be incoherent or even self-contradictory, but others may be consistent, coherent, and metaphysically plausible. However, I do not want to bore the reader with such a systematic exercise. Instead, I will present the combination of claims that I find most promising. It is summarized in table 14.2.

Determinism: I think that a viable interpretation of evolutionary models that use probabilities should at least be consistent with determinism. To the extent that biological variability that is relevant for evolution is caused by hidden variables, the evolutionary process *is* deterministic, even if this does not necessarily exclude the occasional intrusion of objective chance events. As we have seen in the previous section, Brandon and Carson's account fails because it cannot adequately explicate the relationship between such hidden variables, which are likely to exist, and evolutionary probabilities. Therefore, an adequate interpretation of probability in evolutionary theory should proceed *as if* evolution were a deterministic process.

Eliminability: This is the question of whether an omniscient being (e.g., Laplace's demon) could produce a theory of evolutionary change that did not require any probabilistic reasoning. All we can say about this question is that, presumably, the statistical character of evolutionary theory is eliminable in principle if the evolutionary process is fully deterministic. By contrast, such elimination is clearly impossible if the evolutionary process is indeterministic. However, such considerations are strictly counterfactual since, to my knowledge, there are no omniscient beings. The crucial point here is that in principle eliminability does not imply instrumentalism. It is not incoherent to believe that evolutionary theory is eliminable in principle (i.e., for an omniscient being) and, at the same time, to accept a realist interpretation of current evolutionary theory. All we have to assume is that special sciences like biology can have some access to reality even if they fail to produce complete accounts of the processes in their domain. The scientific realist is not committed to the thesis that true theories are complete in the sense that they incorporate the maximally possible causal information concerning the phenomena they treat. Thus the question of eliminability can be dissociated from the realism issue and pushed into the realm of science fiction.

Source of probability: The main issue in this debate so far has been whether the source of probability—that is, the reason(s) why evolutionary models use probabilities—must be sought in the nature of the process itself (as indeterminists hold) or in our ignorance of the real causes of evolutionary change (as some determinists maintain). On my account, the source of probability is located in the hidden variables. To be precise, the source of probability lies in the fact that the hidden variables *are* hidden. Basically, this amounts to the same as saying that the source of probability is ignorance or cognitive limitation, but I prefer this way of speaking because it is more precise.

Interpretation of probability: This is clearly the crux of the matter and will therefore be discussed separately (see below).

Realism: The position that I am advocating is a realist one. Obviously, a defense of scientific realism in general or of realism about evolutionary theory specifically is beyond the scope of this essay. All I would like to claim here is that realism about current evolutionary theory is at least as justified as realism about other mature scientific theories. If it is at all justifiable, on metaphysical and epistemological grounds, to be a realist about the entities and causal processes postulated by well-tested scientific theories, then it is reasonable to assume that current models of natural selection and genetic drift represent the causal structure of evolving populations of organisms at least approximately. The issue at stake here is how to maintain this realism in the face of the challenge leveled specifically at evolutionary theory by Rosenberg and Horan, namely that determinism implies that statistical evolutionary models are mere predictive devices and not representations of biological reality. In the previous sections, I have shown that they have not established this conclusion and that Brandon and Carson's attempt to rescue realism with the help of a thoroughgoing indeterminism fails. The question, then, is how realism about statistical evolutionary models can be defended on the basis of a positive account.

The centerpiece of such a positive account, clearly, will have to be an objective interpretation of probability. Is there an interpretation of probability that is compatible with—but not necessarily committed to—determinism and that is objective and therefore able to sustain a realist stance? Such an interpretation does exist for some domains of the physical world—for example, the domain of statistical mechanics. We have seen (section 3) that the probabilities that appear there can be interpreted as objective properties of so-called ensembles—that is, fictional sets of systems that differ only in initial conditions. Clearly, this interpretation cannot be uncritically transferred to biological contexts because biological systems differ considerably from the idealized physical systems treated in statistical mechanics.

But perhaps there are some interesting analogies to statistical mechanics that could be fruitfully explored. Indeed, one of the pioneers of population genetics, R. A. Fisher (1930, 36), was quite fond of such analogies and likened his "fundamental theorem of natural selection" to the second law of thermodynamics, though I think this is mostly of historical interest. What I would like to suggest is that certain models of evolutionary change in finite populations show an interesting analogy to the ensemble approach in statistical mechanics.

Population geneticists, when treating populations of finite size, are unable to predict exactly what will happen to a particular population. The reason is the intrusion of chance factors like random sampling of gametes in small populations. The first quantitative theory of evolutionary change to systematically take such effects into account was developed by Sewall Wright. He wrote in 1931:

> The gene frequencies of one generation may be expected to differ a little from those of the preceding merely by chance. In the course of generations this may bring about important changes, although the farther the drift from equilibrium the greater will be the pressure to return. The resultant of these tendencies is a certain frequency distribution, or probability curve, for gene frequencies in place of a single equilibrium value. (Wright 1986, 93)

Wright derived the following general formula for the probability distribution of gene frequencies at a single genetic locus:

$$y = C e^{4Nsq} q^{4N(mq_m + v) - 1} (1 - q)^{4N[m(1 - q_m) + u] - 1}$$

In this model, y is the probability that an allele will reach a gene frequency of q from an initial frequency of q_m in a population of size N; s represents the selection coefficient of the allele; u and v are the mutation rates from and to the allele, respectively; and m is the rate of migration. This model shows that if N is very large, the population is most likely to end up at the initial frequency if selection, mutation, and migration are negligible. This is explained by the fact that a large number of chance events are likely to cancel each other. By contrast, if N is very small, the population is most likely to settle at either $q = 0$ or $q = 1$: in other words, the allele will either be fixed or go extinct by random genetic drift irrespectively of its selective value (s). Figure 14.1 shows some of the properties of Wright's model graphically.

How should we interpret the probabilities that Sewall Wright calculated for gene frequency change at a single locus in a finite population? According to Rosenberg's subjective interpretation, we should read these probabilities as expressing degrees of belief: for example, belief in the proposition that an allele will be fixed by random drift. Alternatively, if we were to follow Brandon and Carson and several other philosophers of biology, we would interpret these probabilities as irreducibly stochastic dispositions (propensities) that each individual in the population possesses. However, I suggest that there is an interpretation more in line with Sewall Wright's own thinking than either the subjective or the propensity interpretation. Namely, we can interpret these probabilities as measuring the frequency of populations with a given gene frequency in a hypothetical, infinite set of populations that contain the same number of individuals and the same initial gene frequencies. In other words, we can interpret probability as a property of an ensemble. Note that by *ensemble* I mean, not a single population, but an infinitely large set of populations with the same number of individuals but different in other respects.

That Wright conceived of his probability distributions in this manner is supported, for example, by the following passage from the Galton Lecture that Wright gave at University College, London, in 1950 (explaining the results of his theoretical calculations on evolution in natural populations under the combined action of various evolutionary forces including fluctuations due to random sampling): "The resultant is a probability distribution of *frequencies of gene frequencies* which applies to any one strain in the long run, or to an array of strains, subject to the same conditions, at any one time" (Wright 1986, 586, emphasis mine). When Wright says "frequencies of gene frequencies," he means the proportion of populations with a given gene frequency in an ensemble of populations that are identical in all respects except for the unique chance events that cause gene frequencies to fluctuate randomly.

A look at contemporary textbooks in population genetics shows that scientists still think of probability distributions for gene frequencies in this way. For example, the eminent population biologist Joan Roughgarden (1996, 58) writes: "Because any specific population shows a random amount of sampling error, we cannot hope to predict exactly what any given population will do. But we can combine the results of several populations and observe the properties of a group of populations" (emphasis mine). Roughgarden then goes on to calculate probability distributions

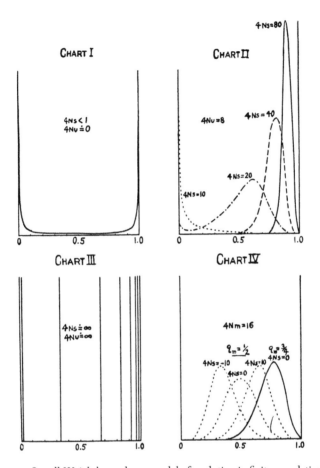

FIGURE 14.1. Sewall Wright's one-locus model of evolution in finite populations (1931). The four graphs show the distribution of probability that a population of size N will reach a gene frequency of q. Gene frequencies are shown on the abscissa. Chart I shows the result for a population in which the product of size and selection coefficient ($4Ns$) is small. Such populations are most likely to either lose ($q = 0$) or fix ($q = 1$) the allele by random drift. Chart III shows the extreme case of an infinite population. In this case, the variance of the distribution approaches zero: that is, the population will reach an equilibrium state determined by the selection coefficients and the mutation rates (u) with a probability of 1. Charts II and IV show intermediate cases with various values for population size, selection coefficients, initial frequencies (q_m), and mutation rates. The probability values (y axis) can be interpreted as frequencies of populations in an ensemble. *Source:* Sewall Wright, "Statistical Theory of Evolution," in *Evolution: Selected Papers*, ed. W. B. Provine (Chicago: University of Chicago Press, 1986), p. 94. Reprinted with permission from *The American Statistician*. Copyright 1931 by the American Statistical Association. All rights reserved.

for gene frequencies under various conditions. The approach taken is still basically Sewall Wright's, except that Roughgarden uses more advanced mathematical theory, namely stochastic theory. In Roughgarden's presentation of the models, the probability that a system is in a certain state at a given time is interpreted as a frequency of systems in that state in an infinite set of systems called the ensemble. She even explicitly draws the analogy to the ensemble approach in statistical mechanics (61).

What I would like to show with this short excursion into population genetic theory is that theoretical biologists, at least when developing certain models of evolutionary change, interpret probability much as physicists do in statistical mechanics. In these models, probability assignments *are* intended to have representational content with respect to real biological systems, contrary to what Rosenberg and Horan think. Probabilities represent frequencies in the limit to infinitely large ensembles of populations. But, as in statistical mechanics, these applications of the concept of probability are fully compatible with determinism. The differences between the individual populations in the ensemble that cause random fluctuations could be entirely due to hidden variables rather than objective chance. Thus, in these models, probability does not necessarily represent irreducibly stochastic propensities, as Brandon and Carson and other philosophers of biology believe.

I think I have at least made it plausible that alternatives to both the Rosenberg/ Horan and the Brandon/Carson account of the statistical character of evolutionary theory exist. I have sketched what an alternative that combines determinism with objective probability and realism about evolutionary theory might look like in the context of some models of evolution in finite populations. However, I am not sure whether the approach I have taken could be transferred to other applications of the concept of probability in evolutionary theory—for instance, macroevolutionary models or the origin of mutations. We must be open to the possibility that no single account of chance and probability can cover all of modern evolutionary theory.[4]

In the next section, I address a question that I have neglected so far: Are biological processes deterministic or not?

7. THE DETERMINISM QUESTION

Brandon and Carson (1996) have produced two independent arguments that are intended to support indeterminism about biological processes. The first argument purports to establish that quantum indeterminism can have population-level effects. Brandon and Carson imagine a population that is at an unstable equilibrium point, such that one newly arising mutation can make the difference as to which of two alleles will be fixed by natural selection. They then argue that, since mutation is a microphysical process that could, in theory, be caused by quantum events, it is at least conceivable for the fate of an entire population to be subject to quantum indeterminism. They put this by saying that quantum events could "percolate up" all the way to the level of biological populations. Another locution they use is that quantum indeterminism could "infect" the population level.

I do not think that, at present, this kind of scenario can be ruled out altogether. Brandon and Carson do admit that their argument does not amount to a strict "no hidden variables-proof" of the kind that has been produced in quantum mechanics. However, I would like to express some caution about such a recourse to the indeterminism of quantum mechanics.

Brandon and Carson's talk about quantum effects "percolating up" or "infecting" the population level is misleading. It does not fit with the orthodox interpretations of quantum mechanics. One of the many remarkable features about quantum mechanics is that, so long as a system is not observed, it evolves according to deterministic laws. Thus there is, in this case, no indeterminism that could "percolate up" or "infect" anything. Quantum indeterminism does not arise simply because systems are very small. It arises only when a microphysical system is subjected to measurement. In other words, it arises when some measurement device is coupled to the system that collapses its wave function. In such situations, the outcome of the measurement can be uncertain in the strong sense of objective chance. To my knowledge, this is the only form of objective chance recognized by modern physics. Note that this indeterminism is a feature that emerges at the macroscopic level under certain conditions — namely measurement conditions — and not something that "percolates up" from the micro level unpredictably. It is a property of a whole quantum system coupled to a measurement apparatus.

The problem, then, is that Brandon and Carson seem to take it for granted that quantum mechanics has shown that we live in a fundamentally indeterministic universe and that the question in biology can only be whether its domain partakes in this indeterminism. However, no such conclusion is warranted. What quantum mechanics shows is that the universe is fundamentally deterministic — unless measurements are performed. But does evolution perform measurements? I do not believe that anyone can answer this question at present. This ought to be a question for scientific research rather than metaphysical speculation. At present, no strong claims should be made about the relevance or irrelevance of quantum mechanics to biological processes.

How about Brandon and Carson's second argument for indeterminism? This argument does not make any use of quantum mechanics. In fact, it purports to demonstrate an autonomous form of indeterminism for biology.

Brandon and Carson's autonomous argument is an argument from empirical evidence. In other words, these authors think that experiments in evolutionary biology support indeterminism in the same way that experiments can be said to support a theory. The evidence they have in mind comes from, for example, experiments with cloned plants, which show a considerable amount of variation even if environmental conditions are kept highly homogenous.

Determinists could make an obvious reply to this empirical argument for indeterminism. They could defend their position by arguing that if cloned plants show variation, then the environment was not homogenous after all. Or perhaps there are some internal hidden variables responsible for this variation, something like "developmental noise." Brandon and Carson are aware of this possible refutation. However, they think that they can defuse it with a methodological argument. They ask what theoretical purpose the postulate of deterministic hidden variables serves. In their view, the only purpose that this postulate serves is to rescue determinism from empirical refutation. These variables, therefore, are not theoretically fruitful; they must be removed on methodological grounds. Brandon and Carson thus apply a methodological rule that reads, "Never postulate hidden variables unless they are theoretically fruitful."

Roberta Millstein (2000) has criticized this argument against determinism. According to her, the indeterminist who attributes any variation that is yet unexplained to objective chance runs the risk of overlooking some important causal variables. Far from being theoretically unfruitful, then, the postulate of hidden variables may lead to the discovery of yet unknown biological processes.[5] Therefore, we are once again faced with the question of where the burden of proof lies: Does the determinist have to show that hidden variables do more theoretical work than merely saving determinism, or does the indeterminist have to show that there are no hidden variables? Determinists and indeterminists are equally at risk of committing an error: the former risk the error of dogmatically sticking to a metaphysical doctrine with the help of the ad hoc postulate of hidden variables, while the latter risk missing some important causal information. How should these cognitive risks be balanced?

If this question cannot be rationally answered, Brandon and Carson have not established indeterminism. Millstein concludes that the only rational attitude for a scientific realist to take is agnosticism. In other words, there are, at present, no strong reasons for preferring either determinism or indeterminism about evolutionary processes.

In the final section, I will show that this conclusion limits our ability to understand the metaphysical foundations of Darwinism.

8. AGNOSTICISM AND THE LIMITS OF METAPHYSICS

If we cannot decide, at present, whether evolutionary processes are deterministic or indeterministic, any account of the statistical nature of evolutionary theory is bound to be incomplete. We are here reaching the limits of metaphysical inquiry. Only empirical science can tell us whether quantum mechanics is relevant for evolutionary theory or whether there is an autonomous biological indeterminism, as Brandon and Carson have suggested. But without this knowledge we cannot give a universally valid interpretation of chance and probability: that is, an interpretation that will cover all the instances of probabilistic models of evolutionary change. Clearly, these models work just as well whether the processes they represent are deterministic or indeterministic. For all explanatory purposes of evolutionary theory, and for all current methods of testing evolutionary models in the field, determinism and indeterminism are empirically equivalent.

In conclusion, I suggest that there are at least two senses in which Darwinism (and neo-Darwinism) is indeed a theory for finite beings. If we accept that evolutionary processes are effectively deterministic (in the sense explained in section 6), then we must admit that the statistical nature of evolutionary models is a manifestation of human cognitive limitation. However, I have shown that this conclusion does not commit us to instrumentalism, provided that we can come up with an objective interpretation of probability that is compatible with determinism and, at the same time, applicable to evolutionary theory or parts thereof. I have sketched such an interpretation for a certain class of models of evolution in finite populations by using the ensemble concept. But, on a more skeptical note, a critical examination of Brandon and Carson's argumentation for indeterminism shows that there are no strong reasons for preferring either a determinist or an indeterminist position with regard to the nature of chance in evolution. This means that evolutionary theory reveals the finite nature of human reason by using concepts like chance and probability in a fruitful way, yet without our fully understanding what exactly they mean.

NOTES

Special thanks to Paul Hoyningen-Huene for many helpful suggestions.

1. An excellent analysis of the different roles of chance in evolution has been given by Roberta Millstein (1997).

2. For the following, I am relying on Sklar (1993).

3. Beatty and Finsen (1989) point out that, in some instances, other parameters of the distribution, such as variance or skew, will be relevant for the genotype's fitness. However, this is of no relevance for my thought experiment.

4. I owe this insight to Kenneth Waters.

5. My favorite example is the groundbreaking research of Barbara McClintock. She would never have discovered mobile genetic elements had she attributed the still unexplained variations in her maize plants to an ineliminable indeterminism or objective chance.

REFERENCES

Beatty, J. 1984. Chance and natural selection. *Philosophy of Science* 51:183–211.

———. 1987. Natural selection and the null hypothesis. In *The latest on the best: Essays on evolution and optimality,* ed. J. Dupré, 53–75. Cambridge, Mass.: MIT Press.

Beatty, J., and S. Finsen. 1989. Rethinking the propensity interpretation: A peek inside Pandora's box. In *What the philosophy of biology is: Essays dedicated to David Hull,* ed. M. Ruse, 17–30. Dordrecht: Kluwer.

Brandon, R. N., and S. Carson. 1996. The indeterministic character of evolutionary theory: No "No hidden variables proof" but no room for determinism either. *Philosophy of Science* 63:315–37.

Darwin, C. 1859. *On the origin of species by means of natural selection or the preservation of favored races in the struggle for life.* London: Murray.

Fisher, R. A. 1930. *The genetical theory of natural selection.* Oxford: Clarendon.

Hodge, M. J. S. 1987. Natural selection as a causal, empirical and probabilistic theory. In *The probabilistic revolution,* vol. 2, *Ideas in the sciences,* ed. L. Krüger, G. Gigerenzer, and M. S. Morgan, 233–70. Cambridge, Mass.: MIT Press.

Horan, B. L. 1994. The statistical character of evolutionary theory. *Philosophy of Science* 61:76–95.

Mills, S., and J. Beatty. 1979. The propensity interpretation of fitness. *Philosophy of Science* 46: 263–86.

Millstein, R. L. 1996. Random drift and the omniscient viewpoint. *Philosophy of Science* 63, suppl.: S10–S18.

————. 1997. *The chances of evolution: An analysis of the roles of chance in microevolution and macroevolution.* Ph.D. diss., University of Minnesota, Minneapolis.

————. 2000. Is the evolutionary process deterministic or indeterministic? An argument for agnosticism. Paper presented at the 17th biennial meeting of the Philosophy of Science Association, Vancouver, B.C.

Popper, K. 1959. The propensity interpretation of probability. *British Journal for the Philosophy of Science* 10:25–42.

Rosenberg, A. 1994. *Instrumental biology or the disunity of science.* Chicago: University of Chicago Press.

Roughgarden, J. 1996. *Theory of population genetics and evolutionary ecology.* Upper Saddle River, N.J.: Prentice Hall.

Sklar, L. 1993. *Physics and chance: Philosophical issues in the foundations of statistical mechanics.* Cambridge: Cambridge University Press.

Sober, E. 1984. *The nature of selection: Evolutionary theory in philosophical focus.* Cambridge, Mass.: MIT Press.

————. 1988. *Reconstructing the past.* Cambridge, Mass.: MIT Press.

von Plato, J. 1982. Probability and determinism. *Philosophy of Science* 49:51–66.

Weber, M. 2001. Determinism, realism and probability in evolutionary theory. *Philosophy of Science,* suppl. 68: S213–S224.

Wright, S. 1986. *Evolution: Selected papers,* ed. W. B. Provine. Chicago: University of Chicago Press.

PART IV

Darwinism and the Place of the Human

O
ne of the major problems connected with Darwinism concerns the issue
of reductionism. The issue arises on two different levels. On the one
hand, there is the problem of how much of the phenomenon of life is
determined by the genes (and these by molecules); on the other, there is the question of the possible autonomy of human culture from biological determinants. With
regard to the second question, Richard Alexander discusses in his comprehensive
essay the possibility of a biological explanation of uniquely and distinctively human
traits, particularly in the arts. It is very important to recognize that the acknowl-
edgment of behavioral traits peculiar to humans does not entail that there can be
no biological explanation for them—and this means that the search for such a biological explanation is not based on the denial of the uniqueness of humans within
the organic world. What is a biological explanation of such traits? Must they be reduced to the tendency to maximize one's offspring? And even if this is the case, why
should reductionism be wrong—is it not the essence of science to reduce complex
phenomena to more elementary facts? Is not the very human phenomenon of nepotism obviously linked to genetic egoism? And cannot human characteristics such as
concealment of ovulation, biparental care, altriciality, long juvenile life, menopause, complex sociality, and the ability to play find a biological explanation? And
could not phenomena like the arts be explained by endless competitive races, faddish effects, and a runaway social selection analogous to Ronald Fisher's famous
runaway sexual selection?

With regard to the first question, Lenny Moss, aiming at a criticism of what
he calls vulgar Darwinism, distinguishes strictly between two concepts of genes:

genes-P (genes for phenotypes) and genes-D (genes as developmental resources defined by their molecular sequence and always indeterminate with respect to the phenotype to which they contribute). It goes without saying that this discussion is linked to the problem of the relation between genetics and developmental biology, a problem continuing in a modern form the old controversy between preformationism and epigenesis. While Moss's criticism of "Weismannism" is based on purely scientific facts (Weismannian organisms, such as humans, are themselves products of evolution, since all plants and many invertebrates have the capacity to transform tissue involved in somatic functions into germ cells), the underlying ontological question is of great philosophical importance: What or who is the real agent in the organic world? Are organisms the marionettes of genes, or are the genes resources for the organisms? And what do the different answers entail for the interpretation of higher organisms and particularly of humans?

The last essay, by Bernd Graefrath, deals, among other things, with the importance of sociobiology for ethics. Graefrath suggests that the categories of functional change and accidental surplus may explain the evolution of complex organs and functions such as the human brain and human culture. With regard to sociobiology's explanatory achievements, he asks: Does the description of animal behavior not often introduce surreptitiously a normative language appropriate only for humans? Even if sociobiology succeeds in explaining human behavior, may it also justify moral norms? Do we not have to distinguish between reasons and causes? But if we accept an appeal to intuition as justification of our most basic moral judgments, is it not tempting to offer a causal explanation of them? And may not the Darwinian theory of evolution lead to a correction of our norms in dealing with animals and thus, to quote Peter Singer, "expand the circle"?

CHAPTER 15

Evolutionary Selection and
the Nature of Humanity

Richard D. Alexander

Although the essentials of organic evolution have been understood for almost 150 years, evolution is still controversial as a general explanation of life and all its traits. It is particularly difficult for people to see evolutionary selection as giving rise to the more complicated aspects of life such as learned and other modifiable traits of humans. Evolution as a process is simple enough that many doubt it can be related in any explanatory way to human culture and history, regardless of the time available for it to operate. I consider this general problem and suggest possible connections between aspects of evolutionary selection and human performances in some difficult areas such as the arts.

CONNECTING LIFE'S SIMPLICITIES AND ITS COMPLEXITIES

More than two-thirds of a century ago, in *The Causes of Evolution,* J. B. S. Haldane (1932/1966) remarked that he would find it easier to be in sympathy with those who believed that consciousness could not arise from a nonconscious state were it not for the fact that this transformation happens in front of us all the time, in the development (ontogeny) of every normal child. The alternative is to imagine some kind of homunculus that could be carried in the sperm or the egg so that it would arrive fully formed in the newly formed single-celled zygote or would somehow be suddenly created there.

Others assume the problem to which Haldane alluded when they argue that humans are too complex to have evolved gradually from the simplicity of single-celled organisms. Of course, complex human individuals arise from single cells every time a fertilized zygote is successful. If it can happen during ontogeny, we may, like Haldane, be sympathetic to the idea that it can happen during evolution. Doubts paralleling those responsible for Haldane's observation continue to be raised regarding the background of the complexity of human culture and sociality, even — or maybe especially — among intelligent and educated people who may be intoxicated with the splendor of the arts, and of culture in general, and perhaps reluctant to entertain the possibility that reduction is a suitable tool for entering into the analysis of such grand human enterprises.

The process of organic evolution that underlies all life is dominated by the uncomplicated subprocess of selection, or the differential survival and reproduction of any and all heritably different and persistent units. Examples of such units are genes, chromosomes (which occasionally recombine), and asexual microorganisms that fission without genetic recombination. These units all persist for extremely long periods relative to the rates at which they are altered by mutation. Because the effects of mutations are random with respect to their usefulness for the organism, in the absence of selection mutations have extremely low likelihoods of generating complexity. Even that unique concentration of function, the sexual organism, can be viewed as a stable unit since, though by definition it lasts on average but a single generation, it still undergoes a vast number of cell divisions and thus many replications of genes (most of which, of course, are destined not to continue, dying with the individual as part of a disposable soma).

The claim by biologists that evolutionary selection per se can yield only traits that maximize the genetic reproduction of their bearers — and that therefore this must be the underlying basis for all human traits and tendencies — jarred sensibilities on a wide scale because it was unexpected and for many seemed from the start to be unlikely and even insulting. How could the massive and variable patterns of culture across the globe and the overall history of humanity, not to say the multiplicity of motives and meanings in the makeup of each of us as individuals, arise out of such a simple and ignoble process as maximizing the reproduction of our genes, more explicitly the reproduction of a few copies of those genes hidden away in our sex cells?

Objections to the claim take at least four forms. First there are those who seek to deny potentially unpleasant outcomes by denying evolution itself. But the process of organic evolution has become an easily observable fact; it is no longer theory. The theoretical nature of evolution today, the challenge for those who seek to falsify, is whether any traits of life — any at all — can be shown not to be outcomes of the process of evolution (Alexander 1978, 1979a).

Second, until recently there could be no broad evolved tendency (not necessarily conscious) to behave as if all human activities — and indeed all activities of all life — might exist because in some way they have served the survival (through reproduction) of unrecognized entities (genes), too small to be viewed even with the magnifying and analyzing devices we began to improve rapidly two or three centuries ago. The human opportunity to evolve to generate particular views on such matters is at least tens of thousands of years old; therefore, whatever views humans in general have held arose in the absence of an understanding of the reasons for those views existing, or any way of approaching that question. This situation is unlike that giving rise to the broad, correct, and apparently evolved intuition that individual interests tend to conflict, evidently arising out of the fact that sexual recombination has across our entire evolutionary history rendered genotypes unique. The enormous difficulty we (including those of us in science and medicine) have experienced in trying to understand the finiteness of individual lifetimes supports the notion that we have not evolved an intuitive understanding of the primacy of particles too small to observe directly.

Third, we typically recoil intuitively from the seeming banality that reproduction is the ultimate function of human activities. This reason is of special importance because, oddly, we are probably evolved to reject reproduction as an explanation for our own tendencies and capabilities, even though we are somewhat willing to accept it for other organisms (see below).

Fourth, evolutionists have not yet been able to describe in detail how evolutionary selection could take us, step by step, from its effects on different genes to the appearance of cultural complexities such as the arts, without needing to claim, at one point or another, "Now here a small miracle takes place" (but see Alexander 1979a, 1989, 1993). Some biologists won't like my using the word *miracle,* but until we bridge the gaps to our own satisfaction, the problem remains.

For some reason it seems to be more difficult for us to visualize a transition, via organic evolution, from the simplicity of differential genetic reproduction to the grandeur of the human scene in all its personal intellectual profundity and collective social drama, than to visualize a transition from absence of consciousness to presence of consciousness or from zygote to adult organism during ontogeny. At one time or another anyone will sympathize with the attitude that evolution simply cannot explain humans; such gloom has even descended on an occasional world-class evolutionary biologist.

During a recent discussion of the basis for the arts, one participant commented, "Maybe a person can have too much education." He meant that it may be possible to know so much about human culture, in the various ways we have thought about it outside an evolutionary framework, that we reduce our likelihood of accepting a different (evolutionary) approach to the problem of its underlying basis or origins. Have human culture, knowledge, history, motivation, and thought been so elabo- rate and diverse—and have humans been oblivious to their probable basis and history for so long—that to become absorbed in the marvels of their manifestations is to become immune to recognizing or accepting their raison d'etre? Can one in- deed have "too much education" in the sense of being overwhelmed by the details of our collective history and achievements because of interpreting them without the guidance of potential causes so fundamental as to require stretching our knowledge and imagination into the logic of biology? Too much education to accept the re- ductionism necessary for explanation, for rebuilding and comprehending the story sensibly? Too much education in the arts themselves to allow sufficient effort to grasp the methods of science as a means of establishing facts, of reliably and de- monstrably "getting it straight"? These questions are relevant to the extent that learning about culture or the arts occurs in the absence of a basic understanding of organic evolution as the only reasonable background of human activities and gener- ates an attitude opposing the incorporation of any possible answers from the science of biology.

In the year 2000, Jacques Barzun published a remarkable 802-page volume titled *From Dawn to Decadence: 500 Years of Western Civilization*. A magisterial display of comprehension of many fields, including history, art, and literature, it deservedly won the National Book Award for criticism. Reading it through in admiration, one cannot doubt that some mentionable portion of Barzun's intent was to understand and explain the basis for the arts. Yet he falters in all his brief mentions of the only likely ultimate and general underpinning of culture: evolutionary selection. It is as if Barzun had written a book on sex without knowing about or accepting the exis- tence of genes and meiosis—or perhaps anything at all below the level of the or- ganism, even the genome as a whole. But the entire point of sex is to recombine those genes—and indeed to recombine only the particular copies that have a chance to be passed to the next generation, meaning the special ones residing in certain places in the testes and ovaries, and scheduled to pass through the reduction division of meiosis and turn into sperm and eggs. It is fair to ask Barzun—and everyone engaged in parallel enterprises—what might be the corresponding "entire point" of culture, or whether there might be one, and how it might affect anyone's interpretations of the arts. In 1979 I wrote on this issue as follows (Alexander 1979a, 68):

> If human evolution, like that of other organisms, has significantly involved selection effective at genic levels, realized through the reproductive strivings of individuals, neither humans as individuals nor the human species as a whole have had a single course to chart in the development of culture but rather [because of the ubiquity and consistency of genomic differences among individuals as a result of sexual recombination] a very large number of slightly different and potentially conflicting courses. In this case it would indeed be difficult to locate "a function for" or even "the functions of" culture. Instead, culture would be, as Sahlins['s] [1976] view may be slightly modified to mean, the central aspect of the environment into which every person is born and . . . must succeed or fail, developed gradually by the collections of humans that have preceded us in history, and with an inertia refractory to the wishes of individuals, and even of small and large groups. [Even if accumulated via cooperative efforts of members of human groups,] culture would represent the cumulative effects of what Hamilton (1964) called inclusive-fitness-maximizing behavior (i.e., reproductive maximization via all socially available descendant and nondescendant relatives) by all humans who have lived. I regard this as a reasonable theory to explain the existence and nature of culture, and the rates and directions of its change.

Flinn and Alexander (1982) pointed out that the adaptiveness of culture, as cumulative social learning, can be maintained by judicious imitation and anti-imitation of the behavior of others and decision making from direct observations of adaptive outcomes, including evolved proximate mechanisms such as pleasure and pain. Such evidence can also come from discussion with others about either proximate effects or adaptive outcomes. Eventually it can include analyses of long-term effects of actions: for example, painful or strenuous acts that nevertheless lead much later to positive effects can be interpreted accurately (adaptively). We are also capable of working out entirely in our own heads the likely outcomes of observed or even imagined actions through scenario building. (For discussions and examples, see Alexander 1979a, 131–33, and 1990b, 7–8.)

On the surface there seem to be many problems with the argument that humans, and other forms of life, are evolved to maximize inclusive fitness: adoption, asymmetrical treatment and marriage of cousins, the so-called avunculate (or "mother's brother" phenomenon), and the "demographic transition" are examples. A substantial literature shows that most of these seeming problems have been misinterpreted (e.g., Alexander 1979a, 1988, 1990b). Thus across history nonrelatives were rarely adopted except as workers or slaves; cousins identified as such by ethnographers are not necessarily cousins; asymmetries in arranged marriages between actual cousins tend to reflect the interests of powerful parents; there are reasons from cousin marriage patterns and confidence of paternity issues to expect men sometimes to tend a sister's offspring more than a spouse's offspring and to expect uncles to tend their future sons-in-law in order to control their behavior toward the uncle's daughters later; and lowered rates of child production are not evolutionarily antireproductive when there are climbable ladders of affluence that can influence offspring success enough to offset using the resources to increase numbers of offspring less well tended (Alexander 1988).

Incidentally, I am deliberately vague or imprecise in my usage of terms like *culture* and *the arts* because I would rather err on the side of inclusiveness. I am not using either an anthropological view of culture or an impression from the humanities. Rather, I wish to assume the problem of somehow rebuilding the explanation of it all because, even if I cannot clarify all of the underlying forces, I think there are such forces general enough to apply to culture and the arts, however defined.

The problem we face in reconciling ourselves to reductionistic biology—and then mastering the synthetic chore of reexamining culture all the way from meme to magnificence—parallels the problem faced by molecular geneticists now that they have described the human genome. To understand the organism—the most important focus of our attention—molecular geneticists now must begin to comprehend ontogeny, progressive changes in the individual organism as it grows and develops—changes that are evolutionarily justifiable only if they increase the reproduction of the organism's component genes sufficiently to overcompensate the delay and risk entailed by evolving a juvenile life. We onlookers tend not to deny that the molecular geneticists will eventually understand ontogeny; at least we don't deny it in the way that we may tend to deny the prospect of getting from evolutionary selection to the arts. But we may doubt that anyone yet comprehends the complexity of the ontogenetic proposition. As organismal biologists, some of us may also wonder if molecular biologists are yet on the right track, one reason being the misleading aspects of early concentration on rare gene effects correlated with rare disorders. Such rare gene effects may have immediate medical significance, but because they have had little chance to be integrated into the genome they probably have little general significance for understanding either the complexities of ontogeny or the human organism as a composite of adaptive (and incidental and not-so-adaptive) traits.

The genome biologists, in this new and obligate challenge of synthesis, must find out how to get from a paltry thirty-one thousand functional genes (or, controversially, a few thousand more) to an organism of many trillions of complexly divergent and organized cells—a organism that may be able to carry out many times thirty-one thousand different actions (physiologically and behaviorally) during almost any split second of its life. Further, whatever the mechanisms, all those actions have presumably been directed by evolution to serve a single function—a function brought into exquisite coordination by the unmatched and unified complexity of the organism, and commonsensical because of our realization of the organism's existence and nature (Alexander 1990a, 1993). By every logic the organism's singular function has to be an on-average equal reproduction of the genes that compose the genome and are variously and predictably represented in the genomes of its relatives. Molecular biologists have started the task—for example, working out the "proteome"—but the distance from proteins to the arts appears almost as incomprehensible as that from genes to the arts.

Given that this whole proposition appears to be so difficult, or even ridiculous, one might ask: Why even think about it? Why argue that humans, modular or not in their individual makeup, are evolved to do but one thing, to reproduce their genes? The underlying and seemingly irrefutable logic is that the genes of any unitary organism acquiring any traits that did anything else at all could not have competed across a single breath of history with others that did no such frivolous thing. In terms of everyday organismal mechanisms there may be no universal adaptation of the organism, only lots of flexible and not-so-flexible evolved and developed mechanisms or modules: lungs, heart, kidneys, liver—circulatory, alimentary, excretory, neural, and muscular systems—and a brain full of geographically localizable modules with separate although coordinated effects. It is worth remembering, however, that none of the individual adaptive mechanisms of the whole organism can be optimized in its particular subfunction. All are necessarily compromised to serve the unitary function of maximizing the success of transferring those paltry few germ cells into the next generation, in packages that, with parental assistance, will be able to do it all over again when their times come. Nothing else works. So it is worth trying to find out if the reproductive-end-all hypothesis has something to offer the students of meaning, culture, the arts, and all the rest.

Dobzhansky (1961, 111) said (so long ago!) that "[h]eredity is particulate, but development is unitary. Everything in the organism is the result of the interactions of all genes, subject to the environment to which they are exposed. What genes determine is not characters, but rather the ways in which the developing organism responds to the environment it encounters." Every part of this thoughtful statement returns us to the suggestion of singular function, and simultaneously to the realization that ordinary genes (i.e., excluding effects of rare genes coincident with rare disorders) cannot be linked in any simple way to singular subfunctions of the organism. Dobzhansky's statement challenges biologists with the staggering proposition that all genes well integrated into the sets of diverse genomes appearing in their population are somehow both pleiotropic and epistatic.

We are continually reminded of the enormity of the problem Haldane mentioned and Dobzhansky specified, that of ontogeny, of development, of understanding the whole organism and its fabulous phenotypic flowerings. Conrad Waddington (1956) called the break of understanding between zygote and organism the "Great Gap" in biology, and it still is that (Alexander 1990a). Ontogeny is not merely less trivial than distinguishing all the genes: its consequences make up the most astounding temporal and spatial concentration of complexity known in our universe. In terms of the size of the paper map of human genomic alternatives mailed out not so long ago by *Science Magazine,* we might wonder if a comparable map of normal ontogenetic alternatives in our species across any brief time period would, say, cover the surface of the earth.

UNDERSTANDING TENDENCIES TO DENY THE SIGNIFICANCE OF REPRODUCTION

Suppose we return to the question of skepticism. If we are evolved to reproduce and nothing else, why can't we just accept it, even if we did not actually evolve to do such accepting? Why should our imaginations seem incapable of grasping the reproductive significance of every one of our acts, or at least be reluctant about it? Such feeling does not arise out of mere disgruntlement over reductionism—at least reductionism any more worrisome than that essential to all analysis. The most devoted student of meaning would not argue that difficult analytical problems can be solved without being temporarily reduced to manageable components (Alexander 1987).

Reproduction is a selfish act, meaning only that it serves the life interests of the reproducer. Surprisingly to us organism-centered observers, because reproduction refers to genes rather than organisms, reproduction (leading to survival of the genes) rather than mere survival (of the organism) is the ultimately selfish act.

Despite our human powers of observation—of our fellows if not ourselves—would any of us wish to be found guilty of a selfish thought or act—let alone a basically selfish nature? Who wishes to live or cooperate with either an avowedly or a sneakily selfish person? The very concept is intensely negative. Hence, perhaps, the occasional effort, still, to dismiss the messenger with the epithet "vulgar sociobiologist." But why, after all, do humans have those keen powers of social observation that we know can ferret out the most minuscule evidence of selfishness, later to be exposed and brandished? Is it possible that we intuitively reject the reproductive end-all because we just don't want to be the ones bringing up the possibility of pervasive selfishness? That we don't want any soul imagining that we personally might be condoning selfishness in any way, shape, or form? Yet selfishness begins with the observable and entirely acceptable everyday process of simply staying alive and healthy. The incredible expense of medical science and practice, insurance, and all the rest attests to the magnitude of our concern with this problem. In terms of the reproduction of genes, selfishness also includes all the beneficence we extend judiciously (and comfortably and righteously) to our collections of relatives—beneficence that sometimes is virtually demanded by both morality and law for the selfish reason that the rest of us do not wish our tax monies or other resources to be utilized for the job.

As has been explained at length since Trivers (1971), reproductive selfishness also includes every act of beneficence to even nonrelatives if such beneficence has a sufficient likelihood of being reciprocated with interest, whether the return comes directly from the benefited parties or indirectly—for example, via the establishment of a reputation for beneficence that brings hopeful returns from observers not previously involved (Alexander 1975a, 1979a, 1993). It is difficult to imagine that anyone invests in the stock market for altruistic reasons; perhaps no one invests in anything at all (except relatives) without expecting (not necessarily consciously) phenotypic rewards that include some kind of interest on the investment. Perhaps, because reproductive selfishness in others is likely to be contrary to our own reproductive interests, and efforts at influencing others sometimes work, we could actually be evolved to reject reproductive explanations of ourselves (unconsciously), without knowing enough about such a possibility to accept its reasonableness consciously. The reason is that such rejection is one way of convincing others that we are good candidates as partners in social reciprocity—in other words, because such pretensions are a way of influencing the behavior of others in directions that serve our own reproductively selfish interests. All of these considerations must be judged with the knowledge that humans have, for some reason, evolved to live in groups, within which they both cooperate and compete, and outside of which they presumably failed consistently. They must also be judged by the realization that some acts of seemingly costly beneficence within human social groups may correlate with total reliance of individuals for their own reproductive success on the continued success of the group: that now and then individuals' interests correspond almost exactly with those of the group.

In the wake of having specified the human genome, the molecular biologists and their universities and the associated manufacturers of profitable medicinal molecules are currently constructing multi-billion-dollar research pathways called "life science initiatives." Their enterprise is the self-centered one of gathering the dollars to be derived from exploiting self-centered human urges to increase our individual and personal human life spans and our physical and emotional comfort. This is true even if we say that the entire enterprise arises out of the organizers' stated intents to help others (too). Without an underlying possibility that the research serves the interests of the (willing) sources of the funds, and the assenting taxpayers and voters, who can believe it likely that these projects would be so focused and so amply funded? As noted, these interests are defined around the concepts of personal survival and the right to generate and assist family. When happiness and good feeling are included in the formula, we have to consider that pleasure does not associate accidentally with actions that typically serve the above aims; neither pleasure nor pain has reason for having evolved to be expressed in any circumstance except when, in the environments of history, they lead to efforts to bring about acts that ultimately are reproductive (Alexander 1987).

If molecular biologists and universities can doggedly pursue their problems of synthesis, arising out of the reduction of the organism to a set of identifiable DNA molecules, then the rest of us—whether artists, humanists, economists, politicians, or social and biological scientists—can surely accept our own problem of synthesizing questions about meaning. Knowing our capabilities, we can surely be comfortable seeking the threads of connection between a fundamental reproductive motivation and the cathedrals of cultural splendor that have been generated across human history. At least we can accept others doing this, even if the effort seems to us unpleasant.

Science is for the most part the pursuit of broadly important facts, meaning facts that have the same or a similar meaning for large numbers of people—or all people. In contrast, the arts are a pursuit of meaning, often in the sense of significance specific to individuals or small groups. As Barzun implied, poetry can be so personal as to be incomprehensible to virtually everyone but the poet (which, as we shall see, need not eliminate its importance for the status of the poet, hence his or her desirability as social or sexual partner). As already implied, there are good biological reasons for meanings often to be individual. In a sexual organism, every individual has a unique genome, therefore a unique composite of life interests (Alexander 1979a). That uniqueness of life pursuits happens to hold even for genetically identical twins only underscores the mechanism: because sexual organisms are overwhelmingly likely to have unique genomes, they can only evolve to behave as if their life interests are unique—even on the rare occasions when they are not. Disagreements between identical twins with different environments and environmental histories reflect that twins have been so rare, and so rarely successful, as to have little or no history of opportunity to respond to their genetic identity. Circumstances that are rare enough across history, such as successful twinning in humans, do not yield the ploddingly directional repetitions of selection that can elaborate distinctive traits.

In the end, do we not take whatever is known or assumed about every circumstance that can affect us and turn it to our own interests, however we interpret those interests? Sometimes, we accept that others (at least) even do it without conscious awareness.

To think about uniqueness of life interests, or of meaning, after accepting the framework of life interests being reproductive, think at first only of the unique collection of genetic relatives in the social realm of each individual. For humans, these collections of known relatives (kindreds) are huge. It is surely no accident that all human societies studied everywhere on earth are centered on extensive kinship systems that operate so as to match their circumstances to reproductive interests, no accident that humans everywhere know in amazing detail who their relatives are and how related they are and that they also have strong opinions about how much help they should be receiving—or giving (Alexander 1979a). The mechanisms of such knowing, when the knowing is coupled to differential nepotism, are in all cases known so far (in all animals) dependent on background patterns of social learning. There apparently are no magic cryptic mechanisms of differential nepotism, at least in humans, except those we can fool by modifying the situations of social learning (Alexander 1990a, 1991)—most of which are probably never conscious. That we use and depend on social learning to know and treat different kin appropriately in reproductive terms, and also the evolved finiteness of our individual lifetimes through senescence (e.g., Alexander 1987), should be every person's introduction to the evidence that genetic reproduction really is the evolutionary end-all.

Only humans distinguish a wide array of relatives; in the context of differential nepotism, most organisms appear to distinguish only relatives and non- (or distant) relatives, such as offspring and nonoffspring. Presumably, humans engage in extensive differential nepotism because (1) relatives are easily identified individually; (2) they remain near for life, or at least for long periods (at least across most of human history); (3) humans can use the kinds of beneficence relatives can supply; and (4) we have generated the mental mechanisms by which differential nepotism can be practiced in reproductively appropriate ways. The question is why this combination of features seems to exist in humans alone (see below).

What, then, can we do about the apparent absence of a satisfactory mechanism linking the lowest and the highest levels of function in human activities—linking, say, the operation of genes to the historical flow of the arts, and perhaps even Barzun's perception of cultural decadence? Our continuing ignorance of ontogeny means that we still have to start with the traits themselves rather than with the genes.

Supposed evolutionary analyses of human traits most often take one or both of two different forms: (1) efforts to explain the functional or phylogenetic origins of a trait (e.g., how—and perhaps when—did art or music or humor begin) and (2) efforts to explain how a trait functioned across some part of evolutionary history—so as to assume its present expression—as in the supposed "environments of evolutionary adaptedness," sometimes curiously described for humans as (restricted to?) "the Pleistocene."

A third perspective, assumed here and in most of my writing (e.g., Alexander 1979a, 1987, 1989), is to examine the current expressions and uses of a trait and from this effort attempt to postulate the function of the trait (in "the environments of history") and relate the trait and the postulated function to parallel traits of other organisms by asking if a set of other traits or conditions correlate with the trait in a way that supports the presumed function. For an example including all three of these kinds of effort, see Alexander (1986, 1989) on humor.

A fourth approach, the one assumed by Alexander (1990b) and in the passage just below, is a comparative functional (reproductive) analysis of different and possibly related traits of the human organism—a kind of jigsaw puzzle or experimental-fit analysis of different unique or distinctive traits of a single species, all of which traits, because they are products of a natural selection elaborating a unified organism with the singular function of reproducing, are expected to work in concert and harmoniously (e.g., Alexander 1990b). This approach considers the human species—and in some ways its representative, the individual organism—as an N of 1 and in this sense parallels the efforts of theoretical physicists to identify and comprehend the nature and interplay of all the components of the single universe so far available to them for analysis. Students of the human species have two advantages over theoretical physicists: (1) occasionally homology with other organisms, because of phylogenetic relationships, remains useful (although the defining attributes of humans sometimes are so functionally distinctive that such comparisons lead instead to error); and (2) the immediacy, complexity, and predictability of evolutionary selection, absent in some sense from the physical universe, and leading inevitably to a kind of harmonious intricacy furthering the singular function of reproduction, yield a special kind of predictability for biologists. Biologists also have the disadvantages, compared to theoretical physicists, that (1) the human species, as with life in general, is immensely more complex than the physical universe as it is currently known; and (2) we are personally involved (thus different individuals and groups of humans sometimes have different and conflicting interests that are affected by particular findings or particular interpretations of the nature of the human species).

This fourth approach is also similar to an effort to understand any complicated novel machine, such as an internal combustion engine, which is also extensively modular and also has a single function—to produce certain torques at certain speeds on the drive shaft. With both the machine and the organism, if we can identify the function we can explore how each part contributes to that function and how the different parts work together to maximize it (Alexander 1993). As noted above, such an "experimental-fit analysis" of the traits of a single species can be combined with the better-known interspecific comparative approach using divergences or parallelisms (by phylogeneticists nowadays called homoplasies, as opposed to homologies) to discover the underlying selective reasons for traits (known as Darwin's comparative method) (Ghiselin 1969; Alexander 1979a). This more usual interspecific comparative approach is difficult in analyzing humans because human distinctiveness—our singularity as a species and our phenotypic distance from our closest relatives—sometimes affords us few relevant comparisons. The second option, or approach, used in the final section of this chapter, calls for explanation via particular aspects or kinds of evolutionary selection, and that means we will need to classify different aspects or kinds of selection more effectively than before.

No one should imagine that any effort to analyze human traits in evolutionary terms represents some kind of ideal cut-and-dried way to identify the truth, or, indeed, anything but a desperate effort. But desperate efforts can work, and if they are all we can muster we have no alternative but to use them whenever we think the questions are important enough to justify it.

TESTING FOR FITS IN REPRODUCTIVE SIGNIFICANCE AMONG UNIQUELY OR DISTINCTIVELY HUMAN TRAITS

If all of life function is understandable as reproductive, how might we discover that? Given the unique problems of verification in self-analysis and the impossibility of the usual kinds of repeatable, controlled experiments when humans are the test animal, one way is by first creating a list of uniquely and distinctively human traits, then seeking to describe, define, and connect these traits so as to demonstrate, if possible, that the only descriptions, definitions, and connections that make any sense whatever link their combined grandeur—which is us—inexorably and irrefutably to the genetic reproduction that modern evolutionary biology has helped us to despise. This is the kind of test attempted in this section. Some arguments that may seem skeletal here are more fully discussed in Alexander (1990b, 1996–97) and Alexander, Richards, and Lahti (in prep.); some will have to be filled in later by others.

We can begin with concealed ovulation (Alexander 1990b). If we have already postulated a plausible scenario for the reproductive function of this astonishing trait of the human female and then are able to show that under that explanation it connects reproductively to one after another of the other traits that distinguish the human species, so as eventually to create a network of interrelated traits that function in harmony with one another as reproductive adaptations, we will have provided a composite hypothesis that at least cannot be summarily dismissed. So long as everything continues to fit, and no other hypothesis has that same feature, we have generated the hypothesis (prediction) that still other traits of humans will fit the same construction. The more traits for which we find unexpectedly good fits, without ad hoc modification, the more likely we are on the right track. If the effort seems to fail with one or more traits, then one or more of our just-so stories is wrong and must be reconsidered, or else the entire proposition is in jeopardy. As long as we continue to succeed, however, we will encounter opportunities to falsify alternatives, particularly if the propositions that appear to combine harmoniously around a reproductive function have a strong likelihood of being evolved adaptations.

Among mammals, concealed ovulation is either unique to the human female or extremely rare among other species. Most mammalian females do the opposite, advertising ovulation in what seem extreme ways. Some investigators, who believe that many mammal females conceal ovulation, are talking about something different from what is being discussed here. They are talking about ovulation concealed only incidentally from other species (such as human observers), as with ovulation advertised by odor alone, "concealed" from species like the olfactorily deprived human one, or about ovulation designed to be advertised across only short distances but nevertheless advertised, not concealed.

In undertaking this discussion there is no reason to worry whether concealment of ovulation ever became perfect: any evolution that seems to be in the direction of concealment is enough to startle the evolutionist. The human female conceals ovulation not only from human males but also largely from herself. If both facts were not true, human males — maybe even those merely "standing on the corner watching all the girls go by" — could sort ovulators from nonovulators; if that were possible, and if forced copulations could be thwarted, birth control by an ovulation-recognition method would be relatively easy and inexpensive.

Because ovulation is displayed elaborately in nearly all mammals, forms of it seen as cryptic may be evolving in the direction of concealment, perhaps in the sense of restricting its evidence to local males only. To show that evolution has achieved complete concealment one need only show that no one within the species is using ovulation as a social or sexual signal; that ovulation can be detected in test situations, or using modern technology — by either observers or by ovulating women themselves — is insufficient. If ovulation is not actually used in normal life situations, then, like a recessive allele in the heterozygous state, it can be assumed to be hidden from participants and thus from the action of selection.

What could possibly be the point of concealing ovulation? For starters, it disenfranchises the males whose strategy is to monopolize females only during ovulation and then to abandon them to look for still others that are also ovulating. A male's only insurance of paternity is to stay long enough to ensure the female's pregnancy. If she conceals her ovulation, this time may incidentally be long enough to establish at least the beginnings of a bond. If a male is remaining near a female, who for that reason is likely to produce his offspring, it may behoove him to protect her — at first, maybe, only from other males so as to give himself paternity, but eventually from all sources of danger because she, as the bearer of his children, has some likelihood of becoming for him the most important female—indeed, the most important other individual—in his species. Moreover, if there is any seasonality at all to ovulation, the longer he protects her the less his chance of accomplishing the same with another female. The benefits of staying and the costs of leaving each continue to grow. The male is increasingly placed in a position to be behaviorally as well as genetically parental. The "spousal" pair has an increasing chance of effective social cooperation, which increasingly raises the value of the female to the male, compared to other females. The opportunity thereby generates for a trend toward offspring having two tending parents. As more parental attention becomes increasingly important to the offspring's success, for whatever reason, not only will males looking merely to identify ovulation lose in the race for successful reproduction, but so will females who advertise ovulation.

Does all of this fit with anything? Consider the human baby. It is by an astonishing degree the physically most helpless (or altricial) of all primate babies. Altriciality is one of its two most distinctive features, the other being the rapid expansion of its brain function. What is the point of the special human trait of infant altriciality? In other mammals and in birds it goes with either being hidden or being protected; in many birds and a few other mammals it goes with having two tending or protecting parents. An altricial baby is one that can afford to put all its calories into being a better individual later on—a better late juvenile or a better adult. By being altricial, a juvenile of any animal is telling us that throughout its evolutionary history it has not gained sufficiently by putting its calories into the alternative strategy of personal protection: it has not gained by being precocial. Something else has been taking care of that: two parents, an inaccessible or hidden nest (or both), a pouch on mama's belly. Altricial babies grow faster (Ricklefs 1983, and discussion in Alexander 1990b). In particular the human baby, because of its altriciality, is able to grow an enormous and calorically expensive brain very rapidly. Consequently, and because of parental protection and teaching, it is able to turn a higher proportion of its expanding attention to learning about the social issues that some biologists believe responsible for the evolution of the uniquely complex human brain and therefore the unique human repertoire of social complexity. Perhaps we have gained a clue about the reason parental care became so important, therefore why concealment of ovulation evolved.

Human female and male parents not only share parental care, one way or another, but also tend the baby until it is an adult—today for life or longer sometimes (e.g., via inheritance). Correspondingly—and unlike altricial songbirds in dangerous nests and altricial parasites in temporary habitats (which use their altriciality to grow rapidly and escape the danger)—the human baby has evolved a *long* juvenile period, arguably with aspects extending beyond the second decade of life. Because of the expense of delaying reproduction, the consequent necessity of an overcompensating increase in reproductive likelihood, and the benefits of extended and multifarious social experiences for all young humans, the long juvenile life of humans fits easily into our chain of probable connections. Already we have causally linked—seemingly in a remarkably plausible way—several unique or distinctive human attributes: concealment of ovulation, biparental care, long-term spousal bonds, infant altriciality, rapid brain evolution, long juvenile life, and social complexity. Perhaps we are getting somewhere.

Another distinctive human trait is worthy of mention in this same context: menopause. Originally menopause was viewed as simple aging, as an antireproductive refraining from adding mutated offspring into the population, or as something that happened after the age when most women would have died. These arguments have been refuted or cast in doubt (Alexander 1990b), though the medical profession for the most part continues to view menopause as a pathological condition to be remedied by drugs, even when those drugs have deleterious side effects. In the human female, menopause is a programmed event that occurs almost midway during the average maximum adult lifetime. If it were incidental or accidental we should see its counterparts everywhere, and we do not. The most interesting postulated reason for menopause, which may also be the one most likely to succeed, is that it optimizes not merely parental care but the tending of the entire local collective of genetic relatives, which can profit in unusual ways from the wisdom and skills of an astute and diligent middle-aged female relative. Given intense and long-term maternal care, there will come a time in her life when the human female can profit by taking care of the variously dependent descendants she already has, in place of producing still more. At minimum she should stop producing children after producing the last one that she has a good chance of rearing to successful independence. At most she might be expected to use her postmenopausal period to tend all of her existing offspring, as well as other kin that need help she can give, to gain and use all the influence she can on who gets what from whom and why. In either case, the apparently unique human attribute of menopause, if indeed evolved, appears explainable only as a reproductive trait, closely connected to extensive nepotistic care. The greater the amount of parental care and other kin care given by a human female, and the longer it is important to the offspring, the more likely is the evolution of a cessation of ovulation long before the average maximum age of death.

Kin care can be important only when sociality causes individuals to live in extended families in which close genetic relatives profit from continuing to associate throughout their lives. As noted, that is surely the situation in which complex kinship systems have always become established and maintained. It also provides a special situation demanding special parental and (other) kin care: complex social competitiveness within the social group, the force described earlier as possibly accounting for the evolution of a rapidly developing brain as a consequence of altriciality.

So to the above list of seven traits we can add menopause, extensive and complex kinship systems (meaning knowledge of them and actions with respect to them), and probably consciousness and extensive learning and social memory. This set of a dozen or so traits places us well along on the road to understanding the human species, assuming the traits are all functionally connected as implied. If they are functionally connected, in the way here suggested, then it is not surprising that we seem to understand each of them better than before. They not only start a description of the human species but are consistent with the idea with which we have conducted the analysis: that reproduction is the evolutionary end-all. The functional connections that seem to have dropped into place were generated from that beginning.

WHAT IS SPECIAL ABOUT HUMAN SOCIALITY?

The essential and underlying question keeps re-emerging: Why the complex human kind of sociality? Thus why a big social brain—including the elaborateness of this thing we call consciousness—and all the possible connections to it that have just been hypothesized? Complex sociality is the central question for the explanatory efforts of not only a Jacques Barzun but every student of human biology. Whatever causes it, or enables it, distinguishes our species and apparently has driven the evolution of many of our unique and distinctive features. What might have taken us in this particular direction? What has been different about the effects of selection on our ancestors—about the reproductive history of our species?

Selection cannot favor social abilities in the absence of social life. In all those life forms in which social organization resembles that of the genes in the genome, and in which the social group approaches the nature of the organism as a group of cells, sociality is relatively simple (e.g., all eusocial forms, with queens and workers typically in huge nuclear families). Conflicts of interest have been dramatically reduced. In genomes, each gene (allele) almost always has an equal chance of being represented in a sperm or egg cell; in a eusocial form each worker usually has an equal chance of having its genes represented in the queen's eggs, and, as with the somatic cells in an organism, evidently a better chance than in having them represented in its own offspring. As with the incredibly numerous and divergent clonal cells of the body—each with the same complement of genes, and the same one as in the testicular and ovarian cells—agreement and cooperation are efficiently blind to details that might involve complicated decisions, conscious contemplation, and the building of alternative possible scenarios for future use. As human organisms, however, we are not blind to the details of complex and sometimes rapid decision making because they are central in our form of sociality. But why are humans so distinctively social, among all organisms, as to evolve a brain apparently uniquely capable of complex social learning and extensive scenario building, in the company of all of the other special, often related human features?

We can begin with the fact that the complexly social wasps, bees, ants, termites, naked mole rats, and all the other species that biologists call "eusocial" (Alexander, Noonan, and Crespi 1991) live in large social groups that are actually nuclear families: mother and offspring or both parents and their offspring. The large social groups of humans are not nuclear families—or faint divergences from such. Though we have generated huge governed units displaying various degrees of unity (patriotism), our nuclear families remain small within them. And we do not parallel the eusocial forms, or the genome, in allowing a few of our fellows (or cells) to do all the reproducing. Instead, as individuals we all begin our lives expecting to reproduce, and, significantly, we seem to treasure that right beyond any other. We have to be impressed by this often overlooked fact about human reproduction, which is responsible for maintenance of a high level of competitiveness within the group, no matter how unified it may be on occasions when the entire group is threatened.

But this fact alone describes nothing in human sociality that is even remotely close to being unique. The same is true of large numbers of social mammals and birds: they live in complexly social, often multimale groups in which all of the individuals "expect" to reproduce. So how are we different?

Consider again our first collection of apparently connected attributes—all the way back to concealment of ovulation and biparental care. Here is another human feature that is unique among mammals: we are the only mammal species that lives in more or less permanent multimale social groups in which considerable paternal care is directed at particular offspring of females. Said differently, we are the only such mammals in which males have a reasonable confidence of paternity and exploit it. So now we have targeted a new question: why we should have come to live in groups so potentially disruptive—thus costly to us—and how we manage it (cf. Alexander 1990b).

There is an enlightening side question. Many bird species—but not mammals—live as humans do, meaning in multimale groups in which both sexes give parental care to their particular shared offspring. How do they do it without being like us in other regards? Birds are unlike mammals in a central feature that is also revealing with respect to the original contention of the centrality of reproduction. In mammals, the female retains the egg after its fertilization and tends the embryo internally. As a result, fertilization takes place "high up" in the oviduct inside the mammal female and then moves farther "up" (meaning farther from the body opening that is the source of sperm) into the uterus, which implants the fertilized ovum and subsequently nourishes the embryo. In such cases we can speculate that, in mammals, compared to most birds and most arthropods, it is more likely that the first sperm (or ejaculate) in wins the race to the egg: first sperm in, that is, at particular times with respect to ovulation.

Birds, arthropods, and other egg-laying organisms are variously different, even though birds are so poorly studied in this regard as to leave open many questions about the nature and degree of their difference from most mammals (possibly excepting those, like bats, that store sperm). Arthropod eggs are typically fertilized just before being laid, and if sperm are stored, they are stored near the external entrance to the female's genital tube; the last sperm deposited thus tend to be closest to the egg that is to be fertilized as it is being laid. In birds the fertilized egg acquires a hard shell and is deposited externally before much development of the embryo has taken place. It seems less likely to be the case in birds (generally) that, as in mammals, first sperm wins the egg. Instead, as with perhaps most arthropods, it must often be virtually the opposite. With the expectation of many qualifications— and the certainty that there are significant unanswered questions about precise mechanisms—the last sperm (ejaculate) in can be more likely to win the egg. Apparently, a prospectively parental male bird can ensure paternity with more effectiveness by frequent copulation at the right times just prior to egg laying (or shelling), if another male has recently copulated with her, than is true for mammals. Males of monogamous songbirds, in which both sexes sometimes cheat, in fact tend to mate frequently with a female that has been out of contact even briefly; by being parental later on, males act as if they retain high confidence of paternity; male mammals, including humans, may often copulate with a mate that has just strayed, but they are probably less likely than male birds to be devoted parents afterward. Whatever details are eventually worked out, we have good reason to believe that mammals and biparental social birds differ in ways like these, important to the difference between them with respect to paternal care in multimale groups. In this connection it is worth considering that in many birds the intromittent organ is very small or virtually absent.

So we are back to concealed ovulation, and it is appropriate to add one more to our list of distinctive human traits related to both reproduction and one another. Humans are also said to be the sexiest of primates—or perhaps mammals— because they copulate so frequently and seem almost constantly preoccupied with sex. One might say, "How else to ensure fertilization if ovulation is concealed?" That might be a suitable explanation early in the history of human evolution. But for modern humans we need to put a different slant on it: given the importance of a long-term pair bond in any species in which both parents tend mutual offspring with long juvenile lives to near adulthood, rates and patterns of copulation in modern humans may have evolved so as to have little to do with the actual pro- duction of babies—at least within bonded pairs. Instead, copulatory rate and pat- tern may have been shaped to initiate, elaborate, and maintain the parental bond. We can hypothesize that the human marriage, spousal, or mating bond is evolved as a parental bond, cemented and sustained—at least during the years of offspring production—to a significant extent by sexual activities. That a parental bond also sets up opportunities for cuckoldry, which lead to special aspects of copulation in social groups, can be left aside for the moment because we at least know that cuck- oldry has not yet led to the disappearance of paternal care in multimale groups of either the human species or many bird species.

The parental bond function for copulation can work if such rates and pat- terns as sustain the pair bond incidentally cause fertilization and the production of babies at a maximal rate or in an optimal pattern. Certain trends and traits may not be explainable except from this hypothesis: examples are copulation during pregnancy and after menopause; ultimate replacement of copulation with other bonding mechanisms such as affection and social and emotional compatibility after the end of baby production; and the linking of copulation more to intimacy and bonding and less to orgasm, especially by the sex (female) that is more vulnerable with respect to receiving parental care from the other. Notice that these arguments assume that human life length is set so as not to interfere with the continuation of parental care and therefore also assume a continuing value for effects of bonding between the parents of any individual. It may at first seem paradoxical that the above argument is supported by parental behavior—and a parental bond—remaining important, and continuing, beyond active sexual behavior of the parents, either with each other or with other partners.

We can briefly address here the question of morality and contributions to its origin. If morality consists of somehow agreed-upon rules about how far any of us can go in serving our own interests within the social group (Alexander 1987, 1993), then we have to ask ourselves how and when the tendencies of human males to respect the pair bonds of others came about (Alexander 1990b). Because males are the sex more likely to be represented in defense of groups consisting of multiple families with dependent offspring, and because of the apparent urgency of paternal care within groups, it is appropriate to ask whether intermale honoring of within-group pair bonds arose in the context of group defense. The question has relevance here because it is clear that, despite concealment of ovulation, extra-pair copulations can lead to losses of paternity confidence and thereby disrupt the parental bond and reduce paternal care. The prominence of such issues in modern life, added to the ceremonial publicity given to weddings, attitudes toward children born out of wedlock, and actions that break up marriages in which there are children, cause us to wonder about the nature and antiquity of morality—rule-dominated patterns of social reciprocity—among potentially disruptive males. This question, as we shall see, is connected to a fundamental one that still remains.

WHAT UNDERLIES THE DISTINCTIVELY HUMAN KIND OF GROUP LIVING?

We are returned to what has steadily become identifiable as a central question: Why should humans have evolved to live in their kinds of social groups when so many traits and tendencies apparently underwent such extraordinary evolutionary changes as a result—meaning, when the trade-offs involved were apparently extremely expensive? We need to discover an overriding importance for human social life that involved proximity of multiple breeding males but fostered a cooperativeness that allowed (or caused) extensive biparental care and all that has become associated with it.

Group living is always expensive, especially when all individuals retain the tendency to seek to reproduce, because it places individuals in more direct competition, engendering costly conflicts. There are not a great many different explanations for group living (Alexander 1974, 1975b, 1990b). Indeed, thwarting of predators in one way or another has gradually moved to the forefront among likely explanations for many cases. There is little doubt that human social groups have been able to thwart predators that individual humans could not; they also hunted cooperatively early on. But, again, humans are in no way unique in these regards. Something more has to be involved.

We can approach the "something more" by thinking again about organisms as genes bundled into groups (genomes) that compete via their collectively produced phenotypes. Organisms compete with other organisms because they are sufficiently alike to use the same resources; as Darwin noted, the more similar two organisms are, the more directly they compete for resources. The same might be true of groups of organisms when they operate to gain resources as units, and it obviously is true of different groups of human organisms today. The reason for competition is that accessible resources will always be limiting in quantity, quality, or cost of acquisition, if for no other reason than because of the ways living creatures exploit them so as to maximize reproduction.

Suppose we start from the present and think backward through history. We can certainly say that the principal hostile force of nature for human groups today is other human groups. In various ways we humans have become "ecologically dominant" (Alexander 1989, 1990b), meaning that we have found ways to reduce the significance of the predators, food problems, and extremes of climate and weather that originally plagued us—at least in ways that affect the modes of interactions of different human groups with one another. Ecological dominance is not merely a way of repeating that we have been able to become the most ubiquitous and influential species across the globe: it also carries in it the explanation for that fact. For our arguments, the main question becomes when and how this situation arose, what form it took, and whether it happened far enough in the past to influence the combinations of human traits we have been discussing. To the extent that evidence can be garnered to prove the existence of multiple-male human groups with altricial babies in the oldest structurally modern humans, we can assume that neighboring human groups have treated each other as major competitors, thereby creating the necessary and sufficient conditions for our composite hypothesis (see also Alexander 1990b). In effect, humans have taken up a unique kind of within-species, group-against-group competition—an ongoing and perhaps endless balance-of-power race—that has led to a distinctive kind of sociality and huge nations. One serious question, then, is not whether, but when, this situation arose, with respect to all the causally intertwined human traits just discussed. The other is the extent to which competing subgroups and within-group alliances contributed to the effect we are considering, as opposed to adversarial relations between geographically contiguous groups.

We can finish this section with a human tendency that may have received too little attention. Alexander (1979a, 1979b, 1987) described a seemingly unique feature of human societies, expecting that if he was wrong in calling it unique this would immediately be pointed out. It was simply that humans alone appear to compete in play, group against group. Other species either compete in play as individuals or, as in chimpanzees, compete as groups sometimes against a single individual. Group-against-group competition in play is evidently still viewed as unique to humans. It is potentially a very significant observation because play is generally interpreted as a way of practicing in low-cost situations for later full-cost social interactions. Thus predators such as cats play as if with prey, and all socially competitive animals that live in groups play, typically one on one, as they will have to compete later on. It is as if playing animals in general are building strategies of later responses in social situations, whether with members of their own species or members of prey species. These strategies, interestingly, parallel our own social scenario building, typically done consciously and later employed both consciously and unconsciously. The difference is that most animals that learn how to compete or deal with prey by playing are probably learning in the way humans learn to type, ride a bicycle, or shoot free throws. The repetition in the play renders the eventual response reflexive, thus quicker (and less open with respect to possible alternatives) than if it were conscious. We humans have had to play, with mental scenarios, in ways that enable us sometimes to use our consciousness later to compete. This situation (or race)—either novel in humans or at least uniquely complex—exists because alternatives are also dreamed up consciously and continually by our scenario-building human adversaries. (Alexander 1989 argued that humor, almost nonexistent in nonhuman species, evolved as social-intellectual play—see also Alexander 1986.) The multiplicity and novelty of ploys would devastate the organism with a repertoire of strategies involving little or no consciousness. Apparently uniquely, humans do all of this, including both physical and social-intellectual play, in preparation for both one-on-one competitions and group-against-group competitions, and including team play in preparation for the latter. We can interpret the intensity and complexity of intergroup competition in play as supporting evidence that humans evolved to compete in social groups within the species, almost as if the different groups were different species. Again, the only serious question about this unusual or unique tendency is when it became prominent, because we know it has been devastatingly prevalent across all of recorded human history. The existence, nature, and intensity of intergroup competition in play tend to support the argument that humans have been involved in their current distinctive kinds of social activities long enough to account for the combinations of traits examined here.

GENETIC SELECTION, PHENOTYPIC FLEXIBILITIES, AND CULTURAL RACES

Connecting genes to culture and its variations is a difficult exercise in terms of kinds and effects of selection, the nature and extent of heritability in trait differences, the functions of phenotypic plasticity, and the overall nature of the organism. At this point the most useful question seems to be whether we can identify aspects of selection that could be responsible for the human traits resulting in culture and the arts.

Some Features of Different Kinds of Selection

As already noted, differential reproduction of genetically different forms, or evolutionary selection, is the only candidate for principal guiding force of evolutionary change. Since Darwin, biologists have generally recognized two major forms of evolutionary selection: natural and sexual. Darwin described the causes, or hostile forces, of natural selection as climate, weather, food shortages, predators, parasites, and diseases. Members of a species—and sometimes members of different species—compete for limited resources affording relief from Darwin's hostile forces. Alleles residing in and transmitted by the particular (and relatively short-lived) individuals that happen to fare better in this competition outreproduce, hence outlast, other alleles located in individuals that fare worse. Faring worse or better can be owing to a single allelic difference. Sexual selection results from differential choice of mates by one or both sexes and from competition among members of either sex for the best partners of the other sex (cf. Bell 1997).

In *natural* selection we can begin with two kinds of interactions between living forms, predatory and competitive. In *predatory* interactions one participant (predator, parasite, disease) uses the other (prey, food, host) as a resource, and the other participant can only lose. In *competitive* interactions—competition for protection from Darwin's hostile forces and high-quality mates and other social partners—two parties strive for the same resource. As a result of a competitive interaction, which is expensive to both parties compared to its absence, one party typically wins and the other loses.

In the third kind of interaction, involving *mutualisms,* both participants can gain, and on average (in the population or species) must do so or the interaction will eventually disappear. Most if not all mutualistic interactions also include some competitive interactions. Thus, even in the sexual interaction, members of one sex compete for access to members of the other sex, or access to the better (more reproductive) sexual partners, and individuals less successful in this effort suffer in comparison to more successful individuals (Fisher 1958; Trivers 1972; Miller 2000). Of course, the interests of individual males and females probably never coincide exactly, so that one sex or the other can suffer reproductively compared to its mating partner or other reproductive competitors in the species. Nevertheless, both parties to a sexual interaction win in the sense that, when sex is the only vehicle of reproduction, successfully mating individuals will outreproduce any individuals failing to mate (excepting those with extensive knowledge of genetic kin and ability to help them). Even in a nonobligate social interaction, such as communication, it is also necessarily true that on average the individuals that communicate—both signalers and receivers—outreproduce those that do not. When this is not true, in both obligate and nonobligate interactions, the system is expected to vanish (Otte 1974): sexual organisms would become parthenogenetic or asexual, and efforts to communicate (or socialize in particular ways) would be dropped. Mutualistic selection thus differs from selection based on predatory or competitive interactions in that both parties in mutualistic selection can gain and must, on average, gain over nonparticipants.

Another way of expressing these ideas may be illustrative (modified from Queller 1994): whenever one organism begins to obtain costly and thus limited benefits (such as food) from another, at a net cost to the benefit giver, the benefit giver will evolve to reduce and eliminate the benefit. This is an effect of natural selection, the benefit takers being Darwin's hostile forces of nature. On the other hand, whenever benefit givers obtain a net reward for giving a benefit, they will evolve to give more of the benefits, and to do so at less expense to themselves, thereby gaining greater rewards. Benefit receivers will evolve to extract more of the benefits, and to use them more effectively, thus favoring the givers of more and greater benefits. This is mutualistic selection. The benefit givers in such selection, which can be either intra- or interspecific, might be labeled beneficent forces of nature. Beneficence need not be genetically altruistic because it can be extended at a reproductive profit to (1) genetic relatives, with the return solely genetic, in terms of the success of the relatives; and (2) nonrelatives, with the return (reciprocity) in beneficence with interest, which can then be used reproductively—for example, to produce offspring or help relatives.

Because of the ubiquity of the organism as evolutionary unit, we can hypothesize that on the whole the organism is an unusually good vehicle for maximizing reproduction among alternatives that might have been produced by all the different and conflicting forces of evolutionary selection in the history of life and also that each particular kind of organism is the best vehicle for maximizing reproduction that could have been produced during its particular population's history, given its nature upon entry into that particular historical sequence.

Social Selection and Social Reciprocity

I will now focus largely on mutualistic and competitive aspects of selection. Further, I will ignore interspecific mutualism and restrict the discussion to social selection, which I identify as a form of mutualistic selection differing from most sexual selection (excepting hermaphroditic species) in that the partners in an interaction may alternate roles or assume the same roles in relation to one another. I will also restrict the discussion to those particular interactions that take the form of social reciprocity (Trivers 1971). Because social reciprocity is pervasive only in the interactions of humans, and because human cultural variations are our specific target, I will restrict the discussion even further to social reciprocity in the human species. My aim is to understand how evolutionary selection on genetic variations could account for the complexity and diversity of human culture—again, defined broadly, so that I am actually asking two questions: (1) how selection could have produced human tendencies to generate such a wide variety of activities, talents, competitions, and ways of life; and (2) the effects of all these activities, competitive races, and life patterns on the continuing genetic makeup of humans—how they affect evolutionary selection. It will be necessary to develop the arguments by considering various aspects of an evolutionary explanation.

Evolutionary selection on any trait can be directional, stabilizing, oscillating, or disruptive. Here I am concerned primarily with directional and stabilizing social selection. In directional social selection, traits exaggerated in a particular direction tend to be favored by prospective social partners (in the cooperativeness of reproduction or reciprocity). Although the resulting directional change may not have an obvious culmination, the amounts of genetic variation that continue to correlate with the trait variations will continually be reduced because selection tends to operate too fast for recurrent mutations to provide continuing new directional possibilities. In stabilizing social selection, a particular expression (or culmination) is favored, such that selection will tend to "fix" the trait at some point on its usual range of variation, removing from the population alleles that yield other expressions of the trait (or removing the relevant effect).

As noted earlier, *social reciprocity* can be said to take two forms, direct and indirect (Alexander 1977, 1979a, 1987). *Direct reciprocity* consists of exchanges of costly benefits of any kind, between parties, with no necessary involvement of third parties. Examples are all kinds of trading, barter, and directly reciprocal investments. *Indirect reciprocity* consists of benefit giving, or investment, with expectation of returns from parties other than the original recipient, as when benefits are given to relatives or partners of other persons, and the reciprocity is expected from those other persons — or even their relatives and associates in reciprocity. In both direct and indirect reciprocity, returns may be extensively delayed and still yield profit to the original donor. The currencies of giving and receiving are also frequently different.

Indirect reciprocity involves reputation and status, and when it is pervasive everyone in a social group is continually being assessed and reassessed by others. Examples of indirect reciprocity are donations to charity, giving of blood, heroism in defensive — or even expansive — wars, and expensive or risky assistance to any citizen of a close-knit group — all of these when there is a strong chance of the act becoming public. The return benefit to the originally beneficent party may come from individuals or groups that expect (not necessarily consciously) either subsequent benefits themselves from the original donor or benefits to others such as relatives or friends who will then have an increased likelihood of providing some benefit (genetic or phenotypic) to the giver of the return benefit. Eventually, merely enhancing the reputation of a benefit giver through public praise can be such a return benefit because of its likelihood of bringing further benefits to the benefit giver from still others. Indirect reciprocity thus contributes to selection of partners for direct reciprocity and binds together aspects of the sociality of an entire group. Thus a war hero who risked his life in defense of a group could receive many kinds of benefits, from increased access to potential mates to prestige and accompanying benefits to offspring and other relatives. Those who provide these benefits are more likely in the future to receive benefits from others who, observing the prestige afforded heroes, may more readily accept the challenge of being a protector of the group.

Humans introduce on a grand scale a special feature of social reciprocity that has given our intellectual race part of its nature: *risk taking,* or increasingly longer-term investments and increasingly greater willingness to invest extensively in the absence of immediate returns. Modern humans invest this way almost continually, often not expecting returns for decades, or even expecting returns only for their descendants after they themselves are dead. Such risky tendencies are likely to persist only as long as they continue to yield greater returns than less risky investments. They call for the evolution of increasing abilities to assess the social future accurately, thus continually reducing the risks of long-term investments.

Long-term and large investments provide countless opportunities for *deception and cheating* and create both evolutionary races and lifetime learning races among reciprocating individuals in abilities to recognize and deal with cheaters and otherwise to maximize investment effectively under increasing risks of being duped or cheated.

Since Trivers (1971) we have gradually learned that only in humans is there strong evidence of extensive evolution of complicated and multitiered potentials for *detecting and responding to deceptions and cheating.* Only in humans is *reputation* (status) likely to be of paramount importance, and, of course, we often say that "reputation is everything." Improving ways to detect and respond to cheating, and to use reputations effectively, may have driven the evolutionary ratchet that created our enormous brains and our massive cleverness in social matters (Alexander 1989). One of the greatest advantages a human can have in social matters is superior ability at building mental scenarios that describe most or all possibilities in upcoming or ongoing interactions with parties having somewhat different life interests. Jack London said it pays to stay one idea ahead of the other fellow. This can be done by the equivalent of looking into other people's minds not only by keen social observation and exceptional memory but also by emotional skills such as empathy and sympathy, thereby anticipating the possible and likely actions of important others in every circumstance that may arise. Many traits useful in human sociality—such as consciousness—reflect broad abilities in mental scenario building.

Consciousness can be defined as knowing what you are thinking about and being able to tell others about it and act on it as a matter of self-understood choice among envisioned alternatives in subsequent social or other situations. It implies the ability to think about times, places, and events separated from your immediate circumstances and the ability to use the understanding so gained to anticipate and alter the future, build further scenarios, plan and think ahead, anticipate different possible outcomes, and retain the potential to act in several alternative ways, depending on circumstances that can be only imperfectly represented at the time the plans or scenarios are being made. Language is a concomitant of consciousness, characterized by features that make communication of useful information about mental scenarios possible: signs, symbols, and displacement in time or space (Hockett 1960; Alexander 1979a, 1983; Pinker 1994, 1997).

How then do we model the probable manner by which evolutionary social selection might have resulted in the arts—or culture as a whole—as we know these features of modern human society? What do we need to say about social selection that has not been said before?

Heritability and Phenotypic Flexibility

Perhaps we need to talk first about heritability of differences in traits and how heritability changes and sometimes virtually disappears. Selection cannot cause evolutionary change unless there are heritable differences in the traits involved in the selection. We need to understand how seeming flexibility in phenotypic expressions — what we oppose to heritable differences as "lifetime learning differences"—connects to evolutionary selection and heritability. Without this connection it is difficult for nonevolutionists to imagine how the learned variations of culture can be related to a history of genetic variations.

Differences in some traits of organisms are owing to differences in alleles, with little ability by the organism to modify the trait using environmental stimuli. Coat color in some mammals, such as horses, is a good example. In other traits, or species, nonheritable changes in coat color also occur; thus some Arctic mammals regularly change coat color between seasons. This ability presumably has to come about as a result of evolutionary selection in an environment that fluctuates with high predictability on a long-term basis. The heritability that allows evolution to occur results in an effect on the animal of a different kind with respect to the observable trait: alleles are favored that give the organism the ability to respond to environmental features such as day length, temperature, or substrate color, which predict the relevant environmental change, by altering coat color — probably by turning certain genes on or off, thus triggering changes in physiology. The genetic bases of all these abilities are most likely adaptive modifications from those yielding a single coat color for life.

In still another kind of coat color change, animals like the chameleon can change color quickly in response to differing substrate colors. This kind of change need not involve turning genes on and off; it could be mediated, for example, through the vision of the animal: what it sees is what it comes to resemble. This change in physiology is also most likely adaptive. All three of the abilities described so far have evolved to be present in the animal when it first encounters the relevant situations, probably effective when it is born or hatched.

The situation of most interest in trying to explain human social or cultural complexity and diversity is slightly different. Here the organism comes into the world unable to make certain responses and acquires the ability to make the responses during its lifetime, as by learning. It begins with abilities to start and continue acquiring responses, the nature of which depend on particular features of the conditioning or "teaching" environment. For example, blushing in humans might be seen as a fourth kind of "coat" color change, which occurs more quickly than even that of a chameleon, has social significance, is modified by social experience, and also must have a basis in evolutionary selection (Alexander 1993).

Sometimes we forget that learning is not a mere relaxation of the developmental or physiological circumstances of the animal. Each species learns particular kinds of things quickly and easily and other things with great difficulty or not at all. Ranges of learning as well as directions of learning also are more or less specified. The way experience affects the learning organism is therefore also surely a result of evolutionary selection. This means that when learned behavior is transmitted between generations as a result of learning and teaching, which is the way culture changes cumulatively, evolved capacities and tendencies are responsible (Flinn and Alexander 1982; Alexander 1990a).

When selection begins on a directionally changing social trait, variations in choosers and chosen may both be heritable, meaning they correlate with genetic differences. In this situation, both choosers and chosen will change evolutionarily (genetically): as Fisher (1958) noted (see also Lewontin 1970), the strength and rapidity of selection on any trait depend on intensity of the selection, amount of variation in heritable traits (meaning the differences are owing to genetic differences), and generation or cycle time. In reciprocity selection we can be referring to both (1) variations in choosing abilities and tendencies and (2) variations in the trait being chosen.

As with the coat color examples used above, selection changes heritability in traits that in various ways underlie the traits originally under selection. If the environment is like the social expressions of humans—that is, able to change very quickly, yet somewhat predictable—then selection is likely to favor quick abilities to alter responses, as in one-lesson learning, preparation through mental scenarios, and the use of rapidly changing behaviors in others to alter one's own behavior profitably.

Phenotypic flexibilities can thus make us think that heritability of trait variations has disappeared, when it has actually become cryptic without disappearing, by influencing underlying physiological variations that are less apparent to us. They can make it too easy for us to accept oversimple interpretations of (supposed lack of) genetic contributions to complex trait variations in organisms. This error is made more likely by one of the simple-to-complex problems discussed in the introduction, that of understanding the ontogeny of the organism: the relationships between the particulateness of the genes and the unity of the organism to which the genes give rise via complexities of coordination and interaction that we as yet have no way of understanding sufficiently. This problem is exacerbated, but also sometimes clarified, by diversity in the ways in which choosers in mutualistic selection function. For example, Zahavi (1975) began the development of the so-called handicap principle, which describes how organisms choosing mates in mutualistic selection can use the amount and kind of expensive striving by potential mates as indicators of their overall good health and ability.

In other words, to some extent the heritability important for assessing effectiveness of evolutionary selection is transferred from the expression of the centrally important trait—which acquires flexibility in responses to environmental variations—to the effort exerted in its expression. Thus, in any race to be first—cultural or otherwise—heritable variations in, say, morphological or physiological traits affecting chances of winning can be outweighed by (potentially heritable) variations in determination or persistence. Only a broadly healthy individual may be able to win by increased determination or persistence, so that evolutionary selection may be transferred to heritable differences in these tendencies, and as well overall quality of the genome, which can be indicated by general health and thus immunity to diseases, ability to obtain high-quality food inexpensively, and many other traits. Included, for example, could be heritable variations in abilities to assess not only which races to enter but how hard to strive in particular kinds of races and in races with particular other kinds of individuals. Such shifts in the "locus" of heritability, significant for the changes that we observe occurring, can be virtually endless, as we can glimpse by thinking again of Dobzhansky's implication of a universality of pleiotropy and epistasis. Giving short shrift to such potentials for continuing heritability in trait variations can make us think, too easily, that many of the traits we observe could not have (or may not have) assumed their present expressions as a result of evolutionary selection. The more complex the interactions of the genes in a genome, the more likely that some kind of significant heritability affecting any particular trait will crop up—a truism to be understood better by considering effects of the genetic recombinations of sexuality, including interbreeding between individuals from demes that have been separated for varying lengths of time in slightly different environments. Such interbreeding produces *novel combinations*— therefore *novel interactions*—of genes within genomes and must lead frequently to changes in selection with regard to epistasis and pleiotropy. This effect is not directly a function of relative amounts of variation within and between demes but of the production of novel genomes that include genes that may never have interacted before. Much of the heritability of trait variations in the world population of humans today is surely owing to increased interbreeding among individuals from demes or populations that have lived in different places long enough to have genotypes that have integrated somewhat different sets of alternative alleles. Thus Dobzhansky (1961) and Roff (1994) may together have given much of the answer to why so much genetic variation remains in populations.

The point of this section has been to emphasize two things. First, there are many avenues by which significant heritability can arise, persist, or emerge in human populations. Second, there are so many levels and kinds of heritable variation in the physiological and behavioral hierarchies of animals like ourselves that it is risky to conclude from simple models of heritability that any particular trait is ever immune to evolutionary selection.

STATUS AND THE ARTS

Status is central in human society (cf. Barkow 1989; Dickemann 1979; Henrich and Gil-White 2001; Hill 1984). It is important to every individual in society, hardly ever approaches being "complete," and by its nature is always in short supply. For humans, unlike any other organisms, the numbers and kinds of differences in status in a complex society can be almost endless and constantly changing; in today's societies, one individual can have different kinds of status in numerous different social circles. These facts create endless and elaborate performances and audiences, as well as endless opportunities and competitive races leading to narrow trends and fads about which little more can be said beyond "Second place is the first loser." These things happen because, for various reasons (see below), audiences to artistic performance and competitive races can sometimes gain as much as or more than the participants. I would be surprised if any species has anywhere near as many performance-audience interactions as humans.

If status is, as seems evident, a measure of access to the resources of reproduction, we can grasp the importance of understanding it, and how to gain it, in a society like our own. How and why status has become so important in human society is another question. It can be complex and dynamic only when indirect reciprocity has become important. Reciprocity and nepotism are both investments that entail risk. Only a genetic return is required to make nepotism profitable, and unlike social reciprocity, much nepotism can occur regularly with a minimum of cognitive skill. Differential nepotism frequently takes the form of reciprocity, however, when two relatives reciprocally assist one other in times of special need for first one, then the other (see discussion in Alexander 1979a, 53–56). Success in reciprocity, with

CONNECTING THE ARTS TO EVOLUTION

People who are successful in the arts possess a set of basic skills that they use in relation to the social scene: observation, perception, appreciation, interpretation, imagination, prognostication, translation, and communication. They depict the social scene through faithful representations mixed with manipulation, exaggeration, and the parading of incongruity—all of which require extraordinary ability to understand the social scene in the first place. They use visual portrayals in art, dance, and drama. They use language in all forms, from oral narration and music to the literature of poetry and fiction. They use metaphor in every sense of the concept, and the race to catch up with all of the ways they do so will never be finished. They use humor, and to those who are sufficiently attentive they demonstrate its deadly seriousness. They combine most or all of these media, sometimes in a single stunning performance that creates novel comparisons and portrays relationships that enlighten and explain the social world as never before.

Audiences of the arts are accepting particular versions of the social scene secondhand from others whom they perceive to have better basic skills or broader experience. To do this effectively they must recognize the relevant talents and abilities in others. Basic to that task are universal and special features of the human brain: social capacity that includes a fertile imagination and all the talents necessary to use it. These traits are the same as those employed by the artists, leading us to wonder if they derive from a single machinery, even if not always comparably.

The result of all this is that the arts are a glorious form of interpretive gossip, multifaceted, multilayered, and altogether still complex beyond anyone's comprehension. People in the arts are by definition the best storytellers among us. What they tell us is never superfluous, impractical, or trivial unless we, the audiences, allow it to be. We gain mightily from knowing how and when to listen, to whom to listen, and what to do with the experience afterward. For the arts are theater, and theater in all of its guises represents the richest, most condensed, and most widely understood of all cultural contributions to our patterns of social scenario building through consciousness and foresight. These scenarios, which we build, review, and revise continually every day of our lives, are obligate passports to social success, and perhaps the central evolved function of the human social brain. We use them to anticipate and manipulate the future—the ever more distant future in ever greater detail.

When Marshall Sahlins (1976) argued that modern hunter-gatherers are models of the original affluent societies because of the surprisingly small amounts of time spent hunting and gathering, he underplayed at least two things. First, time not occupied with securing food and shelter can only be occupied *otherwise;* and this can be done effectively only by engaging in activities that contribute to reproduction. Second, sitting around being social is not necessarily trivial or nonreproductive. Sahlins's comments, however, emphasize that the rise of art in its various expressions need not be restricted to recent societies demonstrating leisure time and affluence in some narrow modern sense.

Even if few could say it as well as Doris Lessing (1992, 35), artists are the people who understand that "[m]yth does not mean something untrue, but a concentration of truths."

the simultaneous retention of social acceptance and harmony, requires the highest levels of cognitive skills (see, e.g., Trivers 1971).

From these assumptions it would appear that the underlying reason for direct reciprocity, followed by indirect reciprocity and an importance for status shifts in human society, is actually group living of the sort discussed earlier, which allows and promotes the kind of extensive differential nepotism apparently unique to humans and the most likely precursor of reciprocity.

The existence of status as an important variable in human society sets the stage for interest in any kind of contest or performance that assists in determining status accurately and in changing it to the observer's or participant's benefit. In Alexander (1979a), I hypothesized that much of the arts represents surrogate scenario building for audiences. I have also suggested previously that, for performers, all of the arts can provide opportunities for the elevation of status. Status and the ability to convey items of great importance to others go together. They may represent the key to understanding not only the arts but much of morality as well, because lowered status is effective punishment for behavior deemed immoral (Alexander 1987). Part of the reason this is possible, of course, is that coalitions and alliances capable of meting out punishment, as well as delivering rewards, can form within human groups and are aspects of indirect reciprocity. It is a large part of the complexity of human social life that, as part of the group-against-group aspect of human sociality, these coalitions can also shift and change in virtually every way imaginable.

THE CONTRIBUTION OF SOCIAL SELECTION TO HUMAN CULTURE: IS IT A RUNAWAY PROCESS?

We need now to engage more specifically the question of the particular flowering of the collective human phenotype usually called "culture." It is obvious that extensive learning abilities in humans, and the long and intimate overlaps of adults and juveniles in social groups, have created a situation in which learned activities are transmitted by learning, thus providing the background for the rapid accumulation of cultural innovations and change, consequently for rapid cultural divergence among variously isolated human groups. But without additional explication these facts seem frustratingly unable to account for the incredible races toward complexity and diversity of human social and cultural activities across the globe— not only the huge specialized human brain but the features that have led to such complex and diverse enterprises as the arts and, indeed, those leading to Barzun's label of decadence. Something more is needed—another step in our understanding of evolutionary process—to connect everything about the human organism to its complex cultural expressions. I ask now whether this "something more" may be considered appropriately under the label of runaway social selection, already introduced in this context by West-Eberhard (1983) and Alexander (1987, 1989).

Ronald A. Fisher (1930) used the adjective *runaway* to apply to a hypothetical selective process that he assigned to sexual selection (see also Trivers 1972; Miller 2000). I suggest that sexual selection and reciprocity selection are special forms of the broader concept of social selection. Social selection is a consequence of competition among individuals for rewards arising out of various kinds of social or mutualistic beneficence.

Fisher's runaway sexual selection has two special features. First is the tendency of the choosing parties to begin favoring individuals with traits that are extreme on some axis of desirability, rather than favoring some particular condition (other than extremeness) of the individuals to be chosen. This form of choosing can only occur in an organism that is able to compare an array of individuals and identify desirable extremes (e.g., Alexander, Marshall, and Cooley 1997). It can evolve to become the standard method of choice only if selection continues for a long time in one direction and the best choices continue to be beyond the previous expressions of the trait. Thus, if females that choose extreme males outreproduce those that do not, the choosing of extremes will spread. So will extremeness in males; otherwise the females choosing them would not gain. Spreading or fixing the choosing of extremes will inject a certain degree of inertia into the process. For this reason, once an "extreme-choosing" tendency is in place, extremes in the traits of the chosen individuals can pass beyond the form in which they are adaptive in any other sense and indeed can become disadvantageous in other contexts. Some such conflict actually exists among all the compromises of selection on different traits of the organism because the effect of selection can only be enhancement of the reproductive integrity of the organism as a whole rather than the state of any of its individual traits. When selection is social, however—when it is a matter of individuals choosing other individuals in a mutualistic or reciprocal social interaction rather than, say, competition to detect or capture food, or to escape enemies more effectively—overshoots in adaptiveness in other respects, because of choosing extremeness, will be more prominent

The second special feature of Fisher's runaway sexual selection is the feedback resulting from the genetic partnership between males and females in jointly produced offspring. Trivers (1972, 166) described it as follows:

[I]f there is a tendency for females to sample the male distribution and to prefer one extreme (for example, the more brightly colored males), then selection will move the male distribution toward the favoured extreme. After a one generation lag, the distribution of female preferences will also move toward a greater percentage of females with extreme desires, because the granddaughters of females preferring the favoured extreme will be more numerous than the granddaughters of females favoring other male attributes. Until countervailing selection intervenes, this female preference will, as first pointed out by Fisher (1958), move both male attributes and female preferences with increasing rapidity in the same direction.

We can note, first, that some aspects of human evolution that intrigue us, such as brain functions that result in cleverness in social interactions, including scenario building and testing the social future by weighing alternatives internally, could have evolved partly through sexual selection. In view of the tendency of the human brain to become larger across history, and of human behavior to become more complex—and seemingly ever more rapidly after these features had exceeded their counterparts in other species—we might be concerned to examine the likelihood that runaway sexual selection is involved. A related question is whether some or all features of runaway selection can also occur in nonsexual social interactions such as the high-risk forms of social reciprocity that are prominent only in the human species.

At first one may imagine that there are no parallels in social selection allowing it to become runaway. But the way an individual gains by selecting its social partners parallels the ways an individual can gain from cooperative interactions with a mate and from a mate's parental care. In both cases there is likely to be genetic change in both the ability to choose good partners and the background of the favored phenotypic attributes because a mutually beneficial interaction can be maintained only if, on average, both interactants gain.

Sexual selection is a distinctive kind of runaway selection because joint production of offspring by the interacting pair causes the process to accelerate (see above Trivers quote). The defining feature of runaway selection is not acceleration, however, but the tendency of the process to go significantly beyond adaptiveness in all contexts except within the particular selective race—much further beyond adaptiveness in other contexts than is ordinarily the case in the myriad compromises among the conflicting adaptive traits that create and maintain the unified organism.

This aspect of runaway selection may hold for reciprocity selection, in which, unlike in sexual selection, both parties can carry tendencies not only to choose extremes but also to display extremes. In social selection, again unlike in sexual selection (except in simultaneous hermaphrodites), an individual can play both roles, of chooser and chosen, with respect to the same traits, and alternations of roles can occur during extended interactions between particular partners. To the extent that social success in ecologically dominant humans (see earlier) becomes the central determinant of reproductive success, runaway reciprocity selection may be a more viable possibility than Fisherian runaway sexual selection.

Fisherian runaway sexual selection presumably begins with a likelihood of heritability (i.e., genetic variability) in both variations in choices and variations in chosen traits. It is easy to see that in this circumstance such competitions, or races, will lead to genetic changes, at first changing both the ability and tendency to choose extremes and the nature of the extremes available for choice. Continuing mutations and genetic recombination (outbreeding among temporarily isolated groups) will tend to offer the choosing parties increasingly extreme possible choices, though to a reduced degree as selection continues to remove heritable variations in trait expressions and the trait becomes sufficiently maladaptive in contexts other than sexual selection. From earlier arguments it is not easy to understand the extent to which relevant heritability is likely to disappear, or even become trivial.

Diminution (or disappearance) of heritability in variations will not necessarily change the tendency to choose extremes: in effect, if an ability and tendency to choose extremes in any of a variety of social races (at least in humans) could become genetically fixed in the population (including the ability and tendency to learn from others, or from observation, the values of such choosing), it could still offer advantages to the chooser of extremes (without evolutionary change in the trait per se) because of the usefulness of even nonheritable trait variations chosen in a social partner. Of course, there may be further evolutionary improvements in the ability to identify and use extreme traits even after all choosers are already choosing extremes.

On the other hand, heritable variations in *what is chosen* will result in a continued march toward greater extremes because this relative quality will not be fixed in the way the ability to identify and favor extremes can be fixed; extremes can be identified only by comparing whatever is available. In Fisher's version of runaway selection, extremes win reproductively at first because they are ecologically superior, meaning functionally superior outside the choice situation itself. Later they continue to win because they are sexually (or, here, socially) superior, even though, as a result of the progress of selection involving choice of extremes, they may have become so extreme as to be otherwise functionally (ecologically) inferior. Here, "sexually or socially superior" means that the choosing parties will have acquired a genetic composition that will cause them to choose the extreme, the resulting choice of the extreme individuals itself causing the chosen individuals to outreproduce.

Extremes, however, will be chosen whether or not, as extremes, they represent heritable variants. If extreme social performances yield special social opportunities to those displaying them (e.g., via their reputations as achievers or winners), then regardless of the basis for the superior performances (i.e., genetic variant or not), winning performances can yield benefits to the kin or other social associates of the individual with the extreme traits. Thus merely joining social competitions or races can pay, though only heritable variations, including the (variant) ability to choose the best from among multiple available races *that might be entered,* will yield evolutionary change. Heritable variations in the ability to choose appropriate races—including not only those generally likely to be profitable but those in which the choosing individual, because of his or her special traits, has a special likelihood of competing successfully—may be all that is needed to drive runaway reciprocity selection. The more different ways that success can be achieved through reciprocity competition, the more robust will be this type of social selection. This is a kind of selection that will yield at least part of the human type of social intelligence, perceptiveness, and perseverance. As we all know, status (or "reputation") can exist independent of the adoption of a *particular* extreme in behavior, so the importance of any behavioral extreme in changing some aspect of culture can also depend on whether a prestigious person adopts, favors, or approves of it.

Since Fisherian runaway sexual selection in nonhuman organisms has remained controversial (or theoretical), but social parallels to it may be robust in human society, one may wonder if Fisher actually derived his idea from observing human social situations rather than from thinking about sexual selection in the birds he ultimately used as his examples. If so, it is likely not the only instance in which an evolutionist was inspired by human traits and tendencies to develop a general evolutionary explanation (e.g., Darwin's observations on human selection of variations in domestic animals and the fact that Hamilton's rule applies in its fullest extent as extensive differential nepotism across multiple levels of relatedness only in humans). This suggestion is ironic in another way, in view of the success with which academic biology departments have managed to exclude the human species from their consideration, leaving its analysis to the almost exclusively nonevolutionary approaches of social science and medical departments, and surely delaying acceptable explanations of the human species in evolutionary terms.

When extremes in particular directions are being chosen in social selection, new extremes never before experienced may be chosen above all other expressions of a trait. This behavior was originally referred to in European ethology as a "superoptimal stimulus" effect and is not specifically related to runaway selection. Whenever superoptimal stimulus effects can be identified, however, they suggest a history of choosing extremes, therefore the possibility of some form of runaway selection. If, for example, we perceive that during the past several centuries art— or even human sociality in general—has flourished and become an enormous and diverse enterprise compared to any "ancestral" condition it might have exhibited, we might suspect that this is evidence of a history of choosing extremes continually in short supply and consequently of some kind of runaway process in however artistry affects the securing of the resources of reproduction. Rather than evidence that evolution cannot be used to explain, say, the arts, explanations of recent flowerings of diversity and complexity in human enterprises can be sought by expanding our understanding of the various subprocesses of evolutionary selection. We should not be discouraged because the pathways to explanation become progressively more difficult and call for expansions of our understanding of the workings of the evolutionary process.

THE ARTS AND COMPETITION

Competitive races among humans, which include performances in the arts, can take either of two directions. They can change in the direction that enables individuals to give performances that are unique, not easily compared to those of others, and highly informative to observers about relatively complicated aspects of sociality. These are the races that we would most often term "artistic." Their participants are likely to have carved out special life niches for themselves. In effect, a flourishing of the arts in its most diverse forms reduces direct and confrontational competitive races because it is more difficult to compare creative artists than to compare, say, athletes in the same sports, chess players who contest one on one, or people who try to best one another's record in any uncomplicated or highly patterned and predictable game or contest.

Trends toward individuality in performance and meaning in the arts—as in poetry—can continue, paradoxically, because the value of gaining insight from the brilliance of others is so great to us. The more profound an insight, as well as the more personal or specific, the more difficult it is likely to be to absorb. This is the reason that a poem can be so personal as to be understandable only to its author, yet the author can be accorded high status because of the poem. Because they expand and enhance our imaginations, poetry and art press continually at the boundaries of mystery and incomprehensibility, so that, as audience, we are always faced with the necessity of deciding whether the artist is incomprehensible because of telling us something unusually profound or because of telling us something trivial or wrong in an obscure way. It is a difficult decision because we know in our souls that the scoffer can lose as profoundly as the gull, whether the poetic message is used or rejected directly or attempts are made to use it indirectly during social interactions with the poet in some other context. The same challenge exists in the assessment and use of humor as insight into the social scene, as all those who have laughed too soon on at least one occasion will understand (Alexander 1986, 1989).

Socially competitive races that parallel those involved in the arts can also take a different form, changing instead in directions that progressively narrow the variations available to the performers and the nature of the prizes and diminish the social messages available to observers. Competitors in such races are forced into increasingly or more directly confrontational races.

Modern humans have generated almost endless possibilities of being "world class" in direct competitions, from championships in golf or boxing to first prize in poetry or painting or musical contests to making the largest pizza or longest hot dog in the world, flagpole sitting for the longest time, or establishing the fastest construction times for field latrines in the history of the military. When reputation comes to be based so strongly on relative achievement compared to other humans that the nature of the competition becomes trivial and the relative ranking of the competitor becomes paramount, extremeness in the competition may be represented by activities or "traits" so radical that, in all regards except the specific competition being considered, they are disadvantageous to the individual exhibiting them. Social messages for audiences, moreover, can be restricted to the identity of contestants that finished in particular rankings. Nevertheless, such rankings represent changes in social status or maintenance of social status.

I can give an example from horse competitions. Judged horse competitions are like other judged contests: they can reach maddening levels of faddishness. These endlessly competitive races involve a wide variety of things, including the details of the clothing worn during different competitive classes; precisely how a rider holds hands, feet, and head; minute details of the equipment worn by the horse; and exactly how the horse holds its head. To take only one example here, in certain breeds, in a competition called "Western Pleasure," a tendency arose to favor horses that moved with their heads held somewhat lower than one is likely to see on the random horse galloping across the prairie or walking in someone's pasture. This favoring evidently began because it was undesirable for horses used in ranch work to hold their heads high enough to be in the way of the rancher's work. But lowered heads became a trend that reached such extremes that horses in these Western Pleasure competitions came to be called "peanut rollers" because their heads were so low they were jokingly regarded as able to roll peanuts as they walked or jogged along. Meantime, in breeds originally bred to move elegantly in parades and the like, and in others to show off unusual gaits, precisely opposite trends were set in motion in similar competitive classes. The horses were encouraged or forced to hold their heads higher and higher until the neck was essentially vertical. While some horse people were drugging their horses, starving them, or bleeding them (yes, I do mean those things) to get their horses to keep their heads down and move ever more slowly (and dejectedly!), those working with other breeds were letting their horses' fore hooves grow long so as to lift their front ends and their heads even higher. Eventually some of the latter people began placing plastic and other kinds of extensions on their horses' front hooves so that they appeared to be walking with their forefeet on boxes. And the riders began to sit farther and farther to the rear on the horse's back so that the horse would have a greater tendency to lift its front end in a fancy way while traveling around the show ring. This position became so extreme that riders began to wear capes that flowed over the horse's rump and to place bulky garlands across the saddle in front of them, both items evidently to conceal how embarrassingly far back the rider was sitting on the horse. Finally, some of the people with a certain breed of horses began to "sore" the tender portions of their horses' forefeet with acid or a knife, and to train the horses with heavy bruising chains around the tops of their hooves, in both instances so that they would lift their feet (and their heads) even higher. It seems sometimes that no judged contest of humans can maintain a happy medium in any one trait or action or fail to become narrow and faddish. Extremes appear, are quickly favored, and eventually become so ridiculously exaggerated as to discourage many people from participating. Perhaps only periodic dramatic changes in direction seriously retard such extreme trends.

Extremes of judging in narrow competitions are not restricted to small audiences or small-time events, as anyone knows who watches events such as figure skating or gymnastics in the Olympic Games. But for individuals or teams representing institutions (or nations), the social message has to do with competition between the institutions, meaning that, because of their symbolism, even narrow and confrontational races with little relationship to social questions or broadly important qualities of the contestant can attract huge audiences and yield high status for excellence in performance.

The incredible tendency of humans to engage in competitive races and carry them to extremes tells us that such competitions must be a very old part of our makeup, and a part that somehow furthered our reproduction. A point with which such races are entirely compatible is that organisms do not evolve just to reproduce successfully but to *outreproduce* everyone like them.

Whenever tendencies exist to recognize and admire "winners," whether they are outstanding artists, musicians, actors, dancers, poets or writers, shamans, humorists, athletes, war heroes, or thinkers, the tendencies may result in prolonged evolution, exaggerating the bases for the skills represented, perhaps producing the extreme skills and capabilities exhibited in the arts today. To the extent that such capacities tend to have a common genetic basis, the selection could be even more effective. Thus a wide variety of artists might achieve their special abilities partly through a common ability to construct and retain mental scenarios. Whether or not the most talented and specialized artists could use their special abilities to succeed in the social scene, their (genetically recombined) offspring might have superior skills in a variety of social situations. The favoring of seemingly different capabilities could have had common ground that led to the exaggeration of rather specific mental capacities, to which we have given labels such as "intelligence," "genius," or "perspicacity." The prediction from this suggestion is that people unusually capable in one intellectual or other endeavor in the arts are likely to be unusually capable in others as well. To the extent that this generalization is accurate, and tends to involve the useful skills of humans very broadly, the concept of "maladaptive in other contexts" because of dramatic overshoot in particular traits is diminished. While reciprocity selection is evidently distinctive and influential in humans, the question of whether it involves a runaway process, at least in the Fisherian sense, remains unclear.

In the 1930s and 1940s, the newsreels that preceded showing of movies in theaters proclaimed that in the twenty-first century people would not have to work more than fifteen or twenty hours a week because technological devices such as robots and calculating or computing machines would take care of most of the mundane tasks humans had to do then. As a specific example, when the labor-saving three-point hitch was developed for small farm tractors (in England by Harry Ferguson), its glory was assumed to be that it would allow the small farmer to do all the necessary work on his 160 or so acres with considerably less time and effort, giving him extra leisure time. In both this special case and in general, the virtual opposite has happened. Today in the United States, in many or most families both parents work even more than one parent worked in the early part of the twentieth century, and children are—seemingly "necessarily"—relegated to day care and preschool. Farmers found that the three-point tractor hitch actually helped set off a new competitive race that required farmers to multiply the number of acres worked in order to make a living wage. Instead of having more leisure time, farmers were able to do more work in the same time, and, like everyone else in the same situation, they seemed to be essentially forced to do that to keep up with or forge ahead of everyone else. More recently, the advent of efficient hydraulic systems on tractors has largely superseded the three-point hitch and set off a new race in which a single family may find itself working one or two thousand acres while selling such products as corn and other grain at prices not too different from those paid when the three-point hitch was invented more than a half-century ago. All of this speaks again to the fact that, like other organisms, humans must be presumed to have evolved not merely to reproduce successfully but to outreproduce their most significant competitors, leading to the proliferation of unending competitive races. As implied in the above discussion of so-called labor-saving devices, we can perhaps be bold enough to ask whether concatenations of some human socially competitive races, because of their combined diversity, narrowness, and runaway and faddish effects, may even have value in comprehending why Barzun (2000)—and others—have felt it necessary to interpret both long- and short-term trends in human sociality as decadent.

Throughout the history of such narrow and frantic competitive races, something more properly called "art" has persisted, apparently because of a second attitude toward competitive races (both strategies can be witnessed being employed by children in that microcosm of the human social world that consists of competing for the approval of a single set of parents). One expression of this second attitude I have earlier described as avoidance of directness in competitiveness rather than the seeking of ever-more confronting forms or expressions of competition. Another similar expression may arise, not through avoidance, but through the incidental perception and adoption of new and appealing avenues of creativity. Whatever their origin, unique life niches are created by some humans for themselves, from which they are able to tell the world things that no one has ever known before. Perhaps, as individuals, these artists can indeed be said to have mounted the "single-minded effort to render the highest kind of justice to the visible universe" alluded to by Joseph Conrad (quoted in Barzun 2000, 791). Incidentally, they also magnify the social diversity that we humans euphemistically refer to as "division of labor."

CONCLUSION

In sum, I have tried to consider every relevant aspect of the topic I originally proposed. At least I don't think I left any gaps that might require miracles. If that is true, perhaps this essay helps make it more likely that some day we will know how to connect the entire array of human activities to a base in reproductive effort. The undertaking is important because it promises dramatic insights into our most precious thoughts and strivings, and those most stubbornly resistant to change. Such human self-understanding may be the most important change imaginable in our universe, not merely to promote happiness and well-being but to lengthen human lifetimes by reducing needless strife and aggression at all levels.

We think of the arts as differing from the sciences in seeking meanings—social facts that are often restricted in their usefulness, or personal rather than general. Part of the reason for the restrictions may be that the uniqueness of individual human genomes has resulted in each of us evolving to behave (in evolutionary terms, appropriately) as if, for each of us, the collection of our personal life interests were unique. We have evolved to use the myriad facts available to us in ways unique to ourselves, which in some sense is the meaning of meaning. Human social races are prevalent because of the importance of the prestige thought to be derived from doing well in almost any such race and the consequences of prestige that bring benefits (power) through the action of direct and indirect reciprocity. Evolutionary social selection, including nepotistic selection and mutualistic and reciprocity selection, may be the last important set of mechanisms required for the connecting of complex aspects of cultural phenomena, such as the arts, to a life basis in evolutionary selection favoring reproductive success. It leads to the exaggeration of all the capabilities that are displayed in artistic works and performances.

I recognize that all of my discussions are imperfect and incomplete. They will be worthwhile, however, if the shortcomings help point the way for those with the next round of insights. I conclude that we seem not to be lacking in mechanisms to approach analysis of our own activities in evolutionary terms, only in knowing how to use them. Perhaps the next generation can make sense of the rudimentary arguments that I and others have generated so far on these difficult topics.

NOTE

I am deeply indebted to four of my former students. Andrew F. Richards has considered most of these issues with me across the past several years and has criticized the manuscript at several stages. Beverly Strassmann, David Marshall, and Mary Jane West-Eberhard provided stimulating criticisms of the chapter on short notice. David Lahti has also helped me immensely by consistently putting forth useful and constructive arguments. Because I did not accept all of the efforts to help me, I stress my own responsibility for any remaining errors or awkwardness.

REFERENCES

Alexander, R. D. 1974. The evolution of social behavior. *Annual Review of Ecology and Systematics* 5:352–83.

———. 1975a. Natural selection and specialized chorusing behavior in acoustical insects. In: *Insects, science, and society: Proceedings of the Comstock Centennial Celebration, Cornell University,* ed. David Pimentel, 35–77. New York: Academic Press.

———. 1975b. The search for a general theory of behavior. *Behavioral Science* 20:77–100.

———. 1977. Natural selection and the analysis of human sociality. In *Changing patterns in the natural sciences,* ed. C. E. Goulden, 283–337. 1776–1976 Bicentennial Symposium Monograph. Philadelphia Academy of Natural Sciences.

———. 1978. Evolution, creation, and biology teaching. *American Biology Teacher* 40:91–104.

———. 1979a. *Darwinism and human affairs.* Seattle: University of Washington Press.

———. 1979b. Evolution and culture. In Chagnon and Irons 1979, 59–78.

———. 1986. Ostracism and indirect reciprocity: The reproductive significance of humor. *Ethology and Sociobiology* 7:253–70.

———. 1987. *The biology of moral systems.* Hawthorne, N.Y.: Aldine de Gruyter.

———. 1988. The evolutionary approach to human behavior: What does the future hold? In *Human reproductive behavior: A Darwinian perspective,* ed. L. L. Betzig, M. Borgerhoff Mulder, and P. W. Turke, 317–41. London: Cambridge University Press.

———. 1989. The evolution of the human psyche. In *The human revolution,* ed. C. Stringer and P. Mellars, 455–513. Edinburgh: University of Edinburgh Press.

———. 1990a. Epigenetic rules and Darwinian algorithms: The adaptive study of learning and development. *Ethology and Sociobiology* 11:241–303.

———. 1990b. *How did humans evolve? Reflections on the uniquely unique species.* Special Publication no. 1. Ann Arbor: University of Michigan Museum of Zoology.

———. 1991. Social learning and kin recognition: An addendum. *Ethology and Sociobiology* 12:387–99.

————. 1993. Biological considerations in the analysis of morality. In *Evolutionary ethics,* ed. M. H. Nitecki and D. V. Nitecki, 163–96. Albany: State University of New York Press.

————. 1996–97. *Understanding humanity: The human species in evolutionary perspective.* Course text for Biology 494, Evolution and Human Behavior, University of Michigan. Revised 1997. Available from Dollar Bill Copying, 611 Church Street, Ann Arbor, Mich. 48109.

Alexander, R. D., David Marshall, and John Cooley. 1997. Evolutionary perspectives on insect mating. In *The evolution of mating systems in insects and arachnids,* ed. J. C. Choe and B. J. Crespi, 4–31. Princeton, N.J.: Princeton University Press.

Alexander, R. D., K. M. Noonan, and B. Crespi. 1991. The evolution of eusociality. In *The biology of the naked mole-rat,* ed. P. W. Sherman, J. U. M. Jarvis, and R. D. Alexander, 3–44. Princeton, N.J.: Princeton University Press.

Alexander, R. D., A. F. Richards, and David Lahti. In prep. *Understanding humanity.*

Barkow, J. 1989. *Darwin, sex, and status: Biological approaches to mind and culture.* Toronto: University of Toronto Press.

Barzun, J. 2000. *From dawn to decadence: 500 years of Western cultural life.* New York: Harper Collins.

Bell, G. 1997. *Selection: The mechanism of evolution.* New York: Chapman and Hall.

Chagnon, N. A., and W. G. Irons, eds. 1979. *Evolutionary biology and human social behavior: An anthropological perspective.* North Scituate, Mass.: Duxbury Press.

Dickemann, M. 1979. The reproductive structure of stratified societies: A preliminary model. In Chagnon and Irons 1979, 331–67.

Dobzhansky, T. 1961. Discussion. In *Insect polymorphism,* ed. J. S. Kennedy, 111. London: Royal Entomological Society.

Fisher, R. A. 1930. *The genetical theory of natural selection.* New York: Dover Publications.

————. 1958. *The genetical theory of natural selection.* 2nd ed. New York: Dover Publications.

Flinn, M., and R. D. Alexander. 1982. Culture theory: The developing synthesis from biology. *Human Ecology* 10:383–400.

Ghiselin, M. T. 1969. *The triumph of the Darwinian method.* Berkeley: University of California Press.

Haldane, J. B. S. 1932/1966. *The causes of evolution.* Repr., Ithaca, N.Y.: Cornell University Press.

Hamilton, W. D. 1964. The genetical evolution of social behaviour, I, II. *Journal of Theoretical Biology* 7:1–52.

Henrich, J., and F. J. Gil-White. 2001. The evolution of prestige: Freely conferred deference as a mechanism for enhancing the benefits of cultural transmission. *Evolution and Human Behavior* 22:165–96.

Hill, J. 1984. Prestige and reproductive success in man. *Ethology and Sociobiology* 5:77–95.

Hockett, C. F. 1960. Logical considerations in the study of animal communication. In *Animal sounds and communication,* ed. W. E. Lanyon and W. N. Tavolga, 392–430. Washington, D.C.: American Institute of Biological Sciences.

Lessing, D. 1992. *African laughter: Four visits to Zimbabwe.* New York: Harper.

Lewontin, R. C. 1970. The units of selection. *Annual Review of Ecology and Systematics* 1:1–18.

Miller, G. F. 2000. *The mating mind: How sexual choice has shaped the evolution of human nature.* New York: Doubleday.

Otte, D. 1974. Effects and functions in the evolution of signaling systems. *Annual Review of Ecological Systems* 5:385–417.

Pinker, S. 1994. *The language instinct.* New York: William Morrow.

———. 1997. *How the mind works.* New York: Norton.

Queller, D. C. 1994. Male-female conflict and parent-offspring conflict. *American Naturalist* 144 Suppl.: 84–99.

Ricklefs, R. E. 1983. Avian postnatal development. In *Avian biology 7,* ed. D. S. Farner, J. R. King, and K. C. Parkes, 1–83. New York: Academic Press.

Roff, D. A. 1994. Evolution of dimorphic traits: Effect of directional selection on heritability. *Heredity* 72:36–41.

Sahlins, M. 1976. *Culture and practical reason.* Chicago: University of Chicago Press.

Trivers, R. L. 1971. The evolution of reciprocal altruism. *Quarterly Review of Biology* 46:35–57.

———. 1972. Parental investment and sexual selection. In *Sexual selection and the descent of man,* ed. B. Campbell, 136–79. Chicago: Aldine.

Waddington, C. H. 1956. *Principles of embryology.* New York: Macmillan.

West-Eberhard, M. J. 1983. Sexual selection, social competition, and evolution. *Quarterly Review of Biology* 58:155–83.

Zahavi, A. 1975. Mate selection: A selection for handicap. *Journal of Theoretical Biology* 53:205–14.

CHAPTER 16

Darwinism, Dualism, and Biological Agency

Lenny Moss

Sometimes the simpler an ontology the greater its imperialistic appetite. Not so long ago the proclivity to explain everything in Marxist terms was at least as prevalent as the propensity to explain everything in some kind of Darwinian terms is today. I once heard it said that the problem with Marxism was that it explained *too* much. What this clever speaker was referring to was not the cross-disciplinary syncretism of the Frankfurt School or similar approaches but rather the straight and narrow style of economistic Marxism, sometimes pejoratively referred to as "vulgar Marxism." It was vulgar Marxism that demoted to secondary status all questions of intersubjectivity, of gender, race, and ethnicity, of culture, psychology, and art, of ethics, human rights, and democracy, in favor of the primacy of class struggle. Who introduced the term *vulgar* and why they chose it I don't know, and I don't doubt that some took it as a compliment, in confirmation of their proletarian mission. But the adjective stuck, and I'm going to borrow it for present purposes because again it seems that the most ontologically narrow style of Darwinism, let us call it vulgar Darwinism, suffers from the same propensity to want to explain too much on the basis of too little. And on further analogy, just as vulgar Marxism denied the status of agency to individuals, I will refer to as vulgar Darwinism the Darwinism that denies agency to living organisms in favor of the abstract dictates of an algorithm or the logic of "the replicator."

Much of the lifework of the philosopher Hans Jonas concerned itself with detailing the career of dualism from Gnosticism onwards and reckoning with the ethical and metaphysical difficulties that it has bequeathed to us. Darwinism upset the trade-off of the specifically Cartesian legacy. Animal life could be countenanced as simple mechanism so long as claims to mind or inwardness could be allocated to a separate substance. Darwinism represented the crowning achievement of the mechanistic wing of dualism but in Jonas's view brought with it a dialectical consequence:

> [F]or if it were no longer possible to regard his [man's] mind as discontinuous with prehuman biological history, then by the same token no excuse was left for denying mind, in proportionate degrees, to the closer or remoter ancestral forms, and hence to any level of animality. . . . In the hue and cry over the indignity done to man's metaphysical status in the doctrine of his animal descent, it was overlooked that by the same token some dignity had been restored to the realm of life as a whole. If man was the relative of animals, then animals were the relative of man and in degrees bearer of that inwardness of which man, the most advanced of their kin, is conscious in himself. (Jonas 1966, 57)

But Jonas's dialectical prediction has yet to come to pass. A mediation of empirical and phenomenological insights that can do justice to both the inner and outer horizons of living nature, a kind of soft naturalism, despite some fits and starts, has largely been the path *not* taken. Nor has this failure lacked severe ramifications for the understanding of man. Post-Darwinian man can be "naturalized" only according to the terms of the "nature" that the prevailing naturalism has provided. If man's animal kin do not already express at least the rudiments of agency and interiority, then how can human evolution be understood as anything other than a "quantum" jump into some form of bloodless cognitivism? Yet a philosophical anthropology that sought to mediate the body and spirit in a flesh-and-blood account of the evolutionary emergence of human particularity began to flourish due to the efforts of thinkers such as Helmuth Plessner, Arnold Gehlen, and Kurt Goldstein, only to become stalled and sidelined by the middle of the century. But why? Why haven't the postdualist benefits of a true Darwinian revolution come to full fruition?

Darwinism, or at least vulgar Darwinism, displaced the locus of agency from living organisms toward something called a gene. Darwinism, as Depew and Weber (1995) have done well in showing us, has itself been a theory in transition. Darwin did not have an adequate account of heredity, although Darwinism is dependent upon some model of heredity. Darwinism has thus been sequentially shaped by what is taken to be the basis of heritable variation. Evelyn Fox Keller (2000) has nominated the twentieth century as the century of the gene, and I second her motion. Certainly a kind of gene-Darwinism has eviscerated the living organism of its agency. I also share Fox Keller's view that the age of genomics will likely lead in a different direction. The nature of genomic Darwinism, however, is yet to be determined.

So long as the agency of organisms is eclipsed and displaced, the inner horizon of organic life, its affective state, will be of little interest to biological thinking and likewise for its metaphysical implications. If life is all about the genes, questions of interiority can be relegated to a purely epiphenomenal status. Random variation and natural selection are the two pillars of Darwinism, and this is where we must look to see if indeed organisms are properly denied a status as agents, and if so on what basis. Is it indeed the case that the two central processes of Darwinian evolution, variation and selection, take place, as it were, behind the backs of actual living organisms? To address this question we must first consider the concept or concepts of the gene that have come to structure the Darwinist, or at least vulgar Darwinist, outlook.

The distinction of having coined a new scientific term does not confer proprietary rights over its future use, but it can be the source of a valuable, and often critical, perspective on the subsequent history of its use. In 1909, Wilhelm Johannsen coined the term *gene* and introduced the distinction between the genotype (which is physically transmitted from one generation to the next) and the phenotype (the ensemble of characteristics that appears in the mature organism).

When he did so, he was keenly aware of the dangers and shortcomings of renewed "preformationism." Since midway through the seventeenth century, when nature became divested of its inherent purposes, naturalistic thinking about the "generation" of life had oscillated between the poles of preformationism and epigenesis.

While by 1900 the idea that miniature organisms were contained in eggs or sperm had long since been discarded, the "rediscovery" of the mid-nineteenth-century experimental work of Gregor Mendel indicated that something "particulate" was transmitted between generations. Purebred red- and white-flowering pea plants could be crossbred to produce all red-flowering pea plants, but if these second-generation red plants were bred with each other then white flowering would reappear in 25 percent of the third-generation plants. The propensity for producing white (as well as that for producing red) flower color thus appeared to be transmitted to progeny in particulate form in such a way that it remained intact even when it was not "dominant." Mendel referred to these transmitted propensities as "unit-characters," not distinguishing between the actual flower color and some chemical agent that, directly or indirectly, caused the plant to produce white or red flowers at some stage of its life cycle. Mendel spoke as if it were simply a piece of the morphology, a piece of the mature form, in this case flower color, that was transmitted between generations.

Johannsen introduced the term *gene,* and the genotype-phenotype distinction, precisely to lead the new science of genetics away from the temptations of preformationism. "Genes," Johannsen (1923) asserted, are not pieces of morphology but chemicals whose influence on the "phenotype" is determined by their "norms of reaction"—that is, by a complex range of developmental patterns realized in different environmental contexts. Recoiling at the thoroughgoing preformationism of his peers in the genetics community of the 1920s, Johannsen pronounced that differences between alternative genes, while often medically and economically significant, never get at the core of an organism and further that these differences do not represent alternative qualities but only deviations from a norm.

I will argue that with regard to genes for phenotypes Johannsen's two intuitions are exactly right and that only on the basis of conflating this sense of *gene* with a second distinctly different sense of *gene* did the truth of these insights lamentably become obscured.

GENE-P AND GENE-D

When scientists and clinicians speak of genes for breast cancer, or genes for cystic fibrosis, or genes for blue eyes, they are referring to a sense of *gene* defined by its relationship to a phenotype (i.e., the characteristics of the person or whole organism) and not to a molecular sequence. The condition for having a gene for blue eyes or a gene for cystic fibrosis entails, not a specific nucleic acid (DNA) sequence, but rather an ability to predict, within certain contextual limits, the likelihood of some phenotypic trait. Molecular studies have revealed that these phenotypic differences are due not to the *presence* of two qualitatively different capabilities but rather to the *absence* of the ability to make the "normal" protein. There thus is no specific structure for the gene for white flowers or the gene for blue eyes or the gene for many diseases because there are many structural ways to be *lacking* the normal resource. The white flower, the blue eye, the albino skin, the cystic fibrosis lung are all highly complex results of what an organism will do in the absence of a certain "normal" molecular structure.

It continues to be useful, in certain contexts, to employ this usage of "the gene." To speak of a gene for a phenotype is to speak *as if,* but only as if, it directly determined the phenotype. It is a sort of preformationism, but one deployed for the sake of instrumental utility. I call this sense of *gene* "Gene-P," with the "P" standing for "preformationist" (see table 16.1). Genes for phenotypes (i.e., Genes-P) can be found where some deviation from a "normal" sequence results, with some predictability, in a phenotypic difference. In the absence of the normal sequence necessary for making brown eye pigment, blue eyes result. Any absence of this brown eye–making resource will thus count as a "gene for blue eyes." Blue eyes are not made according to the directions of the Gene-P for blue eyes; rather, blue eyes are the result of what organisms do in the absence of the brown eye pigment. Reference to the "gene for blue eyes" is a kind of instrumental shorthand with some predictive utility.

Likewise, molecular probes can now be used to identify cystic fibrosis genes, breast cancer genes, and the like, but what these really do is identify one or more of the possible deviations from the norm, the presence of which is correlated with the likelihood of the disease phenotype. These are thus probes for Gene-P.

When biologists focus upon a sequence of DNA that provides the template for making a protein, or a family of related proteins, and examine the interactions in which it participates, then the other sense of *gene* comes into view. Gene-D, "D" for "developmental resource," is defined by its molecular sequence but, in contrast to Gene-P, is always indeterminate with respect to that phenotype toward which and in which it participates.

According to revised findings of the International Human Genome Sequencing Consortium (2004), there are only twenty thousand to twenty-five thousand Genes-D in the human genome. One of them is called NCAM. Although originally named for its role in helping to form attachments between adjacent neural cells (hence the name *neural cell adhesion molecule*), it is in fact expressed in many different tissues, at different developmental stages, and to different ends. As is the case for almost all Genes-D, NCAM has a modular structure, and the template for making any particular NCAM protein consists of some subset of these modules (exons). The Gene-D for NCAM provides templates for molecules that may participate in many different phenotypes, in different ways, in different locations at different times, embryonic or adult, healthy or presumably also aberrant. But it does not determine any of them; it has only the potential for participating in all.

The explanatory story in which Genes-D play a role is *not* one of preformationism but one of epigenesis. Phenotypes are achieved through the complex

TABLE 16.1. Gene-P versus Gene-D

Gene Concept	Examples	Explanatory Model	Ontological Status
Gene-P			
Defined with respect to phenotype but indeterminate with respect to DNA sequence	Gene for breast cancer Gene for blue eyes Gene for cystic fibrosis	Preformationist (instrumental)	Conceptual tool
Gene-D			
Defined with respect to DNA sequence but indeterminate with respect to phenotype	NCAM, actin fibronectin, tubulin 2,000 kinases (20,000 other examples)	Epigenesis	Developmental resource (one kind of molecule among many)
Conflated GeneP / GeneD	???	Preformationist (constitutive)	Virus that invents its own host "the replicator"

interactions of many factors; the role of each being contingent upon the larger context to which it also contributes. What is true for NCAM is true for the Gene-D associated with the cystic fibrosis "locus," with the breast cancer (BRCA1 and BRCA2) loci, and in fact with all of the genes (Genes-D) being identified at the level of specific molecular sequence by the Human Genome Project. The Gene-D or normal molecular resource at the cystic fibrosis locus is not a gene for healthy lungs but a genetic resource that provides template information for a transmembrane chloride ion channel: that is, a protein that may be woven into cellular membranes and play a functional role in the transport of chloride ions into and out of the cell. Similarly, the normal resource at the breast cancer locus (BRCA1) is not a gene for healthy breasts but a template for a large and complex protein that is present in many different cell types and tissues, in many different developmental stages, and that appears to be capable of binding to DNA and influencing cell division in a context-specific way. The crux of the story is this: neither chloride channels nor DNA-binding proteins dictate or determine phenotypes, as their biological significance is always mediated by and contingent upon the complex associations with myriad other factors with which they interact.

To study the biology of a Gene-D is to play one kind of explanatory game, an epigenesis one. To use a Gene-P—that is, the absence of a normal genetic resource—as predictor of a phenotype is to play a different kind of explanatory game, an "as-if" preformationist one. Johannsen was not privy to Genes-D, and his injunctions do not pertain to them. He predicted that the entirety of the organism would not be decomposed down to genes, and he was right. Genes-D are molecular sequences along the chromosome, not pieces of the phenotype. Genes-P are spoken of as if they were pieces of the phenotype, but, as Johannsen predicted, they pertain only to a limited and in some sense superficial set of traits and then only for practical purposes. Now the meanings "Gene-D" and "Gene-P" can both be used responsibly within their proper domains—genetics counselors, for example, use "Gene-P." But just as the word *bank* can be properly used to mean both the side of a river and a good place to invest money without implying that the side of a river is a good place to invest money, so too the word *gene* should not become simultaneously invested with the meanings of both "Gene-D" and "Gene-P." Genes are not at once both molecular sequences and pieces of the phenotype, yet precisely this conflationary confusion has contributed to a vulgar Darwinism that depicts the processes of evolution as taking place wholly behind the backs of living organisms. The problem of how to naturalistically account for the apparent purposiveness of living organisms is managed by deferring the locus of adaptive agency from organisms to conflated genes.

AGENCY AND VARIATION

Kant, in his time, provided a temporary reprieve from Cartesian constraints on explaining organic purposiveness. Purposiveness could exist within a Cartesian, or really Newtonian, mechanistic framework if purposive organization was taken to be a given. Organismic agency was built into his framework by way of the adaptive potential present in the germ. He referred to this as the *Anlagen*—meaning proclivities or aptitudes (see Sloan 2002). This Kantian view did bring with it the liability that it could not countenance novelty—that is, potential for purposive organization that was not already present in the original stock of *Anlagen*. The Darwinian view, of course, brings novelty into the purview of evolutionary process, but does susceptibility to novelty necessarily do away with the need for, or indeed the possibility of, a *de facto* purposive system that is always already there? Vulgar Darwinism says that it does. If indeed Darwin is to be that Newton of a blade of grass that Kant doubted would ever appear, it must. Vulgar Darwinism holds that heritable variation is exclusively random, that it takes place, to whatever extent to that it does, independent of and unaffected by the life history and adaptive capacities of the organism. Now, for this to be true two things must hold. The substrate, the stuff, of heritable variation must indeed undergo its changes unmediated by the ostensibly purposive, self-sustaining dynamics of the living organism, and these changes must bring with them consequences for the phenotype of the organism that are again unmediated by the ostensibly purposive dynamics of the organism. A failure to meet either of these criteria would constitute a failure to exclude the organism in its own lifetime, by its own adaptive energies, from being an agent in its own evolutionary history. It would constitute, in fact, an inability of the Darwinian model to go beyond the Kantian framework in which something like a purposefully organized system is always already there. And I will argue that not just one but both of these criteria fail to hold.

First Criterion

The idea that heritable changes take place independent of the life history of the organism has a name—Weismannism. On a purely speculative basis, Weismann, toward the end of the nineteenth century, theorized that germ tissue was separated and sequestered from the somatic tissue and so buffered from the influence of an organism's life history. This model, it turns out, accurately describes the developmental pattern of *Drosophilia* and most vertebrates, including ourselves, but in fact it does not describe the vast majority of living taxa. All plants and many invertebrates, for example, can produce germ cells from tissue that was previously involved in somatic function. Any changes, adaptive or otherwise, that result from the life interactions of these tissues will thus become part of the germinal stock. While the Weismannian organism is generally taken as the evolutionary exemplar by Darwinists, it is itself an evolutionarily derived state. One cannot explain evolution on the basis of the Weismannian organism because the Weismannian organism is itself a downstream product of evolution.

The response of vulgar Darwinism to this critique of Weismannism has been to defend the first criterion on the basis of the central dogma of molecular biology, a move that is also known as "molecular Weismannism." DNA makes RNA makes protein. The stuff of evolutionary variation is DNA, which, sequestered or not, cannot be influenced, at least in any instructive way, by the life history, or purposive organization, of the organism. True or false? False. Both genetic stability and genetic mutability, it turns out, are highly regulated, and the more salient the mutation to that which is of evolutionary significance, the more highly regulated is the apparent process by which it was brought about. No less than four enzyme-mediated systems monitor and repair spontaneous changes in DNA sequence. But not only does the cell provide its own regulated gateway for allowing, or not, the passage of spontaneous changes in base pair sequences; the cell is also equipped with the means for inducing mutation (Radman 1988). A so-called SOS system, found even in bacteria, is activated in response to threatening conditions, precisely to elicit *de novo* mutations (Radman, Matic, and Taddei 1999). Where such findings provide synchronic evidence for the active role of organisms in the generation of evolutionary diversity, additional evidence can be found at the level of a diachronic analysis. Whereas most, if not all, enzymatic functions are performed by proteins that have been conserved since the single-cell stage, many of the proteins that appear to be instrumental in mediating complex developmental interactions, the so-called cell-adhesion molecules (i.e., the integrins, the cadherins, and the immunoglobulin superfamily), consist of proteins that appear to be patchworks of peptide modules cut and pasted together. This appears to be the result of so-called exon shuffling at the DNA level, exons being the modules of DNA that constitute templates for functional and/or structural domains of proteins. These innovations, these patchwork connectors that allow the cells of complex organisms to be differentially responsive to subtle differences in the surround, could not have come about through random and stepwise mutation of single nucleic acids but only through the specific excision and ligation of whole segments of DNA, processes that entail the mediation of enzymes. Most, if not all, of what appear to be the molecular-level innovations that have accompanied metazoan evolution fail to follow the pattern expected for random mutations that have been gradually selected for behind the backs of the active organism. Evolvability may well have evolved, as we are now being told. Perhaps our vulgar Darwinists will tell us that "once upon a time" classical passive mechanisms of variation led to the evolution of active mechanisms of variation and innovation, but if "once upon a time" means back to some early single-celled organism, then the whole story about metazoan evolution that Darwinism tells must be radically reconfigured, and with it its metaphysical implications.

Second Criterion

More pertinent and perhaps even more interesting than the activity or the passivity of the mechanisms of genetic variation is the phenotypic significance of genetic variation. As suggested above, two conditions must be met for variation to occur outside the reach of the agency of the organism. The first pertains to the mechanisms of variation and the second to the ability of such variation to *directly* determine its phenotypic consequences. The second condition has a name as well—preformationism. I have suggested above that the idea that specific DNA sequences specify phenotypic outcomes is the result of a conflationary confusion between two legitimate but nonoverlapping gene concepts, Gene-P and Gene-D. Simply stated, if the meaning—that is, the phenotypic significance—of changes in the DNA is determined, not by the DNA sequence as such, but by the adaptive developmental capabilities of the organism—in other words, if the organism can "interpret the sequence" or "deploy its potential use" in a multiplicity of ways—then again the organism and not the molecule is in the evolutionary driver's seat. This would be consistent with the concept of Gene-D, the molecular template gene, as a developmental resource. Perhaps the most striking piece of evidence in support of this view has come from transgenic animal experiments. Recombinant DNA technology allows investigators to manipulate animal embryos at the one-celled stage so as to result in an animal whose every cell has been so affected. A typical such manipulation could be the deletion of a certain gene (a Gene-D) that has been highly correlated with significant biological function. If spontaneous somatic mutations in a certain gene are found to be highly correlated with the loss of growth control or some other pathological effect, then surely, reasoned investigators, the complete absence of said gene will wreak disastrous consequences on a developing organism. The results of such experiments typically proved to be shocking. In a characteristic example, transgenic mice were brought to term that had had all of their copies of the gene p53 knocked out. Lesions in p53 had been more highly correlated with tumors in the human brain, lung, breast, colon liver, bladder, and other organs than any other gene, leading investigators to predict "that any embryo deprived of its p53 genes would either perish in the uterus, or emerge with some grotesque deformity." The fetal mice, however, "develop just fine, and look perfectly normal when born" (Angier 1992). The tendency among investigators faced with these "anomalies" was to appeal to the idea of genetic redundancy—that is, the idea that the genome is provided with fallback genes to cover for when necessary. But even if one entertains this mode of explanation, such a "mechanism" would entail the capacity of a developing organism to pick and choose among some inventory of genetic resources, redundant or otherwise. And in light of the more recent revelation that the human genome contains only around twenty-three thousand genes, enthusiasm for yelling redundancy is likely to be seriously dampened.

Surely some lesson can be learned from the finding that humans do not have significantly more genes than the simple nematode worm. Perhaps I can heighten this tension further by pointing out that enzymes for just two catalytic functions, adding and removing phosphate groups (i.e., kinases and phosphatases), have been known to comprise two thousand and one thousand genes respectively and thus, by current counts, over 13 percent of the human genome. Something on the order of a clue must be present here. Gerhart and Kirschner (1997) have identified the role of most cellular phosphorylation as that of contingency making. When the heavily charged phosphoric acid group is added to a protein, it has the power to change its conformation sufficiently to alter its enzymatic or allosteric properties, in effect pulling it apart from the otherwise routine flow of causal interactions and introducing a contingency condition. What might previously have been part of an unbroken chain of metabolic reactions in the interior of the cell has now been separated and been made subject to, for example, the reception, or not, of signals from the surface of the cell. Metazoan evolution appears to take place, not through the evolution of new core enzymatic functions, but rather through pulling apart the old ones, providing a multiplicity of ways in which they can be linked together, and making them all contingent upon a sensitivity to the changing cellular context in which the cell is a dynamic constituent. The proliferation of kinase and phosphatase genes, in the absence of a general expansion of the enzymatic repertoire, suggests that the fine-tuning of contingency-making sensitivity is what is at issue.

Rather than providing an alteration in a predetermined script, as suggested by the conflationary GeneP/GeneD concept and proclaimed by vulgar Darwinism, genetic variations are indeterminate resources deployed by the developing organisms in different ways, in different places, at different times. Developmental epigenesis consists of the reproduction of contingent cycles of interaction. What makes development a predictable process is the adaptive ability of the developing organism to buffer itself from perturbations both from within and from without. In as much as metazoan development is not the execution of a script but a constructive reenactment of context-sensitive relationships between modular units at many different emerging levels of organization, heritable variation can take place at many different levels. Any change that provokes the organism to alter its developmental trajectory—to change its pattern of gene expression, for example—in such a fashion as to reproduce that change in the developmental trajectory of the next generation is the stuff of evolution. As developmental psychologists such as Piaget (1978) and Gilbert Gottlieb (1992) and more recently the biologist Mary Jane West-Eberhard (2003) have suggested, this means that the phenotype can even lead the genotype—that ultimately, adaptive behavior itself can alter the conditions of subsequent cycles of development, reproducing phenotypic patterns that become progressively, selectively stabilized at the level of the genome over subsequent generations.

AGENCY AND SELECTION

Processes of selective stabilization, and selective enhancement, are legion throughout all levels of biological order, and they are all very natural. Within the cell, microtubule organizing centers (MTOCs) randomly sprout microtubules in all directions, but only those *selectively stabilized* due to contextual cues escape the fate of rapid depolymerization (Gerhart and Kirschner 1997). Immature B cells of the immune system sport *de novo* immunoglobulins on their surface, which may or may not result in selective recognition of antigen and subsequent clonal expansion. Organismic selection in the context of its ecological relationships is not distinctly different from those many contingent cellular interactions that mediate the course of ontogeny. Differential survival and differential fecundity taken as abstract quantities surely have bearing on evolutionary history, as any Darwinism must hold. But selection can be brought down to earth and resituated in actually existing mutually selecting relationships in which the organism is anything but passive. In the absence of the conflationary gene, the distinction between variation and selection itself begins to break down. Consider the case of organisms, let's say microorganisms, that have entered into a *de novo* symbiotic association. To do so successfully entails becoming metabolically coupled in a fashion that is salutary for each. They have in effect selectively stabilized each other. The formation of a new joint metabolic enterprise entails novelty—that is, variation. Where in this case would variation end and selection begin? Does a proper Darwinism require that we must abstract organisms out of their mutually selecting interactive ecological milieu in order to comprehend the meaning of selection? The bias of vulgar Darwinism toward the Hobbesian individual, let alone toward the Hobbesian gene, has obscured much of the panorama of life. Symbiotic associations are the norm, not the exception. We enjoy the milk of the cow that harbors a protozoan, which harbors a bacteria, which can digest cellulose. Trees depend upon symbiotic associations with fungi in their roots, and it may well have been a symbiotic association between autotroph and heterotroph that allowed life to crawl out of the sea and colonize the land. We harbor a greater number of prokaryotic cells in our digestive tracts than the number of eukaryotic cells of which we are composed. The move from genetics to genomics, which has already begun to call attention to the lack of genetic resources in particular species, will likely lead to an appreciation of the distribution of vital genetic resources between different types of organisms that have mutually selected each other. We have no reason to doubt that the proclivity of organisms, from unicellular to gigacellular, to enter into new metabolic enterprises, to select, and selectively induce and selectively stabilize each other, constitutes an active role in evolutionary history.

I have considered the processes of both variation and selection and have argued that in neither case do the claims of vulgar Darwinism to exclude the organism from an active role in evolutionary history hold up to empirical scrutiny. I would now like to briefly turn back to Jonas's desideratum for Darwinism and questions of philosophical anthropology.

ANTHROPOLOGICAL AFTERTHOUGHTS

The challenge for a post-Cartesian philosophical anthropology would be to locate in the evolutionary movement toward the human body — that is, in the physical stuff of which we are made — the emergence of our sociocultural form of life. Vulgar Darwinism has given us eugenics, sociobiology, and now evolutionary psychology, or perhaps it would be at least as accurate to say that eugenics, sociobiology, and evolutionary psychology have given us vulgar Darwinism. These sciences, or pseudosciences as the case may be, have certainly sought to biologize humankind, and biologize the differences between human individuals and groups, but they have offered very little by way of accounting for what is distinctive about the human species, other than perhaps the idea of a largely disembodied language instinct. Within the framework of a new or renewed epigenesis, unconstrained by conflationary confusions, a nonvulgar Darwinism can address the post-Cartesian mission that Jonas foretold, namely returning the horizon of interiority to our cognition of prehuman life and finding our way to an evolutionary philosophical anthropology that locates human particularity in its full-bodied emergence from, yet continuity with, prehuman forms of life.

Perhaps a good way to begin such an enterprise would be to revisit earlier efforts in this direction such as those of Helmuth Plessner (1970, 1975), who, following the likes of Nietzsche, Bergson, and especially Dilthey, attempts to put forward an understanding of "life" that is independent of and irreducible to Cartesian notions of either mind or body. Plessner sought to characterize the active agency of organisms in a concept of positionality that captures both its centripetal and its centrifugal tendencies. The animal lives outward from its center, projecting, reaching,

yet always simultaneously pulling back into its boundary and maintaining centered-
ness. The animal lives at its center, in its here and now, but cannot make an object
of its center to itself. The human form of life for Plessner, by contrast, is characterized by the double perspective of both being a centered body and reflectively
attaining an eccentric positionality from which one sees a world in which one has
a body. The human form of life is such as to disclose a world to itself, a world of
which it is a member, but also to be always a body self-centered within the world.
The human thus has a body and is a body. In his attempts to formulate a methodology suitable for the human sciences, Dilthey endeavored to recover, from what
is textually expressed, originary experience that can be understood only through
some form of reenactment. Following Dilthey, but seeking to radicalize the reach
of interpretation of historical text into the somatic/organismic realm, Plessner's anthropology analyzes human expressivity at the level of the mimetic and gestural—
blushing and turning pale, laughing and crying—the latter two of which Plessner
considers to be uniquely indicative of the tensions of the human doubled standpoint. While understanding of organic expressivity must begin at the level of our
experience, as Jonas intimates, it need not stop there.

Can the origins of human eccentric positionality—that is of the conditions
of world disclosure—be accounted for in terms of the evolution of the animate
human body? Plessner, like Gehlen (1988) after him, looked to the immaturity of
human birth, and what they then referred to as human instinctual underdetermi-
nation, as the biological basis for both the ability and the need of humans to stand
in front of a world that they disclose to themselves, as well as being simultaneously
centered within it. Changes in developmental timing such as immature birth or
neoteny have indeed gained empirical support as significant events in evolutionary
history. But I would suggest that there are other sources of empirically oriented
studies that lend credence to Plessner's model. The neuroscientist Merlin Donald (1991) has proposed an intermediate stage in hominid cognitive/cultural evolution that he refers to as mimetic culture. The enabling transition for Donald
is the attainment of an enhanced control of skeletal muscles that makes possible
the recall and rehearsal of patterns of movement. Rehearsable motor skills would
constitute the basis for a mimetic culture in which "autocued" motor sequences
would be expressively performed as primitive intentional representations, ritual
dance, a mimetic "prototheater" of the everyday, and thereby provide the com-
municative and norm-enforcing means for social actions requiring task differentiation, such as big game hunting. Interestingly, Donald's paleoanthropological
model is further buttressed by studies of the residual cognitive capabilities of

patients who have experienced episodes of extreme dyslexia. The practical and philosophical implications of this kind of anthropology, I would suggest, will be very different from those born of what I have called vulgar Darwinism, but I leave it to future work to detail further just what those implications will be.

REFERENCES

Angier, Natalie. 1992. Proof of genetic redundancy: Mice normal without vital gene. *New York Times,* March 24.

Depew, D., and B. Weber. 1995. *Darwinism: Evolving systems dynamics and the genealogy of natural selection* Cambridge, Mass: MIT Press.

Donald, M. 1991. *Origins of the modern mind: Three stages in the evolution of culture and cognition.* Cambridge, Mass.: Harvard University Press.

Gehlen, A. 1988. *Man: His nature and place in the world.* New York: Columbia University Press.

Gerhart, J., and M. Kirschner. 1997. *Cells, embryos, and evolution.* Malden, Mass: Blackwell Science.

Gottlieb, G. 1992. *Individual development and evolution: The genesis of novel behaviour.* New York: Oxford University Press.

International Human Genome Sequencing Consortium. 2004. Finishing the euchromatic sequence of the human genome. *Nature* 431:931–45.

Jonas, H. 1966. *The phenomenon of life.* Chicago: University of Chicago Press.

Johannsen, W. 1909. *Elemente der Exakten Erblichkeitslehre.* Jena: G. Fisher.

————. 1923. Some remarks about units in heredity. *Hereditas* 4:133–41.

Keller, E. F. 2000. *The century of the gene.* Cambridge, Mass.: Harvard University Press.

Piaget, J. 1978. *Behavior and evolution.* New York: Pantheon Books.

Plessner, H. 1970. *Laughing and crying: A study of the limits of human behavior.* Evanston, Ill.: Northwestern University Press.

————. 1975. *Die Stufen des Organischen und der Mensch.* 3rd ed. Berlin: de Gruyter.

Radman, M. 1988. The high fidelity of DNA duplication. *Scientific American,* August, 40–46.

Radman, M., I. Matic, and F. Taddei. 1999. The evolution of evolvability. In *Molecular strategies in biological evolution,* ed. L. H. Caparole, 146–55. New York: New York Academy of Sciences.

Sloan, P. 2002. Performing the categories: Eighteenth-century generation theory and the biological roots of Kant's a-priori. *Journal of the History of Philosophy* 40:229–53.

West-Eberhard, M. J. 2003. *Developmental plasticity and evolution.* Oxford: Oxford University Press.

CHAPTER 17

Darwinism

Neither Biologistic nor Metaphysical

Bernd Graefrath

n 1860, the year of his death, Arthur Schopenhauer had a chance to become acquainted with Charles Darwin's book *The Origin of Species by Means of Natural Selection, or the Preservation of Favoured Species in the Struggle for Life* (1859). In the London *Times,* which he read nearly every day, he had found an excerpt from it, but his judgment about it was not too favorable: in a letter, Schopenhauer wrote that Darwin's book "is not at all congenial to my theory, but a superficial empiricism, which is not sufficient in this case: it's just a variation of Lamarck's theory."[1] This means: as a biologist, Darwin lacks originality, and as a philosopher, he lacks metaphysical depth. But the first claim is not true, and the second—if true—should not necessarily be regarded as a fault. Although Charles Darwin still shared some Lamarckian hypotheses with earlier evolutionary thinkers (among them his grandfather Erasmus Darwin) that were later discarded by a more strict Darwinism or neo-Darwinism, his theory marked a turning point in the history of biology. In retrospect, it is not very important that Darwin still believed in the hereditary transmission of acquired characteristics. What makes his theory outstanding is his successful approach of a biological explanation that can do without any teleological factors in the evolutionary process of all species, including humankind.

Darwin's biological theory is not without philosophical implications. At least it has undermined some paradigms of traditional metaphysics. David Hume had already done much to destroy the foundations of metaphysical systems that claim that many empirical facts about our world can only—or at least most plausibly—be accounted for by the assumption of a creator who is a rational designer. In his *Dialogues concerning Natural Religion* (1779), he lets Philo say: "The world plainly resembles more an animal or a vegetable, than it does a watch or a knitting-loom. Its cause, therefore, it is more probable, resembles the cause of the former. The cause of the former is generation and vegetation. The cause, therefore, of the world, we may infer to be some thing similar or analogous to generation or vegetation." Demea objects to this with a question: "But what wild, arbitrary suppositions are these? What *data* have you for such extraordinary conclusions?" Philo has a safe, though modest defense: "Right, cries Philo: This is the topic on which I have all along insisted. I have still asserted, that we have no *data* to establish any system of cosmogony."[2] Ever since Darwin, opponents of the argument from design can take their stand more self-assured: instead of pointing out the possibility of many alternative explanations, they can now shift the burden of proof to the philosophical theist.[3] The most acceptable explanation of the biological part of the cosmological process is now one that assumes a natural evolution, developing without purpose.

But is the only impact of Darwinism on philosophy negative? Currently it is popular to make Darwinism take the lead in a broader project of reductive naturalism. This endeavor holds that the natural sciences are the only sciences in the strict sense and that if any knowledge is to be found in the humanities (including philosophy), it has to fulfill the criteria of knowledge set by the empirical sciences. One popular example is sociobiology, which is very prone to a reductionist misunderstanding of its own status. Some sociobiologists claim that their science not only sheds new light on the behavior of animals (including humans) but provides the *only* acceptable perspective for understanding human culture or even every particular human action. The most far-reaching application of this approach can be found in epistemology and ethics, if it is claimed that these philosophical disciplines either are nonsense or have to be reconstructed as a part of scientific biology.[4]

This kind of biologistic naturalism, is, in my view, one example of a false interdisciplinary cooperation. Another is the attempt to base a new metaphysical system on Darwin's biological theory. This project is less popular but must be taken seriously. There it is claimed that Darwin has earned himself a place among the great philosophers and metaphysicians because he gained important ontological insights.[5] I think that in this case as well, Darwinism should be defended as an autonomous biological theory. But biological findings can be relevant for philosophy—which, nevertheless, remains an autonomous discipline too. The interesting task for future research is then to clarify this relevance in more detail. One good example is so-called "evolutionary ethics"—if its range is properly understood.[6] Another example is a project that during the last hundred years has lost more and more appeal among professional philosophers: the task of giving a picture of the world as a whole and of humankind's place in it. One of the founding fathers of contemporary analytic philosophy, G. E. Moore, still thought that this should be a positive concern of philosophy.[7] Today, such an effort is usually[8] rejected as impossible, too difficult, or too uncertain. But, as can be seen from bestseller lists, there is a need for orientation that can be supplied by a comprehensive view of the world, built on a scientific basis. One might call such an undertaking "hypothetical metaphysics,"[9] and Schopenhauer can be counted among its proponents,[10] although he seldom showed the kind of modesty suitable for embarking on the presentation of very generalized theories that must refrain from any sort of dogmatism.

All these points have to be discussed in more detail, and it will be shown that Darwinism, properly understood, is neither biologistic nor metaphysical but should nevertheless be studied by philosophers because it does have important implications for their work.

DARWINISM AS A SCIENTIFIC RESEARCH PROGRAM

The first advice for a scientific research program is: Do not load an unnecessary burden on your shoulders! (And do not let others put a burden on your shoulders—even if they have the best intentions!) This means Darwinism should first be evaluated as a biological theory. In that respect, it can well be summarized as follows: "Genes mutate, organisms compete with each other and are selected, and species evolve."[11] Today it can be said that as a biological theory, Darwinism has been most successful. It has beaten all its rivals if we apply the usual scientific standards, like "accuracy, scope, simplicity, fruitfulness."[12] This does not automatically mean that it is also most popular in the public at large. So-called "creation science" has many followers and is influential in some areas, but this is not the kind of "success" that counts in science. There has been some controversy over whether Genesis should be treated as a respectable biological alternative to Darwinism, and "creation scientists" often claim that Darwinism is "just another theory" or "just another paradigm." Karl Popper has to bear some responsibility for these misunderstandings. His falsificationism can easily be taken to mean that all theories that have not been falsified yet should therefore be regarded as having an equally good standing. Further, Popper has criticized Darwinism's formula of "the survival of the fittest" as tautological and therefore unscientific (as if it postulated only the survival of those that actually survive).[13] But Darwinism should properly be understood as an empirically testable theory predicting, for example, that in a particular geological stratum certain types of fossils will not be found or that for any species that will ever be discovered there will be a plausible explanation about how it could have evolved step by step from previous species.[14]

The reference to Thomas S. Kuhn's concept of incommensurable paradigms in scientific research is more interesting, but it is not of much help for the "creation scientist." Especially in the further development of Kuhn's approach in the writings of Imre Lakatos,[15] it is clear that there still are reliable standards by which competing research programs can be evaluated. And measured by these standards, "creation science" is just a bad scientific alternative. This insight also makes clear which strategy should be chosen to keep Genesis out of biology classes: creation science should be excluded, not because it comes from a religious source, but because, if it is taken to be a biological theory, it is just a bad theory.

Once Darwinism is established as the best theory of biological evolution, diverse attempts can be undertaken to extend its scope to other areas than those to which it was first applied. Already Darwin himself had carefully pointed out a possible direction: "Light will be thrown on the origin of man and his history."[16] But even among evolutionists, there has always been some resistance to the idea that even the evolution of the human mind could be explained in Darwinian terms. Alfred Russel Wallace thought that Charles Darwin was taking his theory too far when he undertook a naturalistic explanation of the origin of the mind,[17] and even today some philosophers share Wallace's reservations.[18] But if Darwinism could explain the origin of all species except for *Homo sapiens* with all its characteristics, it would not be an acceptable biological theory. And indeed, the evolution of the mind can completely be explained by natural causes, and the opposition to such an explanation can often be traced back to a very common misunderstanding of Darwinism, which, interestingly enough, can be found among some of its foes *and* some of its friends. This misunderstanding depends on what Stephen Jay Gould calls the "Panglossian Paradigm." Even for a hard-core Darwinist, there can be excesses of adaptationism.[19] It could even be said that especially for a hard-core Darwinist it is essential to deny what could be called a pan-adaptationism or hyperselectionism, which is rather more congenial to the traditional argument from design. One of Darwin's strongest arguments stresses the existence of rudimentary organs, which require a historical and not a teleological explanation. Consequently, there is room for historical accidents in a Darwinian account of evolutionary history: it does not have to postulate that all characteristics of biological organisms are making a positive contribution to reproductive success. Two standard categories of such an account are *functional change* and *accidental surplus*. The first is sufficient to answer objections of the type "What good is half a wing?" The second is sufficient for a biological explanation of the human mind. In the words of Stephen Jay Gould:

> The earliest Cro-Magnon people, with brains bigger than our own, produced stunning paintings in their caves, but did not write symphonies or build computers. All that we have accomplished since then is the product of cultural evolution based on a brain of unvarying capacity. . . . Natural selection may build an organ "for" a specific function or group of functions. But this "purpose" need not fully specify the capacity of that organ. . . . Our large brains may have originated "for" some set of necessary skills in gathering food, socializing, or whatever; but these skills do not exhaust the limits of what such a complex machine can do.[20]

Human consciousness could thus be an evolutionary by-product of other capacities that were more directly favored by natural selection. Such a biological account of the origin of human consciousness does not imply a reductionist materialism in the philosophy of mind, and it certainly does not imply that because of these humble origins the human mind is less valuable or less important for the picture we form about what it means to be a human being. As far as the so-called mind-body problem is concerned, several nonreductivist theories are compatible with a Darwinian account of the origin of human consciousness. Epiphenomenalism is a serious candidate,[21] as is John Searle's position of a "property pluralism."[22]

Some might still object that a biological explanation is not suitable for an adequate philosophical understanding of subjectivity. This point is well taken, insofar as a biological answer should not be regarded as exhaustive. But this can easily be misunderstood, and it is most important to stress that this holds true for all biological (and even all scientific) theories. We have to be on our guard in order not to commit what I would call the "nothing but" fallacy—that is, the false idea that if an explanation gives a convincing answer, then all other contributions to the understanding of a phenomenon must be false or at least superfluous. For our present context, however, the most important point is that Darwinism gives a most plausible biological explanation of the origins of the human brain with all its capacities.

Today's Darwinists are also exploring the force of a biological explanation of human behavior as we encounter it in our society: Darwinism enhances our understanding by calling our attention to the way in which even complex aspects of human culture could be explained by the mechanisms of an evolution that consists in the differential copying success of genes relative to their alleles.[23] We can accept this extension of the original biological theory as a valuable contribution to our understanding without thereby being committed to a naturalism in the reductionist sense.[24] But one could still go further and claim that scientific biology would give a complete explanation of all human behavior and that no other reasonable questions besides those that could be treated by the empirical methods of the natural sciences would remain. This is a problematic philosophical theory that cannot itself be justified by empirical means. The project of so-called "evolutionary ethics" offers a good illustration of the limits of such a naturalism.

"EVOLUTIONARY ETHICS" AND THE LIMITS OF PHILOSOPHICAL DARWINISM

Can biological knowledge explain and justify morality? A careful philosopher should refuse to answer this question with a clear "yes" or "no." Before clear-cut answers can be given, several distinctions must be drawn. The most important stresses the difference between the categories of *explanation* and *justification:* while explanations (e.g., in evolutionary biology) refer to causes, justifications (e.g., in philosophical ethics) refer to reasons. At a deeper level, the thesis of a complete gap between reasons and causes can be critically discussed. But before we turn to the question about the foundation of justification in ethics, we must first distinguish several levels of explanation. What does it mean to have a morality, and what exactly do sociobiologists try to explain? The focus of their research is the behavior of animals that live in groups and interact with one another. Some kinds of behavior can be described as "altruistic," but this term is already problematic when applied to animals that cannot understand what they are doing. The term does have its proper place in the sphere of human *actions,* and these cannot adequately be described as mere events, like the wind moving the trees. Things become even more problematic when certain kinds of animal behavior are, for example, described as "adultery" or "fulfilling parental duties." It is inappropriate to use the same term for the description of forms of behavior that in important respects do not fall into the same category. When we apply a term like *adultery* to describe human behavior, we presuppose an *understanding* of social practices, and when we talk about rights and duties, we presuppose an understanding of normative discourse. This must not necessarily hinder us from ascribing certain moral rights to certain animals,[25] but it must be stressed that their behavior would not appropriately be described by saying that they were doing something like making a claim.

A further distinction concerns what it means to understand human morality. The sociologist Max Weber stresses that to give an adequate description of human behavior, it is important to understand what people believe themselves to be committed to.[26] This requires more than merely describing what people usually do: the normative element in customs must not be neglected. But the term *morality* also has an even wider meaning, which is somewhat obscured by the English term *ethic*. We understand what it means to be under a moral obligation, even if this obligation is not accepted as an obligation in the community we live in. We have to distinguish between the *ethos* of a community, which tells us what people in this community regard as obligatory, and *morality proper,* which tells us what is obligatory without reference to what a community's customs happen to be. Philosophical *ethics* then is concerned with the justification of this morality in the strict sense.

Sociobiologists can contribute to our understanding of how and why certain types of habits and customs spread while others are no longer practiced. They can furthermore explain the evolutionary roots of our capacity to make moral judgments. But they cannot, on the basis of their scientific knowledge alone, answer the question of what we *ought* to do. This is the basic reason why the political program of social Darwinism fails from a justificatory point of view: even if we do not consider its one-sided view of the evolutionary process, there is always the unsolved problem of answering why we should copy nature (or anything else). The step from *is* to *ought* requires a special justification.

Today, the naturalistic fallacy is accepted as a real fallacy that has to be avoided. Usually, this view is — at least theoretically[27] — shared even by those who regard themselves as proponents of "evolutionary ethics." Properly speaking, the term is then a misnomer because it does not claim to be an ethics at all. But there are several ways in which sociobiological findings can be relevant for ethics in the strict sense. Each of these has to be evaluated on its own terms. Biological knowledge can play an important role in hypothetical imperatives. Even though not all moral obligations can be reduced to this type of prudential reasoning, even the most categorical imperatives have to be applied to particular cases in a hopefully successful way. For example, if there is a kind of "natural" tendency for male domination, but we find it morally obligatory that females have the same moral status as males, biological knowledge can help us to determine what the most promising course of action will be to counter the "natural" tendency that we cannot justify.

Another example is the problem of our dealings with the so-called "third world." If we take seriously the claim that all persons have the same moral status, wherever they live, then people in rich countries will have to make drastic changes in their way of living. But when we are planning a political program that will bring these changes into action, it would be wise to consider certain anthropological limitations. One should not overstrain what humans can bear, and a certain degree of realism is appropriate for success with political reforms. It is true that such a reference to natural limitations can easily be misused by conservative forces who are just looking for an excuse not to change anything. But then it is our task to find out what can really be changed and what cannot be changed very fast.

This kind of relevance of biological facts for ethics might be a good corrective against certain types of traditional ethics that underestimate the problems of application from general insights to particular solutions. But it is not certain that this type of interdisciplinary project justifies a special role for evolutionary knowledge. After all, many types of empirical knowledge are relevant for the application of moral principles. For example, the limits on natural resources of our planet Earth are certainly relevant for the understanding of our responsibility for future generations, but this does not mean that there is a special philosophical place for a project of "geological ethics."

A more particular relevance of evolutionary theory for philosophical ethics has been mentioned above and deserves elaboration. This concerns the distinction between reasons and causes and the foundation of philosophical ethics in general. Not every type of philosophical ethics will regard this as a big problem: if it were possible to justify certain moral principles as categorical imperatives in an aprioristic way, then reasons could never be reduced to mere causes. But if one thinks that this project has not been successful so far, one might consider John Rawls's models of justification as the best thing we currently have. Then basic "intuitions" (in an ontologically harmless sense) play an important justificatory role, and a sociobiologist might ask whether these basic intuitions are, in the final analysis, just some kind of biologically explainable feelings that were caused by evolutionary processes. Jeffrie G. Murphy gives a good presentation of the challenge that philosophical ethics of this type must face:

> According to Rawls, a theory puts us in the desirable and theory-confirming state of reflective equilibrium if it does a better job than any other theory of embracing and ordering the largest possible subset of our pretheoretical convictions about what is good and evil, right and wrong, just and unjust. *But where do these convictions come from and what is their status in justification*—i.e., *why do we place so much confidence in them?* Wilson and other sociobiologists are, I think, dead right in wanting to ask this question and dead right in think-

ing that an honest answer to it will unmask, at least partially, some of the pretensions of moral philosophy. If these convictions can be shown to be simply the result of biological instincts preserved in evolution by natural selection, then their status in moral epistemology will be affected. For moral theory would then be *relativized* to a degree that many of the "New Deontologists" in ethics—Rawls, Nozick, Fried, Dworkin—would no doubt find objectionable.[28]

Murphy does not think that this challenge is fatal. Nevertheless, it could have an influence insofar as a Humean approach in ethics might then be regarded as more acceptable than a Kantian one. Perhaps we just have to accept that at the foundation of our moral discourse we find certain moral sentiments without which we could not form or understand moral judgments but that we share with others and that thus enable us to find intersubjective validity.[29] The main point is that a fatal relativism is not the consequence of such a biologically enlightened philosophical ethics. There is room for argument in normative discourse, and biological knowledge can even contribute to moral progress by helping us to form a more appropriate view of the world and our place in it.

THE PLACE OF HUMANKIND IN A DARWINISTIC WORLD

Nelson Goodman has argued that "facts are small theories, and true theories are big facts."[30] In a similar vein, Imre Lakatos has blurred the distinction between "science" and "metaphysics."[31] From this perspective, metaphysical systems could be regarded as very big theories that give a comprehensive view of the world, based on an interpretation of theories from several sciences. Such an approach is much more hypothetical than traditional metaphysics, but it may be a fruitful field for philosophical research. Wittgenstein postulated in his *Tractatus Logico-Philosophicus* that "Darwin's theory has no more to do with philosophy than any other hypothesis in natural science."[32] This was meant as a rejection of certain misunderstandings about the scope of philosophy. But we can give his sentence a more positive turn if we distinguish between different regions of the traditional philosophical landscape. Perhaps Wittgenstein would claim that certain regions of this landscape should not even be considered to belong to philosophy. But it is an open question whether a more restrictive or a more far-reaching understanding of philosophy is most helpful in answering the questions we are interested in. Of course, this presupposes that we are interested in more topics than logic and philosophy of language can cover. Certainly Wittgenstein's own interests were *not* restricted in this way. And if the idea of an inquiry concerning the place of humankind in the universe is regarded as worthy to be undertaken, the philosophers might still be most qualified to give it a try. The Darwinian theory will then have something to do with philosophy, just like any other theory of natural science.

As a preparation of this project, it is helpful to examine what Quine calls the "ontological commitments" of a theory. This point raises fundamental questions about the nature of scientific research. For example, in the case of Darwinism, the concept of *species* is of central importance. Should we regard this concept as the name of a real category that we found "out there," or is it rather a category that we ourselves created because it is instrumentally useful in the building of a biological theory? The roots of this controversy go back at least to the foundations of Kant's theoretical philosophy, and the case has not been decided yet. Among theorists in biology, a realist interpretation of the concept of species is most common,[33] but the functionalist account also has some adherents.[34] In our context, it is not necessary to take a side in this debate (although it should be mentioned that "Ockham's Razor" would recommend the latter approach). We are in the sphere of hypothetical metaphysics now, and we should be content to place the Darwinian theory in a more general view of the world without deciding whether the Darwinian theory should be based on a realist or an instrumentalist interpretation. It is more important that an unacceptable relativism can still be avoided, because not every scientific theory is equally well qualified to work in the functionalist sense. And it should be remembered that even scientific realists must rely on operationalist tests, so that the burden of proof may be on the realists' side to explain how they can reliably proceed from an instrumentalist to a realist interpretation of their theory.

An important ontological commitment of Darwinism is the rejection of essentialism, which is replaced by explanations in terms of populations (without essences). Thus a Platonistic idea of species is hardly compatible with evolutionary theory. An important concept that is not an ontological commitment of Darwinism is the concept of progress. Darwin's biological theory is not a theory about progress but a theory about "descent with modification."[35] He even set himself the motto: "Never use the words 'higher' or 'lower.'"[36] "Progress" is not a category that can be established by biological research, and working biologists can do without it. This does not mean that the concept is meaningless. On the contrary: without such a concept, we could not even ask which direction new developments *ought* to take. But since *progress* is a normative term, it must first be established which developments should count as progress in a positive sense. Charles Darwin himself saw this quite clearly: for example, he thought that the moral sentiments of humans were in need of cultural refinement but that biological evolution per se could never be presupposed to be destined toward a positive end.

This discussion of progress brings us back to the task of developing a more general view of the place of humankind in the universe. A Darwinian account of the origin of all species should bring us to acceptance of the fact that *chance* plays a vital role in cosmic evolution. The roots of reason are irrational, and the roots of all species, including humankind, are genetic copying errors. There is no guarantee for optimism concerning the future. Does this sound pessimistic? Many people who were exposed to Darwinism for the first time certainly felt this way. For example, George Bernard Shaw wrote: "As compared to the open-eyed intelligent wanting and trying of Lamarck, the Darwinian process may be described as a chapter of accidents. As such, it seems simple, because you do not at first realize all that it involves. But when its whole significance dawns on you, your heart sinks into a heap of sand within you."[37] But we can become adjusted to this view, just as we can become adjusted to the Copernican view of our solar system. Although some sociobiologists think that "the genes hold culture on a leash,"[38] there is still enough room for the hope of progress, and beforehand we cannot even rule out the possibility of, in Francis Bacon's words, "the effecting of all things possible."[39] On the contrary, one of the main problems of the future may be, not too little biological freedom, but too many technological possibilities, so that we have an *embarras de richesse*.

On the moral side, the Darwinian view of the world could lead to a new relationship between humans and (other) animals.[40] This does not mean that because of a mere hereditary relationship we would be under a special obligation of treating all animals or those genetically close to us as equals. But biological knowledge can be relevant for the interpretation of the behavior of animals. For example, a common evolutionary history makes it most plausible that the "pain behavior" of a cat should be regarded as real pain, contrary to the Cartesian view that mere animals can no more feel pain than mere machines can. If we think that human pleasure and pain are relevant in our moral deliberations, then we are under pressure to explain why we exclude animals that also show signs of sentience. It is to be expected that for many of these current exclusions no good justification can be given, and thus we will have to "expand the circle"[41] of the beings with moral status.[42]

Expanding the circle in this way will change our philosophical attitude toward the world we live in. The most general consequences of Darwinism for the way we picture our place in the world can perhaps be explored better by literature than by traditional philosophy. But it would have to be a certain kind of philosophical literature that is rare today. At least one example should be set forth here: the Polish philosopher and science fiction author Stanislaw Lem. Looking back on his work, he writes:

> I was born to be a philosopher at a time when it is no longer possible to build great systems in the realm of philosophy because this kingdom had fallen apart through the invasion of the sciences, so that the philosopher could no longer be the sovereign creator of a worldview. . . . So when I was looking for a position and a subject in which I could do what my intellect was best suited for, I settled down in that junk shop that is called science fiction because I took its wrong and misleading name seriously. Science fiction—for me this meant scientific strictness and, at the same time, the privilege of creative freedom, which is still offered by art. So, in spite of it all, philosophy, although in disguised form.[43]

One example from Lem's work is especially worth mentioning: In "Golem XIV"[44] he presents lectures by a computer of the twenty-first century whose intellect is superior to all human understanding. In the first lecture, it talks about humankind; in the last lecture, it talks about itself. Justice cannot be done to the rich content of this work in just one human lecture, but at least one idea should be mentioned to give food for thought: Is it possible that a Darwinistic process of evolution could be repeated on many other levels, from the evolution of superior machine intelligence to a kind of Darwinistic struggle between competing universes? The latter idea may be too speculative, but it is already seriously discussed among some scientists.[45] The former idea goes back at least to Samuel Butler, who in 1872 published his satire *Erewhon*. In it, the idea of a "Book of the Machines" is developed, the content of which is much more substantial than the satirist may have expected. Thus Butler writes:

> It is said by some with whom I have conversed upon this subject, that the machines can never be developed into animate or *quasi*-animate existences, inasmuch as they have no reproductive system, nor seem ever likely to possess one. If this be taken to mean that they do not marry, and that we are never likely to see a fertile union between two vapour-engines with the young ones playing about the door of the shed, however greatly we might desire to do so, I will readily grant it. But the objection is not a profound one. No one expects that

all the features of the now existing organisations will be absolutely repeated in an entirely new class of life. The reproductive system of animals differs widely from that of plants, but both are reproductive systems.[46]

It might furthermore be objected that machines still need humans for their "procreation," but Butler has a good reply: "The humble bee is a part of the reproductive system of the clover."[47] And humans might be viewed as doing for the machines what the bees are doing for the clover. This idea that the actors themselves are not conscious of all or even the most important functions they are fulfilling is familiar from a piece of philosophy that may at some time have been regarded as the equivalent of science fiction: Schopenhauer's "metaphysics of sexual love,"[48] which today cannot be read without taking Schopenhauer to be the first sociobiologist.

CONCLUSION

Thus, after all, even Schopenhauer, who wrote that Darwin's theory contained only "superficial empiricism," saw that scientific theories can have philosophical implications. He wanted to make sure not only that his philosophical theory was compatible with results from the natural sciences but that it gave the best general explanation of them all. One can use the term *metaphysics* for this project, although it is probably more misleading than helpful. Darwinism makes some metaphysical systems very implausible, but it still is compatible with many philosophical theories. First of all, it is the best biological theory we have. It does have implications for the explanation of human culture, but it does not force us to explain cultural developments in exclusively biologistic terms. Darwinism is neither biologistic nor metaphysical, and when it is freed from these burdens, its relevance for an enlightened philosophy can fruitfully be explored.

NOTES

1. Arthur Schopenhauer, letter to Adam von Doss, March 1, 1860, in *Gesammelte Briefe,* 2nd ed., ed. Arthur Huebscher (Bonn: Bouvier, 1987), 471–72 (my translation).

2. David Hume, *Dialogues concerning Natural Religion,* ed. Norman Kemp Smith (Indianapolis: Bobbs-Merrill, 1947), 176–77.

3. Cf. J. L. Mackie, *The Miracle of Theism: Arguments for and against the Existence of God* (Oxford: Oxford University Press, 1982), esp. 133–49.

4. Cf. Michael Ruse, *Taking Darwin Seriously: A Naturalistic Approach to Philosophy* (Oxford: Blackwell, 1986).

5. Vittorio Hoesle and Christian Illies, *Darwin* (Freiburg: Herder, 1999), 8.

6. Cf. Bernd Graefrath, *Evolutionaere Ethik? Philosophische Programme, Probleme und Perspektiven der Soziobiologie* (Berlin: Walter de Gruyter, 1997).

7. G. E. Moore, *Some Main Problems of Philosophy* (New York: Macmillan, 1953), 1–2.

8. An exception is J. J. C. Smart, *Our Place in the Universe* (Oxford: Blackwell, 1989).

9. Cf. Bernd Graefrath, *Es faellt nicht leicht, ein Gott zu sein: Ethik fuer Weltenschoepfer von Leibniz bis Lem* (Munich: C. H. Beck, 1998), esp. 117–23.

10. Cf. Dieter Birnbacher, "Induktion oder Expression? Zu Schopenhauers Metaphilosophie," *Schopenhauer-Jahrbuch* 69 (1988): 7–19.

11. David L. Hull, "Are Species Really Individuals?" *Systematic Zoology* 25 (1976): 181.

12. Thomas S. Kuhn, "Reflections on My Critics," in *Criticism and the Growth of Knowledge,* ed. Imre Lakatos and Alan Musgrave (Cambridge: Cambridge University Press, 1970), 261.

13. See Karl R. Popper, *Unended Quest: An Intellectual Autobiography* (Glasgow: Fontana/Collins, 1976), 167–80 ("Darwinism as a Metaphysical Research Programme").

14. Cf. Stephen Jay Gould, "Evolution as Fact and Theory," in *Hens' Teeth and Horses' Toes* (New York: W. W. Norton, 1994), 253–62.

15. Imre Lakatos, "Falsification and the Methodology of Scientific Research Programmes," in Lakatos and Musgrave, *Criticism,* 91–195.

16. Charles Darwin, *The Origin of Species* (Harmondsworth: Penguin Books, 1985), 458.

17. Cf. Stephen Jay Gould, "Natural Selection and the Human Brain: Darwin vs. Wallace," in *The Panda's Thumb: More Reflections in Natural History* (New York: W. W. Norton, 1980), 47–58.

18. Hoesle and Illies, *Darwin,* 176–77; Vittorio Hoesle, *Moral und Politik: Grundlagen einer Politischen Ethik fuer das 21. Jahrhundert* (Munich: C. H. Beck, 1997), 263–64.

19. Cf. Daniel C. Dennett, *Darwin's Dangerous Idea: Evolution and the Meanings of Life* (Harmondsworth: Penguin Books, 1996), 239.

20. Gould, "Natural Selection," 56–57.

21. Dieter Birnbacher, "Epiphenomenalism as a Solution to the Ontological Mind-Body Problem," *Ratio,* n.s., 1 (1988): 17–32.

22. John R. Searle, *The Mystery of Consciousness* (London: Granta Books, 1997), 211. See also John R. Searle, *The Rediscovery of the Mind* (Cambridge, Mass.: MIT Press, 1994), esp. 106–9 ("Consciousness and Selectional Advantage").

23. See Richard Dawkins, "In Defence of Selfish Genes," *Philosophy* 56 (1981): 556–73, esp. 571.

24. Cf., e.g., Richard D. Alexander, *Darwinism and Human Affairs* (Seattle: University of Washington Press, 1982).

25. Cf. Joel Feinberg, "Human Duties and Animal Rights," in *On the Fifth Day: Animal Rights and Human Ethics,* ed. Richard K. Morris and Michael W. Fox (Washington, D.C.: Acropolis Books, 1978), 45–69.

26. For Max Weber's term *verstehende Soziologie,* see his *Gesammelte Aufsaetze zur Wissenschaftslehre,* ed. Johannes Winckelmann (Tubingen: Mohr/Siebeck, 1988), 427–74.

27. See Robert J. Richards, "A Defense of Evolutionary Ethics," *Biology and Philosophy* 1 (1986): S265–93.

28. Jeffrie G. Murphy, *Evolution, Morality, and the Meaning of Life* (Totowa, N.J.: Rowman and Littlefield, 1982), 99–100.

29. Cf. Bernd Graefrath, *"Moral Sense" und praktische Vernunft: David Humes Ethik und Rechtsphilosophie* (Stuttgart: J. B. Metzler, 1991).

30. Nelson Goodman, *Ways of Worldmaking* (Indianapolis: Hackett, 1978), 97.

31. Lakatos, "Falsification," 182, 184.

32. Ludwig Wittgenstein, *Tractatus Logico-Philosophicus* (1921), revised English translation (London: Routledge, 1974), para. 4.1122.

33. See Ernst Mayr, *Toward a New Philosophy of Biology* (Cambridge., Mass.: Harvard University Press, 1988), esp. 335–58.

34. See Philip Kitcher, "Species," in *The Units of Selection: Essays on the Nature of Species,* ed. Marc Ereshefsky (Cambridge, Mass.: MIT Press, 1992), 317–41.

35. Darwin, *Origin of Species,* 404.

36. Quoted in Antony Flew, *Darwinian Evolution* (London: Paladin, 1984), 123.

37. George Bernard Shaw, "Preface: The Infidel Half Century," in *Back to Methuselah: A Metabiological Pentateuch,* rev. ed. (London: Oxford University Press, 1945), xliii.

38. Edward O. Wilson, *On Human Nature* (Cambridge, Mass.: Harvard University Press, 1978), 167.

39. Quoted in Peter B. Medawar, *The Hope of Progress* (London: Methuen, 1972), 110.

40. Cf. James Rachels, *Created from Animals: The Moral Implications of Darwinism* (Oxford: Oxford University Press, 1990).

41. Peter Singer, *The Expanding Circle: Ethics and Sociobiology* (Oxford: Oxford University Press, 1981).

42. Cf. Bernd Graefrath, "Zwischen Sachen und Personen: Ueber die Entdeckung des Tieres in der Moralphilosophie der Gegenwart," in *Tiere und Menschen: Geschichte und Aktualitaet eines prekaeren Verhaeltnisses,* ed. Paul Muench and Rainer Walz (Paderborn: F. Schoeningh, 1998), 383–405.

43. Stanislaw Lem, "Nachwort," in Stanislaw Lem, *Dialoge* (Frankfurt: Suhrkamp, 1980), 307–8 (my translation). Cf. Bernd Graefrath, "Taking 'Science Fiction' Seriously: A Bibliographic Introduction to Stanislaw Lem's Philosophy of Technology," *Research in Philosophy and Technology* 15 (1995): 271–85.

44. Stanislaw Lem, *Imaginary Magnitude* (London: Mandarin, 1985), 97–248. Cf. Bernd Graefrath, *Lems Golem: Parerga und Paralipomena* (Frankfurt: Suhrkamp, 1996).

45. See Paul Davies, *The Mind of God: Science and the Search for Ultimate Meaning* (Harmondsworth: Penguin Books, 1992), 221–22 ("Cosmological Darwinism").

46. Samuel Butler, *Erewhon,* ed. Peter Mudford (Harmondsworth: Penguin Books, 1970), 210. Cf. Bernd Graefrath, *Ketzer, Dilettanten und Genies: Grenzgaenger der Philosophie* (Hamburg: Junius, 1993), esp. 185–215 ("Eier, Hennen und Maschinen: Samuel Butlers Philosophie von 'Ergindwon'").

47. Butler, *Erewhon,* 211.

48. Arthur Schopenhauer, *Die Welt als Wille und Vorstellung,* vol. 2, ed. Arthur Huebscher (Zurich: Diogenes, 1977), 621–64 ("Metaphysik der Geschlechtsliebe").

INDEX

381

Milton Keynes UK
Ingram Content Group UK Ltd.
UKHW022151310824
447596UK00021B/438